ASTRONOMY AND
ASTROPHYSICS LIBRARY

Springer
Berlin
Heidelberg
New York
Barcelona
Hong Kong
London
Milan
Paris
Singapore
Tokyo

Physics and Astronomy   ONLINE LIBRARY

http://www.springer.de/phys/

ASTRONOMY AND
ASTROPHYSICS LIBRARY

Series Editors: I. Appenzeller · G. Börner · M. Harwit · R. Kippenhahn
J. Lequeux · P.A. Strittmatter · V. Trimble

Series homepage – http://www.springer.de/phys/books/aal

G. S. Bisnovatyi-Kogan

Stellar Physics

1: Fundamental Concepts and Stellar Equilibrium

Translated from the Russian by A. Y. Blinov and M. Romanova

With 24 Figures and 22 Tables

 Springer

Dr. G. S. Bisnovatyi-Kogan

Space Research Institute
Russian Academy of Sciences
Profsoyuznaya 84/32
117 810 Moscow
Russia

Translators:
Dr. A. Y. Blinov
Dr. M. Romanova

Space Research Institute
Russian Academy of Sciences
Profsoyuznaya 84/32
117 810 Moscow
Russia

Cover picture: Abundance of protoplanetary disks around newborn stars as seen by the Hubble space telescope in the Orion Nebula (STScI-PR94-24 and by courtesy of A. M. Quetz, Max-Planck-Institut für Astronomie, Heidelberg, Germany).

Library of Congress Cataloging-in-Publication Data

Bisnovatyi-Kogan, G. S. (Gennadii Semenovich)
    [Fizicheskie voprosy teorii zvezdnoi evoliutsii. English]
    Stellar physics / G.S. Bisnovatyi-Kogan ; translated from the Russian by A.Y. Blinov
and M. Romanova.
      p. cm. -- (Astronomy and astrophysics library, ISSN 0941-7834)
    Includes bibliographical references and index.
    Contents: v. 1. Fundamental concepts and stellar equilibrium
    ISBN 354063262X (v. 1 : acid-free paper)
    1. Stars--Evolution. I. Title. II. Series.

    QB806 .B5713 2000
    523.8'8--dc21

                                 00-038607

Title of the original Russian edition: Fizicheskie voprosy. Teorii zvezdnoi evoliutsii. © Nauka, Moskva 1989
The Russian edition was published in one volume.

ISSN 0941-7834
ISBN 3-540-63262-X Springer-Verlag Berlin Heidelberg New York

Springer-Verlag Berlin Heidelberg New York
a member of BertelsmannSpringer Science+Business Media GmbH

© Springer-Verlag Berlin Heidelberg 2001
Printed in Germany

The use of general descriptive names, registered names, trademarks, etc. in this publication does not imply, even in the absence of a specific statement, that such names are exempt from the relevant protective laws and regulations and therefore free for general use.

Data conversion by PTP-Berlin, Stefan Sossna
Cover design: *design & production* GmbH, Heidelberg

Printed on acid-free paper   SPIN 10032130   55/3141/mf - 5 4 3 2 1 0

To my teacher, Yakov Borisovich Zeldovich

Preface to the English Edition

The development of the theory of stellar evolution has been relatively rapid since 1989, which is when the Russian edition of this book was published. Progress in the field concerned mainly a better understanding of the physical background of stellar processes, in particular the improvements made in calculating new opacity tables. The latter led to a better description of some observational phenomena, such as Cepheid oscillation models, but otherwise led mainly to quantitative corrections of previously known results. The field that may be strongly influenced by the increase of opacity according to the new tables is the mass loss from massive evolved stars. This latter phenomenon has not yet been investigated, however. Many new results have been obtained in helioseismology, the theory of supernova explosions, accretion theory, and 2-D calculations of different phenomena, such as star formation, explosions of rotating magnetized stars and numerical simulations of stellar convection.

This book has been updated and now includes over 150 new references. New material has also been added to otherwise well established sections. This includes the CAK theory of mass loss from hot luminous stars, the description of the Eggleton method of stellar evolution, and a more detailed consideration of the accretion disk structure around black holes.

The revisions and additions of new material substantially increased the number of pages, making it desirable to produce two essentially self-contained volumes. The first volume, "Fundamental concepts and stellar equilibrium", contains the material related to the first six chapters of the Russian edition. The second volume, "Stellar evolution and stability", includes the material of the other chapters of the Russian edition. While both volumes retain the structure of the Russian edition, each of these two volumes now has a self-contained character and could be interesting for different kinds of readers. The first one contains a detailed description of physical processes in stars and the mathematical methods of evolutionary calculations. Thus this volume will be of interest for physicists and specialists in numerical mathematics regardless of the level of their actual involvement in the work on stellar evolution. The second volume contains both the qualitative and the quantitative descriptions of stellar evolution, explosions, stability and oscillations. This will be of interest for the wider astronomical community, observers and theoreticians alike, working or interested in astrophysical phenomena related to stellar formation and evolution. Finally, those who directly work in the theory of stellar evolution, or want to study this field in depth, will find that both volumes provide them with a comprehensive introduction to and survey of the present state of the art in this field.

I am very grateful to my colleagues from the Cambridge Institute of Astronomy, D.O. Gough, P. Eggleton, and C. Tout, and from Queen Mary and Westfield College, I.W. Roxburgh, A.G. Polnarev, and S.V. Vorontsov, for their hospitality during my stay in these places, many discussions and help with the work on different parts of the book, as well as improvements to the English. I am also grateful to M.M. Romanova, A. Blinov, and S.V. Repin, who translated the Russian edition of the book into English and prepared the TEX file of this translation.

Moscow, August 2000   G.S. Bisnovatyi-Kogan

Preface to the Russian Edition

The desire of astrophysicists to gain insights into the mysteries of the birth and death of stars has required the application of almost all branches of modern physics. The results of atomic physics are necessary for studies of stellar birth out of the interstellar medium; and knowledge of the structure of white dwarfs and neutron stars requires the use of liquid- and solid-state theory, and the theory of phase transitions, superconductivity, and superfluidity. Between these extremes, in the area where stars mostly exist, the laws of nuclear physics and weak interactions, and the theories of matter and radiative transfer, are at work.

The equilibrium of a star is described by the equations of hydrodynamics supplemented by general relativity and electromagnetic field theory. The problem of turbulence and convection, not yet completely resolved in terrestrial applications, is even more important and difficult in problems of stellar evolution. This book deals with many of these problems, trying to develop the theory of stellar evolution from a physical standpoint. In this regard, I have followed D.S. Frank-Kamenetski's excellent *Physical Processes in Stellar Interiors* [215]; however, spectacular achievements in the field during the last 25 years have considerably reduced the overlap with this book. An essential part of the items treated here has been considered more qualitatively in Ya.B. Zel'dovich's lectures published in *Physical Grounds of Stellar Structure and Evolution* [107].

The astrophysical results given in Part II have much of the descriptive character typical of astronomy. The reason for this is that the major results are obtained here through numerical experiments which, just as in a real astronomical situation, can only be described rather than reproduced in a book. I have also tried to clarify, whenever possible, the physical sense of the results.

The material in this book is to some extent presented according to my personal preference, as particular attention is often paid to items connected to my scientific interests. Nonetheless, I have tried to preserve a general understanding of the problems discussed and to give results which are basic, as I see them, to the theory of stellar evolution.

I have tried to select from a large variety of papers those reviews and monographs either representing an important advance in solving some astrophysical problem, or dealing with interesting physical problems which are not necessarily important (or are not regarded as such) for the development of the theory of stellar evolution. I have considerably softened the selection rules for my own results. Some problems in the book remain unanswered; I have included them in the hope that some reader will succeed in finding their solution.

The book is concerned only with the evolutionary paths of single stars. The theory of stellar evolution for close binaries, in which there is a considerable increase in the evolutionary paths, is treated in the recently published monograph by A.G. Masevich and A.V. Tutukov *Stellar Evolution: Theory and Observations* [156]. The relationship between theory and observations is also considered in this book.

I gratefully acknowledge the assistance of and useful discussions with S.I. Blinnikov, S.A. Lamzin and A.F. Illarionov, and express my particular thanks to E.V. Bugaev and D.G. Yakovlev, who have read several chapters of the book and made many helpful remarks.

Moscow, September 1989   G.S. Bisnovatyi-Kogan

Contents

Contents of Volume 2

1 Thermodynamic Properties of Matter

1.1 Perfect Gas with Radiation

Most stellar material is characterized by a high temperature and a relatively low density. Under these conditions the kinetic energy of particles is far above their interaction energy, and a non-relativistic, non-degenerate perfect gas model is a realistic approximation. The thermodynamic properties of planetary material, including those of the earth, have not been studied in such detail. Planets are much cooler at the same density and their matter is therefore condensed into the liquid and solid states; this presents serious difficulties for investigation.

Thermodynamic equilibrium exists between matter and radiation in stellar interiors owing to fast particle collision processes and to photon absorption and emission. The gravity is counterbalanced by both the radiation and gas pressure.

Stellar material consists of different chemical elements, mostly hydrogen and helium. In fact more than 98.5% by mass is accounted for by these two elements alone. The rest comprises a mixture of almost all stable isotopes in the periodic table. The abundances of the most plentiful observable elements in the Sun are given in Table 1.1 [5]. Since the difference between the surface temperature and that at the centre is about three or four orders of magnitude, and the respective difference for the density is about 10 orders, the ionization state of the material also changes.

In the central region of stars with masses $M \geq M_\odot$, where $M_\odot$ denotes the mass of the Sun, all atoms are almost totally ionized. Let i be the atomic number of an element which may have different values of ionization state ranging from neutral ($j = 0$) up to fully ionized ($j = i$). The binding energy of the j-times ionized ion of the i-th element is denoted by ε_{ij}, the energy of a totally ionized atom being equal to zero: $\varepsilon_{ii} = 0$. The specific energy E (erg g^{-1}), pressure P (dyn cm^{-2}), and specific entropy S (erg g^{-1} K^{-1}), of this mixture of atoms, ions, and electrons together with radiation are as follows [145]:

$$E = \frac{3}{2}\frac{kT}{\mu m_\mathrm{u}} + \frac{aT^4}{\varrho} - \sum_i \sum_{j=0}^{i} \frac{x_i}{m_i} y_{ij}\varepsilon_{ij}, \qquad (1.1.1)$$

Table 1.1. The abundance of the most plentiful chemical elements in the Sun [5]

Element	Chemical symbol	Atomic number	Atomic mass	Common logarithm of abundance	
				by number of atom	by mass
Hydrogen	H	1	1.0080	12.00	12.00
Helium	He	2	4.0026	10.93	11.53
Carbon	C	6	12.0111	8.52	9.60
Nitrogen	N	7	14.0067	7.96	9.11
Oxygen	O	8	15.9994	8.82	10.02
Neon	Ne	10	20.179	7.92	9.22
Sodium	Na	11	22.9898	6.25	7.61
Magnesium	Mg	12	24.305	7.42	8.81
Aluminium	Al	13	26.9815	6.39	7.78
Silicon	Si	14	28.086	7.52	8.97
Phosphorus	P	15	30.9738	5.52	7.01
Sulphur	S	16	32.06	7.20	8.71
Chlorine	Cl	17	35.453	5.6	7.2
Argon	Ar	18	39.948	6.8	8.4
Calcium	Ca	20	40.08	6.30	7.90
Chromium	Cr	24	51.996	5.85	7.57
Manganese	Mn	25	54.9380	5.40	7.14
Iron	Fe	26	55.847	7.60	9.35
Nickel	Ni	28	58.71	6.30	8.07

Relative content by mass
Hydrogen $X_{\mathrm{H}} = 0.73$
Helium $X_{\mathrm{He}} = 0.25$
Other elements $\sum x_i = 0.017$

Number of nucleons per nucleus,
$\mu_{\mathrm{n}} = 1.26$
Mean atomic mass at total
ionization $\mu_{\mathrm{M}} = 0.60$.

$$P = \frac{\varrho k T}{\mu m_{\mathrm{u}}} + \frac{1}{3} a T^4, \tag{1.1.2}$$

$$S = \frac{k}{\varrho} \sum_i \sum_{j=0}^{i} n_{ij} \left\{ \frac{5}{2} \ln \left[\left(\frac{m_i k T}{2\pi\hbar^2} \right)^{3/2} \frac{g_{ij}}{n_{ij}} \right] \right\}$$

$$+ \frac{k}{\varrho} n_{\mathrm{e}} \left\{ \frac{5}{2} + \ln \left[\left(\frac{m_{\mathrm{e}} k T}{2\pi\hbar^2} \right)^{3/2} \frac{2}{n_{\mathrm{e}}} \right] \right\} + \frac{4}{3} \frac{a T^3}{\varrho}, \tag{1.1.3}$$

where

ϱ is the density, T is the temperature,
$k = 1.38064 \times 10^{-16}$ erg K^{-1} (Boltzmann's constant),
$\hbar = 1.0546 \times 10^{-27}$ erg s (Planck's constant, divided by 2π),
$a = \pi^2 k^4 / 15\hbar^3 c^3 = 7.565 \times 10^{-15}$ erg cm^{-3} K^{-4} is the radiation energy
  density constant,

$c = 2.9979 \times 10^{10}$ cm s^{-1} is the velocity of light in vacuum,

x_i is the mass fraction of the element with atomic number i,

y_{ij} is the fraction of the i-th element ionized to the j-th state, so that
$\sum_{j=0}^{i} y_{ij} = 1$,

$m_i \approx A_i m_u$ is the nuclear mass of the i-th element with atomic mass $A_i \geq 4$,

$m_u = 1.66057 \times 10^{-24}$ g is the atomic mass unit and is equal to 1/12 of the mass of the isotope carbon-12,

$m_e = 9.10953 \times 10^{-28}$ g is the mass of the electron[1],

$n_{ij} = x_i \varrho y_{ij}/m_i$ cm^{-3} is the number density of ions of type i in the j-th state of ionization, $\qquad(1.1.4)$

g_{ij} is the statistical weight of ion of thype i in the j-th state of ionization,

$n_e = \sum_i \sum_{j=1}^{i} j n_{ij}$ cm^{-3} is the electron number density under conditions of overall charge neutrality, $\qquad(1.1.5)$

$\mu = \left[\sum_i \frac{m_u}{m_i} x_i \sum_{j=0}^{i}(1+j) y_{ij} \right]^{-1}$ is the number of nucleons per gas particle (the mean atomic mass). $\qquad(1.1.6)$

For a fully ionized gas consisting of hydrogen, helium, and other elements with $A_i \approx 2i \gg 1$,

$$\mu \simeq \left[2x_H + \frac{3}{4} x_{He} + \frac{1}{2} x_A \right]^{-1}, \qquad x_A = \sum_{i \geq 6} x_i, \qquad (1.1.7)$$

$m_{He} \approx 4 m_u, \qquad m_H \approx m_u.$

The zero energy in (1.1.1) is the rest energy of totally ionized ions and electrons. The ratio of ionization stages in thermodynamic equilibrium is given by the Saha equations [145]:

$$\frac{y_{i,j-1}}{y_{ij}} = n_e \frac{g_{i,j-1}}{2g_{ij}} \left(\frac{2\pi\hbar^2}{m_e kT} \right)^{3/2} e^{I_{ij}/kT} = n_e K(T). \qquad (1.1.8)$$

Here, $I_{ij} = \varepsilon_{i,j-1} - \varepsilon_{ij}$ is the ionizaton energy (potential) of the j-th electron and $I_{i0} = 0$. The ionization energy of the most abundant elements is given in Table 1.2. After substituting (1.1.4) and (1.1.5) into (1.1.8) and solving this set of equations we obtain the degrees of ionization. In the case of one (i-th) atomic species with one ionization state we find the analytical solution

$$n_e = n_{i1} = \frac{\varrho}{m_i} y_{i1}, \qquad y_{i0} = 1 - y_{i1},$$

$$\frac{1 - y_{i1}}{y_{i1}^2} = \frac{\varrho}{m_i} \frac{g_{i0}}{2g_{i1}} \left(\frac{2\pi\hbar^2}{m_e kT} \right)^{3/2} e^{I_{i1}/kT} = F_{\varrho,T},$$

---

[1] $m_p = 1.67265 \times 10^{-24}$ g is the mass of the proton, $m_n = 1.67495 \times 10^{-24}$ g is the mass of the neutron, $m_D/m_u = 2.01410$, $m_{3H}/m_u = 3.01605$, $m_{3He}/m_u = 3.01603$ are the relative masses of deuterium, tritium and helium-3. Constants are taken from [5, 180].

Table 1.2. Ionization potentials and total momenta of the outer electronic shells of the most abundent elements [180]

Atomic number	Element symbol	Ionization potentials, eV	Total momenta
1	H⁻, H	0.747; 13.5985	0; 1/2
2	He	24.5876; 54.418	0; 1/2; 0
6	C	11.260; 24.284; 47.89; 64.49	0; 1/2; 0; 1/2
7	N	14.534; 29.602; 47.45; 77.47	3/2; 0; 1/2; 0
8	O	13.618; 35.118; 54.94; 77.41	2; 3/2; 0; 1/2
10	Ne	21.565; 40.964; 63.46; 97.12	0; 3/2; 2; 3/2
11	Na	5.1391; 47.287; 71.64; 98.92	1/2; 0; 3/2; 2
12	Mg	7.646; 15.035; 80.15; 109.2	0; 1/2; 0; 3/2
13	Al	5.9858; 18.828; 28.448; 120	1/2; 0; 1/2; 0
14	Si	8.152; 16.346; 33.493; 45.14	0; 1/2; 0; 1/2
15	P	10.49; 19.73; 30.18; 51.47	3/2; 0; 1/2; 0
16	S	10.36; 23.33; 34.83; 47.31	2; 3/2; 0; 1/2
17	Cl	12.968; 23.81; 39.61; 53.47	3/2; 2; 3/2; 0
18	Ar	15.760; 27.63; 40.74; 59.81	0; 3/2; 2; 3/2
20	Ca	6.113; 11.872; 50.91; 67.10	0; 1/2; 0; 3/2
24	Cr	6.766; 16.50; 30.96; 49	3; 5/2; 0; 3/2
25	Mn	7.4368; 15.640; 33.67; 51.2	5/2; 2; 5/2; 0
26	Fe	7.87; 16.18; 30.65; 54.8	4; 9/2; 4; 5/2
28	Ni	7.63; 18.17; 35.2; 54.9	4; 5/2; 4; 9/2

1 eV = 11604 K $\qquad$ $X_0; X_+; X_{++}; X_{+++}$ $\qquad$ $X_0; X_+; X_{++}; X_{+++}$

and so

$$y_{i1} = \left(\frac{1}{4F_{\varrho,T}^2} + \frac{1}{F_{\varrho,T}} \right)^{1/2} - \frac{1}{2F_{\varrho,T}}. \qquad (1.1.9)$$

The study of stellar evolution often requires knowledge of the adiabatic indices

$$\gamma_1 = \left(\frac{\partial \ln P}{\partial \ln \varrho} \right)_S, \quad \gamma_2 = \left(\frac{\partial \ln T}{\partial \ln P} \right)_S, \quad \gamma_3 = \left(\frac{\partial \ln T}{\partial \ln \varrho} \right)_S,$$

and heat capacities

$$c_v = T \left(\frac{\partial S}{\partial T} \right)_\varrho, \quad c_p = T \left(\frac{\partial S}{\partial T} \right)_P.$$

If the ionization is not total, all these magnitudes should be computed numerically. For such a computation it is convenient to rewrite them using partial derivatives:

$$\left(\frac{\partial \ln P}{\partial \ln \varrho} \right)_T, \quad \left(\frac{\partial \ln P}{\partial \ln T} \right)_\varrho, \quad \left(\frac{\partial S}{\partial \ln \varrho} \right)_T, \quad \left(\frac{\partial S}{\partial \ln T} \right)_\varrho = c_v.$$

Next we use the well known properties of Jacobians

$$\frac{\partial(u,v)}{\partial(x,y)} = \frac{\partial(u,v)}{\partial(t,s)}\frac{\partial(t,s)}{\partial(x,y)} \quad ; \quad \frac{\partial(u,v)}{\partial(x,v)} = \left(\frac{\partial u}{\partial x}\right)_v , \tag{1.1.10}$$

to obtain

$$\gamma_1 = \left(\frac{\partial \ln P}{\partial \ln \varrho}\right)_T - \left(\frac{\partial \ln P}{\partial \ln T}\right)_\varrho \left(\frac{\partial S}{\partial \ln \varrho}\right)_T \Big/ \left(\frac{\partial S}{\partial \ln T}\right)_\varrho , \tag{1.1.11}$$

$$\gamma_2 = \left[\left(\frac{\partial \ln P}{\partial \ln T}\right)_\varrho - \left(\frac{\partial \ln P}{\partial \ln \varrho}\right)_T \left(\frac{\partial S}{\partial \ln T}\right)_\varrho \Big/ \left(\frac{\partial S}{\partial \ln \varrho}\right)_T\right]^{-1} , \tag{1.1.12}$$

$$\gamma_3 = -\left(\frac{\partial S}{\partial \ln \varrho}\right)_T \Big/ \left(\frac{\partial S}{\partial \ln T}\right)_\varrho , \tag{1.1.13}$$

$$c_v = (\partial S/\partial \ln T)_\varrho , \tag{1.1.14}$$

$$c_p = c_v - \left(\frac{\partial S}{\partial \ln \varrho}\right)_T \left(\frac{\partial \ln P}{\partial \ln T}\right)_\varrho \Big/ \left(\frac{\partial \ln P}{\partial \ln \varrho}\right)_T , \tag{1.1.15}$$

$$c_p/c_v = \gamma_1 \Big/ \left(\frac{\partial \ln P}{\partial \ln \varrho}\right)_T . \tag{1.1.16}$$

The first law of thermodynamics, and the fact that the differential of the free energy $F = E - TS$ is a total derivative, allows us to express derivatives of the entropy in terms of derivatives of the energy and pressure:

$$\left(\frac{\partial S}{\partial \ln T}\right)_\varrho = \frac{1}{T}\left(\frac{\partial E}{\partial \ln T}\right)_\varrho , \quad \left(\frac{\partial S}{\partial \ln \varrho}\right)_T = -\frac{1}{\varrho T}\left(\frac{\partial P}{\partial \ln T}\right)_\varrho . \tag{1.1.17}$$

For a fixed degree of ionization y_{ij}, (1.1.3–6) give

$$S = \frac{k}{\mu m_u}\ln(T^{3/2}/\varrho) + \frac{4}{3}\frac{aT^3}{\varrho} + \text{const.} \tag{1.1.18}$$

and all the derivatives can then be obtained analytically:

$$\left(\frac{\partial \ln P}{\partial \ln \varrho}\right)_T = \beta_g, \quad \left(\frac{\partial \ln P}{\partial \ln T}\right)_\varrho = 4 - 3\beta_g,$$

$$\left(\frac{\partial S}{\partial \ln \varrho}\right)_T = -\frac{P}{\varrho T}(4 - 3\beta_g), \quad \left(\frac{\partial S}{\partial \ln T}\right)_\varrho = \frac{3}{2}\frac{P}{\varrho T}(8 - 7\beta_g), \tag{1.1.19}$$

where $\beta_g = P_g/P$ is the ratio of the gas pressure to the total pressure. The expressions for the adiabatic indices and heat capacities of a monatomic gas with $\mu = $ const. and radiation take the form [218]

$$\gamma_1 = \beta_g + \frac{2}{3}\frac{(4-3\beta_g)^2}{8-7\beta_g}, \quad \gamma_2 = \left[4 - 3\beta_g + \frac{3}{2}\beta_g\frac{8-7\beta_g}{4-3\beta_g}\right]^{-1},$$

$$\gamma_3 = \frac{2}{3}\frac{4-3\beta_g}{8-7\beta_g},$$

$$c_v = \frac{3}{2}\frac{P}{\varrho T}(8-7\beta_g), \tag{1.1.20}$$

$$c_p = \frac{3}{2}\frac{P}{\varrho T}(8-7\beta_g)\left[1 + \frac{2}{3}\frac{(4-3\beta_g)^2}{\beta_g(8-7\beta_g)}\right],$$

$$\frac{c_p}{c_v} = 1 + \frac{2}{3}\frac{(4-3\beta_g)^2}{\beta_g(8-7\beta_g)} = \gamma_1/\beta_g.$$

Equations (1.1.18–20) are widely used in descriptions of stellar matter, since the major part of a star's mass is fully ionized, so that $\mu = $ const. The ionization of material in stellar shells at lower temperatures is not total and $\mu = \mu(\varrho, T)$.

Problem. Derive the equation for the concentration of electrons in a plasma consisting of $H^0, H^+, H^-, He^0, He^+, He^{++}$ ions and atoms and singly charged ions of k other elements.

Solution. From the Saha equation (1.1.8) and Table 1.2 we have for the hydrogen

$$y_{H^0} = \frac{4}{n_e}\left(\frac{m_e kT}{2\pi\hbar^2}\right)^{3/2}e^{-\frac{0.747}{T_e}}y_{H^-} \equiv \frac{y_{H^-}}{n_e}Q_{H^0}, \quad g_{H^0} = 4,$$

$$y_{H^+} = \frac{1}{n_e}\left(\frac{m_e kT}{2\pi\hbar^2}\right)^{3/2}e^{-\frac{13.6}{T_e}}y_{H^0} \equiv \frac{y_{H^0}}{n_e}Q_{H^+} = \frac{y_{H^-}}{n_e^2}Q_{H^0}Q_{H^+}. \tag{1}$$

Using the relation $y_{H^-} + y_{H^0} + y_{H^+} = 1$ then gives

$$y_{H^-} = n_e^2\left(n_e^2 + n_e Q_{H^0} + Q_{H^0}Q_{H^+}\right)^{-1}. \tag{2}$$

Similarly, for the helium

$$y_{He^0} = n_e^2\left(n_e^2 + n_e Q_{He^+} + Q_{He^+}Q_{He^{++}}\right)^{-1},$$

$$y_{He^+} = \frac{y_{He^0}}{n_e}Q_{He^+}, \quad y_{He^{++}} = \frac{y_{He^0}}{n_e^2}Q_{He^{++}}Q_{He^+}, \tag{3}$$

where

$$Q_{He^+} = 4\left(\frac{m_e kT}{2\pi\hbar^2}\right)^{3/2}e^{-\frac{24.6}{T_e}}, \quad Q_{He^{++}} = \left(\frac{m_e kT}{2\pi\hbar^2}\right)^{3/2}e^{-\frac{54.4}{T_e}}, \tag{4}$$

and for heavy elements

$$y_{j+} = \frac{Q_{j+}}{n_e + Q_{j+}}, \quad Q_{j+} = 2\frac{g_{j+}}{g_{j0}}\left(\frac{m_e kT}{2\pi\hbar^2}\right)^{3/2} e^{-\frac{I_{j1}}{kT}}. \tag{5}$$

Here T_e is the temperature written in electron-volts. Applying (1.1.4) to all elements and using (1.1.5) to determine the electron concentration under conditions of electroneutrality,

$$n_{H^+} + n_{He^+} + 2n_{He^{++}} + \sum_{j=1}^{k} n_{j+} = n_e + n_{H^-},$$

we obtain the equation for the normalized electron concentration x_e

$$x_e = \frac{m_u}{\varrho} n_e$$

$$= x_H \frac{q_{H^0} q_{H^+} - x_e^2}{x_e^2 + x_e q_{H^0} + q_{H^0} q_{H^+}} + \frac{x_{He}}{4} \frac{x_e q_{He^+} + 2q_{He^+} q_{He^{++}}}{x_e^2 + x_e q_{He^+} + q_{He^+} q_{He^{++}}}$$

$$+ \sum_{j=0}^{k} x_j \frac{m_u}{m_j} \frac{q_{j+}}{x_e + q_{j+}}. \tag{6}$$

Here, $q_i = m_u Q_i / \varrho$. All values in (6) are dimensionless and close to unity, which might be useful for numerical calculations.

With the degree of ionization versus ϱ and T in hand we can find the thermodynamic functions and their derivatives. An example of such a calculation is shown in Fig. 1.1, where $\gamma_3(\varrho, T)$ is plotted as a function of temperature for the mixture characterized by $x_H = 0.75, x_{He} = 0.22$ and solar ratios for

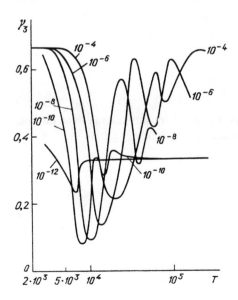

Fig. 1.1. γ_3 as a function of T at $\rho = $ const. indicated at corresponding curves, for the normal composition (Table 1.1) in the region of hydrogen and helium ionization.

the other elements (see Table 1.1). The two minima on the $\gamma_3 |_\varrho (T)$ curves correspond to the ionization of hydrogen and to single ionization of helium. At low densities $\varrho = 10^{-12}$ g cm^{-3} the second minimum lies in the region of radiation pressure dominance and is therefore unimportant.

1.2 Degenerate Relativistic Gas

During the late stages of stellar evolution or in supernova explosions the kinetic energy of electrons in the central regions of stars may approach their rest energy, so their velocity may become close to that of light:

$$kT \sim m_e c^2, \quad \langle v_e \rangle \sim c. \tag{1.2.1}$$

The calculation of thermodynamic functions therefore requires the full relativistic expression for the electron energy and momentum. On the other hand, the density may also be so high that the mean phase space number density is close to unity. It is therefore necessary to take into account Pauli's exclusion principle for electrons (spin = 1/2), which states that their number in a phase space cell must equal either unity or zero. The mean number of electrons with energy ε per cell is defined by the Fermi function

$$f_e = \left[1 + \exp\left(\frac{\varepsilon - \mu_{te}}{kT} \right) \right]^{-1}, \tag{1.2.2}$$

where μ_{te} is the electron chemical potential,

$$\varepsilon = \left(m_e^2 c^4 + p^2 c^2 \right)^{1/2}, \tag{1.2.3a}$$

and

$$p = \frac{m_e v}{\sqrt{1 - \frac{v^2}{c^2}}} \tag{1.2.3b}$$

is the electron momentum.

The thermodynamic functions are obtained by integrating over momentum space, taking into account the statistical weight $g_e = 2$ [145]:

$$n_e = 2 \frac{4\pi}{(2\pi\hbar)^3} \int_0^\infty f_e p^2 dp, \tag{1.2.4}$$

$$E_e = \frac{2}{\varrho} \frac{4\pi}{(2\pi\hbar)^3} \int_0^\infty f_e \varepsilon p^2 dp, \tag{1.2.5}$$

$$P_e = 2 \frac{4\pi}{(2\pi\hbar)^3} \frac{1}{3} \int_0^\infty f_e p v \, p^2 dp, \quad v = \frac{pc}{\sqrt{p^2 + m_e^2 c^2}}, \tag{1.2.6}$$

$$S_e = -\frac{2}{\varrho}\frac{4\pi}{(2\pi\hbar)^3}k\int_0^\infty [f_e \ln f_e + (1 - f_e)\ln(1 - f_e)]\, p^2 dp. \qquad (1.2.7)$$

Transforming the integrals and introducing dimensionless variables

$$x = \frac{pc}{kT}, \quad \alpha = \frac{m_e c^2}{kT} = \frac{5.93013 \times 10^9 K}{T}, \quad \beta = \frac{\mu_{te}}{kT} \qquad (1.2.8)$$

gives

$$n_e = \frac{1}{\pi^2}\left(\frac{kT}{c\hbar}\right)^3 I_{n^-}, \quad E_e = \frac{1}{\pi^2 \varrho}\left(\frac{kT}{c\hbar}\right)^3 kT I_{E^-},$$

$$P_e = \frac{1}{3\pi^2}\left(\frac{kT}{c\hbar}\right)^3 kT I_{P^-}, \qquad (1.2.9)$$

$$S_e = \frac{k}{\pi^2 \varrho}\left(\frac{kT}{c\hbar}\right)^3\left(I_{E^-} - \beta I_{n^-} + \frac{1}{3}I_{P^-}\right),$$

where

$$I_{n^-} = \int_0^\infty \frac{x^2 dx}{1 + \exp(\sqrt{x^2 + \alpha^2} - \beta)},$$

$$I_{P^-} = \int_0^\infty \frac{x^4 dx}{\sqrt{x^2 + \alpha^2}\left[1 + \exp(\sqrt{x^2 + \alpha^2} - \beta)\right]}, \qquad (1.2.10)$$

$$I_{E^-} = \int_0^\infty \frac{\sqrt{x^2 + \alpha^2}\, x^2 dx}{1 + \exp(\sqrt{x^2 + \alpha^2} - \beta)}.$$

When $kT \geq 0.1 m_e c^2$, in thermodynamic equilibrium positrons must also be taken into account. Annihilation of an electron–positron pair $e^- e^+$ results in the creation of photons with zero chemical potential in an equilibrium state, $\mu_{ph} = 0$. The equilibrium annihilation condition $\mu_{te} + \mu_{te+} = \mu_{ph}$ enables us to write:

$$\mu_{te+} = -\mu_{te}. \qquad (1.2.11)$$

Thermodynamic functions for positrons can be derived from (1.2.9), where $-\beta$ should be substituted for β and integrals $I_{i+}, i = n, E, P$ obtained by substitution of $-\beta$ for β in (1.2.10) should be used. Nucleons and nuclei may be considered non-degenerate and non-relativistic, so we can write, including the radiation,

$$E_{N,r} = \frac{3}{2}\frac{kT}{\mu_N m_u} + \frac{aT^4}{\varrho}, \qquad (1.2.12)$$

$$P_{N,r} = \frac{\varrho kT}{\mu_N m_u} + \frac{aT^4}{3}, \qquad (1.2.13)$$

$$S_{N,r} = \frac{k}{\varrho}\sum_i n_i\left\{\frac{5}{2} + \ln\left[\left(\frac{m_i kT}{2\pi\hbar^2}\right)^{3/2}\frac{g_i}{n_i}\right]\right\} + \frac{4}{3}\frac{aT^3}{\varrho}. \qquad (1.2.14)$$

Only totally ionized matter is considered here. If nuclear reactions are absent and the mass fractions of elements are constant (x_i = const.), we have, similarly to (1.1.18)

$$S_N = \frac{k}{\mu_N m_u} \ln \left(\frac{T^{3/2}}{\varrho} \right) + \text{const.} \tag{1.2.15}$$

In (1.2.12–15) the mean number of nucleons per nucleus is denoted

$$\mu_N = \left(\sum_i \frac{x_i}{A_i} \right)^{-1}. \tag{1.2.16}$$

In order to obtain the full thermodynamic functions P, E and S, it is necessary to sum up the corresponding expressions for electrons, positrons, nuclei and radiation. The nuclear charge is connected with an excess of electrons over positrons. We have

$$\frac{\varrho}{\mu_Z m_u} = n_{e^-} - n_{e^+}, \qquad \mu_Z = \left(\sum_i \frac{Z_i x_i}{A_i} \right)^{-1}, \tag{1.2.17}$$

where μ_Z is the number of nucleons per electron. Combining (1.2.17) with (1.2.9) and (1.2.10) gives the $\mu_{te}(\varrho, T)$ dependence. In the case of total ionization $y_{ii} = 1$, $i = Z$, (1.1.6), (1.2.16) and (1.2.17) yield

$$\frac{1}{\mu} = \frac{1}{\mu_N} + \frac{1}{\mu_Z}. \tag{1.2.18}$$

The rest energy of nuclei is constant in the absence of nuclear transformations and is therefore adopted as the nominal zero energy level in this section.

We next consider the limiting cases of (1.2.9).

1.2.1 Strong Degeneracy

At zero temperature, electrons fill phase space up to the Fermi momentum p_{Fe}. Taking account of the statistical weight factor, we find that the density of electrons is twice the number of cells in a phase space sphere of radius p_{Fe}:

$$n_e = \frac{2}{(2\pi\hbar)^3} \frac{4\pi}{3} p_{Fe}^3 = \frac{p_{Fe}^3}{3\pi^2\hbar^3}. \tag{1.2.19}$$

Combining the result with (1.2.17) then gives, in the absence of positrons,

$$p_{Fe} = \left(\frac{3\pi^2 \varrho}{\mu_Z m_u} \right)^{1/3} \hbar = \left(\frac{1.027 \varrho}{10^6 \mu_Z} \right)^{1/3} m_e c. \tag{1.2.20}$$

The kinetic energy of an electron on the surface of the phase space sphere is called the Fermi energy:

$$\varepsilon_{\text{Fe}} = (m_e^2 c^4 + p_{\text{Fe}}^2 c^2)^{1/2} - m_e c^2 = m_e c^2 (\sqrt{y^2 + 1} - 1),$$

$$y = \frac{p_{\text{Fe}}}{m_e c},$$

$$\varrho = \frac{m_e^3 c^3 \mu_Z m_u}{3\pi^2 \hbar^3} y^3 = 9.740 \times 10^5 \mu_Z \left[\left(\frac{\varepsilon_{\text{Fe}}}{m_e c^2} + 1 \right)^2 - 1 \right]^{3/2}. \tag{1.2.21}$$

Recalling that $f_e = 1$ for $p < p_{\text{Fe}}$ and $f_e = 0$ for $p > p_{\text{Fe}}$, we find from (1.2.5) and (1.2.6) that

$$E_e = \frac{2}{\varrho} \frac{4\pi}{(2\pi\hbar)^3} \int_0^{p_{\text{Fe}}} (p^2 c^2 + m_e^2 c^4)^{1/2} p^2 dp$$

$$= \frac{m_e^4 c^5}{24\pi^2 \hbar^3 \varrho} g(y) = \frac{6.002 \times 10^{22}}{\varrho} g(y), \tag{1.2.22}$$

$$P_e = \frac{2}{3} \frac{4\pi c}{(2\pi\hbar)^3} \int_0^{p_{\text{Fe}}} (p^2 + m_e^2 c^2)^{-1/2} p^4 dp = \frac{m_e^4 c^5}{24\pi^2 \hbar^3} f(y),$$

where

$$f(y) = y(2y^2 - 3)\sqrt{y^2 + 1} + 3 \sinh^{-1} y,$$

$$g(y) = 3y(2y^2 + 1)\sqrt{y^2 + 1} - 3 \sinh^{-1} y, \tag{1.2.23}$$

$$g(y) + f(y) = 8y^3 \sqrt{y^2 + 1}.$$

The temperature correction, if the system is strongly degenerate, can be found from expansion of the general formulae with the aid of the relation [145]

$$\int_0^\infty \frac{\phi(u) du}{e^{u - u_0} + 1} = \int_0^{u_0} \phi(u) du + \frac{\pi^2}{6} \phi'(u_0) + \frac{7\pi^4}{360} \phi'''(u_0) + \dots, \tag{1.2.24}$$

which holds for $e^{-u_0} \ll 1$. Writing $u = \sqrt{x^2 + \alpha^2} - \alpha$, $u_0 = \beta - \alpha$, and neglecting the contribution of positrons $\sim e^{-u_0}$, we obtain [166] from (1.2.9) and (1.2.10)

$$n_e = \frac{1}{3\pi^2} \left(\frac{m_e c}{\hbar} \right)^3 \left[y_1^3 + \frac{\pi^2}{\alpha^2 y_1} \left(y_1^2 + \frac{1}{2} \right) + \frac{7\pi^4}{40\alpha^4 y_1^5} + \dots \right], \tag{1.2.25}$$

$$E_e = \frac{m_e^4 c^5}{24\pi^2 \hbar^3 \varrho} \left[g(y_1) + \frac{4\pi^2}{\alpha^2 y_1} (3y_1^2 + 1)\sqrt{y_1^2 + 1} \right.$$

$$\left. + \frac{7\pi^4}{5\alpha^4 y_1^5} (2y_1^4 - y_1^2 + 1)\sqrt{y_1^2 + 1} + \dots \right], \tag{1.2.26}$$

$$P_e = \frac{m_e^4 c^5}{24\pi^2 \hbar^3} \left[f(y_1) + \frac{4\pi^2}{\alpha^2} y_1 \sqrt{y_1^2 + 1} \right.$$

$$\left. + \frac{7\pi^4}{15\alpha^4 y_1^3} (2y_1^2 - 1)\sqrt{y_1^2 + 1} + \dots \right], \qquad (1.2.27)$$

$$S_e = \frac{m_e^2 c}{3\hbar^3 \varrho} k^2 T \left[y_1 \sqrt{y_1^2 + 1} \right.$$

$$\left. + \frac{7\pi^2}{15\alpha^2 y_1^3} \left(y_1^2 - \frac{1}{2} \right) \sqrt{y_1^2 + 1} + \dots \right]. \qquad (1.2.28)$$

Here $y_1 = \sqrt{\beta^2 - \alpha^2}/\alpha$, the expansion parameter $\alpha y_1 \gg 1$, and upon transforming integrals (1.2.10) into (1.2.24) the functions $\phi(u)$ take the form $\phi_n = (u + \alpha)\sqrt{u^2 + 2u\alpha}, \phi_E = (u + \alpha)^2 \sqrt{u^2 + 2u\alpha}, \phi_P = (u^2 + 2u\alpha)^{3/2}$. To find the explicit form of the E_e, P_e and S_e dependence on ϱ and T, retaining only the terms $\sim \alpha^{-2}$, we use y, found from (1.2.20) and (1.2.21) together with (1.2.25) and obtain the relations between y, y_1 and α:

$$y = \left(\frac{3\pi^2 \varrho}{\mu z m_u} \right)^{1/3} \frac{\hbar}{m_e c}, \quad y^3 = y_1^3 + \frac{\pi^2}{\alpha^2 y_1} \left(y_1^2 + \frac{1}{2} \right). \qquad (1.2.29)$$

For small values of $(y^3 - y_1^3)$ we may write

$$y_1^3 = y^3 - \frac{\pi^2}{\alpha^2 y} \left(y^2 + \frac{1}{2} \right).$$

Substituting $y_1(y)$ into (1.2.23, 25–28) gives

$$g(y_1) = g(y) - \frac{8\pi^2}{\alpha^2 y} \left(y^2 + \frac{1}{2} \right) \sqrt{y^2 + 1},$$

$$f(y_1) = f(y) - \frac{8\pi^2}{3\alpha^2} \left(y^2 + \frac{1}{2} \right) \frac{y}{\sqrt{y^2 + 1}},$$

and the explicit expressions for the thermodynamic functions

$$E_e = \frac{m_e^4 c^5}{24\pi^2 \hbar^3 \varrho} \left[g(y) + \frac{4\pi^2}{\alpha^2} y \sqrt{y^2 + 1} \right],$$

$$P_e = \frac{m_e^4 c^5}{24\pi^2 \hbar^3} \left[f(y) + \frac{4\pi^2}{3\alpha^2} y \frac{y^2 + 2}{\sqrt{y^2 + 1}} \right], \qquad (1.2.30)$$

$$S_e = \frac{m_e^2 c}{3\hbar^3 \varrho} k^2 T y \sqrt{y^2 + 1}.$$

In the limiting cases functions f and g take the form

$$f(y) \approx \frac{8}{5}y^5 \left(1 - \frac{5}{14}y^2\right),$$

$$g(y) \approx 8y^3 + \frac{12}{5}y^5 \left(1 - \frac{5}{28}y^2\right) \quad \text{for } y \ll 1,$$

(1.2.31)

and

$$f(y) \approx 2y^4 \left(1 - \frac{1}{y^2}\right),$$

$$g(y) \approx 6y^4 \left(1 + \frac{1}{y^2}\right) \quad \text{for } y \gg 1.$$

Using (1.2.31) we obtain from (1.2.30) in the non-relativistic limit $y \ll 1$

$$E_e = \frac{m_e c^2}{\mu_Z m_u} + \frac{m_e^4 c^5}{10\pi^2 \hbar^3 \varrho} y^5 \left(1 - \frac{5}{28}y^2 + \frac{5\pi^2}{3\alpha^2 y^4}\right),$$

$$P_e = \frac{m_e^4 c^5}{15\pi^2 \hbar^3} y^5 \left(1 - \frac{5}{14}y^2 + \frac{5\pi^2}{3\alpha^2 y^4}\right),$$

(1.2.32)

and in the ultrarelativistic limit $y \gg 1$

$$E_e = \frac{m_e^4 c^5}{4\pi^2 \hbar^3 \varrho} y^4 \left(1 + \frac{1}{y^2} + \frac{2\pi^2}{3\alpha^2 y^2}\right),$$

$$P_e = \frac{m_e^4 c^5}{12\pi^2 \hbar^3} y^4 \left(1 - \frac{1}{y^2} + \frac{2\pi^2}{3\alpha^2 y^2}\right),$$

(1.2.33)

$$S_e = \frac{m_e^2 c}{3\hbar^3 \varrho} k^2 T y^2.$$

1.2.2 Very Low Matter Density

The matter density may become so low that the concentration of pairs will exceed the initial concentration of electrons. In this case $\beta \ll 1$ may be used as a small parameter; if $\beta = 0$, then $n_e = n_{e+}$. Expanding (1.2.10) in powers of β and integrating by parts gives

$$I_{n\mp} = I_2 \pm \beta I_1, \quad I_{P\mp} = I_3 \pm 3\beta I_2 + \frac{3}{2}\beta^2 I_1,$$

$$I_{E\mp} = I_4 \pm \beta(\alpha^2 I_0 + 3I_2) + \frac{\beta^2}{2} I_5,$$

(1.2.34)

where

$$I_0(\alpha) = \int_0^\infty \frac{dx}{1 + \exp\sqrt{x^2 + \alpha^2}},$$

$$I_1(\alpha) = \int_0^\infty \frac{(2x^2 + \alpha^2)dx}{\sqrt{x^2 + \alpha^2}(1 + \exp\sqrt{x^2 + \alpha^2})},$$

$$I_2(\alpha) = \int_0^\infty \frac{x^2 dx}{1 + \exp\sqrt{x^2 + \alpha^2}},$$

$$I_3(\alpha) = \int_0^\infty \frac{x^4 dx}{\sqrt{x^2 + \alpha^2}(1 + \exp\sqrt{x^2 + \alpha^2})}, \qquad (1.2.35)$$

$$I_4(\alpha) = \int_0^\infty \frac{x^2\sqrt{x^2 + \alpha^2}dx}{1 + \exp\sqrt{x^2 + \alpha^2}},$$

$$I_5(\alpha) = \int_0^\infty \frac{(3x^2 + \alpha^2)\exp\sqrt{x^2 + \alpha^2}}{(1 + \exp\sqrt{x^2 + \alpha^2})^2} dx.$$

For $\alpha = 0$, we may rewrite [145] the integrals (1.2.35) in terms of Riemann's Γ function and ζ function using the relation

$$F_\nu(0) = \int_0^\infty \frac{x^{\nu-1}dx}{1 + e^x} = (1 - 2^{1-\nu})\Gamma(\nu)\zeta(\nu), \quad \nu > 0 \qquad (1.2.36)$$

and the formula

$$\int_0^\infty \frac{dx}{1 + e^x} = \ln 2.$$

Taking from [145] the values of $\zeta(n)$ for integer $\nu = n$ and substituting $\Gamma(n) = (n-1)!$ gives

$$I_0(0) = \ln 2, \qquad I_1(0) = \frac{\pi^2}{6}, \qquad I_2(0) = \frac{3}{2}\zeta(3) = 1.80308;$$

$$I_3(0) = I_4(0) = \frac{7\pi^4}{120}, \qquad I_5(0) = 3I_1(0) = \frac{\pi^2}{2}. \qquad (1.2.37)$$

Using (1.2.34–37), the definition of y (1.2.29) and recalling (1.2.9) and (1.2.17), we can rewrite the thermodynamic functions as

$$\beta = \frac{y^3\alpha^3}{\pi^2 A_2}, \qquad n_{e\mp} = \frac{15}{\pi^4}I_2(0)\frac{T^3}{k}A_1\left(1 \pm \frac{1}{6I_2(0)}\frac{y^3\alpha^3}{A_1}\right),$$

$$E_{e-} + E_{e+} = \frac{7}{4}\frac{T^4}{\varrho}B_0\left(1 + \frac{30}{7\pi^6}\frac{B_2}{B_0 A_2^2}y^6\alpha^6\right),$$

$$P_{e-} + P_{e+} = \frac{7}{12}T^4 A_0\left(1 + \frac{30}{7\pi^6}\frac{y^6\alpha^6}{A_0 A_2}\right), \qquad (1.2.38)$$

$$S_{e-} + S_{e+} = \frac{7}{3}\frac{T^3}{\varrho}\frac{3B_0 + A_0}{4}\left(1 + \frac{15}{7\pi^6}\frac{6B_2 - 2A_2}{3B_0 + A_0}\frac{y^6\alpha^6}{A_2^2}\right),$$

where the new functions

$$A_0(\alpha) = \frac{120}{7\pi^4} I_3(\alpha), \quad A_2(\alpha) = \frac{6}{\pi^2} I_1(\alpha),$$

$$B_0(\alpha) = \frac{120}{7\pi^4} I_4(\alpha), \quad B_2(\alpha) = \frac{2}{\pi^2} I_5(\alpha), \tag{1.2.39}$$

$$A_1(\alpha) = \frac{I_2(\alpha)}{I_2(0)}.$$

have been introduced. In the case of ultrarelativistic pairs, for $\alpha \ll 1$, these functions can be represented asymptotically [166]

$$A_0 = 1 - \frac{15}{7\pi^2} \alpha^2, \quad A_2 = 1 - \frac{3}{2\pi^2} \alpha^2,$$

$$B_0 = 1 - \frac{5}{7\pi^2} \alpha^2, \quad B_2 = 1 - \frac{1}{2\pi^2} \alpha^2, \tag{1.2.40}$$

$$A_1 = 1 - \frac{\ln 2}{2 I_2(0)} \alpha^2.$$

The thermodynamic functions in a gas of low density, when the pairs are nearly ultrarelativistic, are obtained from (1.2.38–40):

$$n_{e\mp} = \frac{15}{\pi^4} I_2(0) \frac{aT^3}{k} \left(1 - \frac{\ln 2}{2 I_2(0)} \alpha^2 \pm \frac{1}{6 I_2(0)} y^3 \alpha^3 \right),$$

$$E_{e^-} + E_{e^+} = \frac{7}{4} \frac{aT^4}{\varrho} \left(1 - \frac{5}{7\pi^4} \alpha^2 + \frac{30}{7\pi^6} y^6 \alpha^6 \right),$$

$$P_{e^-} + P_{e^+} = \frac{7}{12} aT^4 \left(1 - \frac{15}{7\pi^4} \alpha^2 + \frac{30}{7\pi^6} y^6 \alpha^6 \right), \tag{1.2.41}$$

$$S_{e^-} + S_{e^+} = \frac{7}{3} \frac{aT^3}{\varrho} \left(1 - \frac{15}{14\pi^4} \alpha^2 + \frac{15}{7\pi^6} y^6 \alpha^6 \right).$$

In the non-relativistic limit, for $\alpha \gg 1$, retaining two terms in the expansion of the denominator in (1.2.35) gives [93]

$$A_0 = \frac{360}{7\pi^4} \alpha^2 \left[K_2(\alpha) - \frac{1}{4} K_2(2\alpha) \right],$$

$$B_0 = \frac{120}{7\pi^4} \alpha^2 \left[\alpha K_1(\alpha) + 3 K_2(\alpha) - \frac{\alpha}{2} K_1(2\alpha) - \frac{3}{4} K_2(2\alpha) \right],$$

$$A_1 = \frac{\alpha^2}{I_2(0)} \left[K_2(\alpha) - \frac{1}{2} K_2(2\alpha) \right], \quad A_2 = \frac{6\alpha^2}{\pi^4} \left[K_2(\alpha) - K_2(2\alpha) \right], \tag{1.2.42}$$

$$B_2 = \frac{2\alpha^2}{\pi^2} \left[\alpha K_1(\alpha) + 3 K_2(\alpha) - 2\alpha K_1(2\alpha) - 3 K_2(2\alpha) \right],$$

Table 1.3. Values of the functions $A_i(\alpha), B_i(\alpha)$ for $0 \leq \alpha \leq 10$

α	A_0	A_1	A_2	B_0	B_1	B_2
0.00	1.0000	1.0000	1.0000	1.0000	1.0000	1.0000
0.50	9.4989(-1)	9.5476(-1)	9.6299(-1)	9.8119(-1)	9.8342(-1)	9.8702(-1)
1.00	8.2749(-1)	8.4020(-1)	8.6278(-1)	9.2303(-1)	9.3130(-1)	9.4529(-1)
1.50	6.7622(-1)	6.9345(-1)	7.2532(-1)	8.3028(-1)	8.4519(-1)	8.7168(-1)
2.00	5.2709(-1)	5.4480(-1)	5.7846(-1)	7.1580(-1)	7.3497(-1)	7.7039(-1)
2.50	3.9653(-1)	4.1217(-1)	4.4246(-1)	5.9438(-1)	6.1464(-1)	6.5311(-1)
3.00	2.9030(-1)	3.0290(-1)	3.2762(-1)	4.7800(-1)	4.9689(-1)	5.3345(-1)
3.50	2.0806(-1)	2.1764(-1)	2.3656(-1)	3.7418(-1)	3.9040(-1)	4.2216(-1)
4.00	1.4661(-1)	1.5360(-1)	1.6748(-1)	2.8635(-1)	2.9949(-1)	3.2544(-1)
4.50	1.0189(-1)	1.0685(-1)	1.1675(-1)	2.1497(-1)	2.2520(-1)	2.4549(-1)
5.00	7.0003(-2)	7.3461(-2)	8.0361(-2)	1.5877(-1)	1.6650(-1)	1.8188(-1)
5.50	4.7634(-2)	5.0006(-2)	5.4746(-2)	1.1563(-1)	1.2133(-1)	1.3271(-1)
6.00	3.2147(-2)	3.3756(-2)	3.6973(-2)	8.3190(-2)	8.7329(-2)	9.5597(-2)
7.00	1.4345(-2)	1.5066(-2)	1.6510(-2)	4.1752(-2)	4.3848(-2)	4.8039(-2)
8.00	6.2613(-3)	6.5769(-3)	7.2085(-3)	2.0259(-2)	2.1280(-2)	2.3321(-2)
9.00	2.6856(-3)	2.8211(-3)	3.0922(-3)	9.5667(-3)	1.0049(-2)	1.1014(-2)
10.0	1.1356(-3)	1.1929(-3)	1.3076(-3)	4.4175(-3)	4.6404(-3)	5.0864(-3)

Here and in other tables of this book the value in parenthesis is the order of magnitude

where $K_n(\alpha)$ is the modified third-order Bessel function (McDonald's), expanded for $\alpha \gg 1$ as follows [93]:

$$K_1(\alpha) \approx \sqrt{\frac{\pi}{2\alpha}}e^{-\alpha}\left(1 + \frac{3}{8\alpha} - \frac{15}{128\alpha^2}\right),$$

$$K_2(\alpha) \approx \sqrt{\frac{\pi}{2\alpha}}e^{-\alpha}\left(1 + \frac{15}{8\alpha} + \frac{105}{128\alpha^2}\right).$$

(1.2.43)

Table 1.3 represents the functions $A_i(\alpha), B_i(\alpha)$, found by numerical integration in [167] for $0 \leq \alpha \leq 10$.

1.2.3 Weak Degeneracy

The case of weak degeneracy corresponds to $f_e \ll 1$ in (1.2.2). We can therefore expand the integrals (1.2.10) in series, making use of the large size of the exponential in the denominator. Using the two-term approximation for these integrals gives [218, 166, 363, 93]:

$$I_{n\mp} = \alpha^2\left[K_2(\alpha)e^{\pm\beta} - \frac{1}{2}K_2(2\alpha)e^{\pm2\beta}\right],$$

$$I_{P\mp} = 3\alpha^2\left[K_2(\alpha)e^{\pm\beta} - \frac{1}{4}K_2(2\alpha)e^{\pm2\beta}\right],$$

$$I_{E\mp} = \alpha^3 \left[K_1(\alpha)e^{\pm\beta} + \frac{3}{\alpha}K_2(\alpha)e^{\pm\beta} \right.$$

$$\left. - \frac{1}{2}K_1(2\alpha)e^{\pm 2\beta} - \frac{3}{4\alpha}K_2(2\alpha)e^{\pm 2\beta} \right]. \tag{1.2.44}$$

Using (1.2.44) and the relation (1.2.29) for y we can derive from (1.2.17) with sufficient accuracy

$$\sinh\beta = \frac{\alpha y^3}{6K_2(\alpha)}\left[1 + \frac{K_2(2\alpha)}{K_2(\alpha)}\sqrt{1 + \frac{\alpha^2 y^6}{36K_2^2(\alpha)}} \right],$$

$$\cosh\beta = \sqrt{1 + \frac{\alpha^2 y^6}{36K_2^2(\alpha)}} + \frac{K_2(2\alpha)}{K_2(\alpha)}\frac{\alpha^2 y^6}{36K_2^2(\alpha)},$$

$$\cosh 2\beta = 1 + 2\frac{\alpha^2 y^6}{36K_2^2(\alpha)}, \quad \sinh 2\beta = \frac{\alpha y^3}{3K_2(\alpha)}\sqrt{1 + \frac{\alpha^2 y^6}{36K_2^2(\alpha)}}, \tag{1.2.45}$$

$$\beta = \ln\left[\frac{\alpha y^3}{6K_2(\alpha)} + \sqrt{1 + \frac{\alpha^2 y^6}{36K_2^2(\alpha)}} \right] + \frac{K_2(2\alpha)}{K_2(\alpha)}\frac{\alpha y^3}{6K_2(\alpha)}.$$

We have derived (1.2.45) making use of a small value of the terms which contain $K_2(2\alpha)$, corresponding to the case of weak degeneracy[2]. Using (1.2.44) and (1.2.45) we find from (1.2.9)

$$n_{e^-} + n_{e^+} = \frac{6\varrho}{\mu_Z m_u y^3}\left[\frac{K_2(\alpha)}{\alpha}\sqrt{1 + \frac{\alpha^2 y^6}{36K_2^2(\alpha)}} - \frac{K_2(2\alpha)}{2\alpha} \right],$$

$$E_{e^-} + E_{e^+} = \frac{6kT}{\mu_Z m_u y^3}\left\{ \left[K_1(\alpha) + \frac{3}{\alpha}K_2(\alpha) \right]\sqrt{1 + \frac{\alpha^2 y^6}{36K_2^2(\alpha)}} \right.$$

$$+ \left[\frac{3}{2\alpha}K_2(2\alpha) - K_1(2\alpha) + \frac{K_1(\alpha)K_2(2\alpha)}{K_2(\alpha)} \right]$$

$$\left. \times \frac{\alpha^2 y^6}{36K_2^2(\alpha)} - \frac{1}{2}K_1(2\alpha) - \frac{3}{4\alpha}K_2(2\alpha) \right\},$$

$$P_{e^-} + P_{e^+} = \frac{6\varrho kT}{\mu_Z m_u y^3}\left[\frac{K_2(\alpha)}{\alpha}\sqrt{1 + \frac{\alpha^2 y^6}{36K_2^2(\alpha)}} \right.$$

$$\left. + \frac{1}{2\alpha}K_2(2\alpha)\frac{\alpha^2 y^6}{36K_2^2(\alpha)} - \frac{1}{4}\frac{K_2(2\alpha)}{\alpha} \right],$$

---

[2] We use here $\sinh^{-1} x = \ln(x + \sqrt{x^2 + 1})$.

$$S_{e^-} + S_{e^+} = \frac{6k}{\mu_Z m_u y^3} \left\{ \left[K_1(\alpha) + \frac{4}{\alpha} K_2(\alpha) \right] \sqrt{1 + \frac{\alpha^2 y^6}{36 K_2^2(\alpha)}} \right.$$

$$- \frac{y^3}{6} \ln \left[\frac{\alpha y^3}{6 K_2(\alpha)} + \sqrt{1 + \frac{\alpha^2 y^6}{36 K_2^2(\alpha)}} \right]$$

$$+ \left[\frac{K_2(2\alpha)}{\alpha} - K_1(2\alpha) + \frac{K_1(\alpha) K_2(2\alpha)}{K_2(\alpha)} \right] \frac{\alpha^2 y^6}{36 K_2^2(\alpha)}$$

$$\left. - \frac{K_1(2\alpha)}{2} - \frac{K_2(2\alpha)}{\alpha} \right\}. \tag{1.2.46}$$

Equations (1.2.46) are valid for a weakly degenerate gas of arbitrary density, including a very low one where the number of pairs greatly exceeds the initial number of electrons and $\lambda = \alpha^2 y^6 / 36 K_2^2(\alpha) \ll 1$. It is also important that the gas should be non-relativistic because for $\alpha \ll 1$ the pairs produced fill phase space even at a very low density. Thus, applying (1.2.46) requires that $\alpha \geq 1$; when $\alpha \gg 1$ expansion (1.2.43) is correct[3]. For $\lambda \gg 1$, retaining two terms in the expansion in powers of $1/\lambda$, we find from (1.2.46) and (1.2.43) the perfect gas thermodynamic functions corrected with respect to degeneracy, relativistic effects and pair production (see also [166])

$$n_{e^-} + n_{e^+} = \frac{\varrho}{\mu_Z m_u} \left\{ 1 + \frac{9\pi}{\alpha^3 y^6} e^{-2\alpha} \left[1 + \frac{15}{4\alpha} \right. \right.$$

$$\left. \left. - \frac{\alpha^{3/2} y^3}{6\sqrt{\pi}} \left(1 + \frac{15}{16\alpha} \right) \right] \right\},$$

$$E_{e^-} + E_{e^+} = \frac{m_e c^2}{\mu_Z m_u} + \frac{3}{2} \frac{kT}{\mu_Z m_u} \left\{ 1 + \frac{5}{4\alpha} \right.$$

$$+ \frac{\alpha^{3/2} y^3}{12\sqrt{\pi}} \left(1 - \frac{15}{16\alpha} \right) + \frac{6\pi}{\alpha^2 y^6} e^{-2\alpha}$$

$$\left. \times \left[1 + \frac{21}{4\alpha} - \frac{\alpha^{3/2} y^3}{6\sqrt{\pi}} \left(1 + \frac{27}{16\alpha} \right) \right] \right\},$$

$$P_{e^-} + P_{e^+} = \frac{\varrho kT}{\mu_Z m_u} \left\{ 1 + \frac{\alpha^{3/2} y^3}{12\sqrt{\pi}} \left(1 - \frac{45}{16\alpha} \right) \right.$$

$$\left. + \frac{9\pi}{\alpha^3 y^6} e^{-2\alpha} \left[1 + \frac{15}{4\alpha} - \frac{\alpha^{3/2} y^3}{12\sqrt{\pi}} \left(1 + \frac{15}{16\alpha} \right) \right] \right\},$$

---

[3] Neglecting weak degeneracy in (1.2.46) gives $P_{e^-} + P_{e^+} = kT(n_{e^-} + n_{e^+})$.

$$S_{e^-} + S_{e^+} = \frac{k}{\mu_Z m_u} \left\{ \frac{5}{2} - \ln\left(\sqrt{\frac{2}{\pi}} \frac{\alpha^{3/2} y^3}{3} \right) + \frac{15}{4\alpha} \right.$$

$$+ \frac{\alpha^{3/2} y^3}{24\sqrt{\pi}} \left(1 + \frac{45}{16\alpha} \right) + \frac{9\pi}{\alpha^2 y^6} e^{-2\alpha}$$

$$\left. \times \left[1 + \frac{23}{4\alpha} - \frac{\alpha^{3/2} y^3}{6\sqrt{\pi}} \left(1 + \frac{35}{16\alpha} \right) \right] \right\}. \tag{1.2.47}$$

In (1.2.47) $E_{e^-} + E_{e^+}$ includes the rest energy of produced pairs and their kinetic energy without relativistic corrections, while $P_{e^-} + P_{e^+}$ includes relativistic corrections to the pairs' pressure. In the limit of a very low density $\lambda \ll 1$, retaining the first two terms in the expansion in powers of λ, we find from (1.2.46) relations identical to the non-relativistic limit of equations (1.2.38) combined with (1.2.42).

1.2.4 Non-relativistic Gas

In this case, $\alpha \sim \beta \gg 1$ and the contribution of positrons is negligible. Equations (1.2.9) and (1.2.10) become

$$\frac{\varrho}{\mu_Z m_u} = \frac{\sqrt{2}}{\pi^2} \left(\frac{m_e kT}{\hbar^2} \right)^{3/2} F_{1/2}(\beta - \alpha),$$

$$E_e = \frac{\sqrt{2}}{\pi^2 \varrho} \left(\frac{m_e kT}{\hbar^2} \right)^{3/2} \left[kT F_{3/2}(\beta - \alpha) + m_e c^2 F_{1/2}(\beta - \alpha) \right]$$

$$= \frac{m_e c^2}{\mu_Z m_u} + \frac{kT}{\mu_Z m_u} \frac{F_{3/2}(\beta - \alpha)}{F_{1/2}(\beta - \alpha)},$$

$$P_e = \frac{2\sqrt{2}}{3\pi^2} \left(\frac{m_e kT}{\hbar^2} \right)^{3/2} kT F_{3/2}(\beta - \alpha) \tag{1.2.48}$$

$$= \frac{2}{3} \frac{\varrho kT}{\mu_Z m_u} \frac{F_{3/2}(\beta - \alpha)}{F_{1/2}(\beta - \alpha)},$$

$$S_e = \frac{\sqrt{2}}{\pi^2} \frac{k}{\varrho} \left(\frac{m_e kT}{\hbar^2} \right)^{3/2} \left[\frac{5}{3} F_{3/2}(\beta - \alpha) - (\beta - \alpha) F_{1/2}(\beta - \alpha) \right]$$

$$= \frac{k}{\mu_Z m_u} \left[\frac{5}{3} \frac{F_{3/2}(\beta - \alpha)}{F_{1/2}(\beta - \alpha)} - (\beta - \alpha) \right],$$

where $F_\nu(\xi)$ are the Fermi integrals

$$F_\nu(\xi) = \int_0^\infty \frac{y^\nu \, dy}{1 + \exp(y - \xi)},$$

$$y = \frac{x^2}{2\alpha} = \frac{p^2}{2 m_e kT}, \quad \xi = \beta - \alpha. \tag{1.2.49}$$

In the non-relativistic limit the kinetic energy of electrons is separated from their rest energy. If $e^{-\beta+\alpha} \gg 1$, then $f_e \ll 1$, and degeneracy is not important. In this limit we have

$$F_{1/2}(\beta - \alpha) = e^{\beta-\alpha}\Gamma\left(\frac{3}{2}\right)\left[1 - 2^{-3/2}e^{\beta-\alpha}\right],$$

$$F_{3/2}(\beta - \alpha) = e^{\beta-\alpha}\Gamma\left(\frac{5}{2}\right)\left[1 - 2^{-5/2}e^{\beta-\alpha}\right].$$

(1.2.50)

The first terms in integrals (1.2.50) yield the ordinary gas thermodynamic functions (see Sect. 1.1). Taking into account the first relation (1.2.48) and (1.2.49) gives

$$
\begin{aligned}
e^{\beta-\alpha} &= \frac{\varrho}{\mu_Z m_u}\pi^{3/2}\sqrt{2}\left(\frac{\hbar^2}{m_e kT}\right)^{3/2} \\
&\quad \times \left[1 + \frac{\varrho}{\mu_Z m_u}\frac{\pi^{3/2}}{2}\left(\frac{\hbar^2}{m_e kT}\right)^{3/2}\right] \\
&= \sqrt{\frac{2}{\pi}}\frac{\alpha^{3/2}y^3}{3}\left(1 + \frac{\alpha^{3/2}y^3}{6\sqrt{\pi}}\right).
\end{aligned}
$$

This relation provides the same thermodynamic functions as yielded by neglecting corrections for relativistic effects ($\sim 1/\alpha$) and pair production ($\sim e^{-2\alpha}$) in (1.2.47). In order to determine the Fermi integrals (1.2.49) in the limit of strong degeneracy $\beta - \alpha \gg 1$, we use (1.2.24). Retaining the first two terms of the expansion gives

$$F_{1/2}(\beta - \alpha) = \frac{2}{3}(\beta - \alpha)^{3/2} + \frac{\pi^2}{12}\frac{1}{(\beta - \alpha)^{1/2}},$$

$$F_{3/2}(\beta - \alpha) = \frac{2}{5}(\beta - \alpha)^{5/2} + \frac{\pi^2}{4}(\beta - \alpha)^{1/2}.$$

(1.2.51)

Deriving from the first relation (1.2.48)

$$
\begin{aligned}
\beta - \alpha &= \frac{1}{2}\left(\frac{3\pi^2\varrho}{\mu_Z m_u}\right)^{2/3}\frac{\hbar^2}{m_e kT} \\
&\quad \times \left[1 - \frac{\pi^2}{3}\left(\frac{\mu_Z m_u}{3\pi^2\varrho}\right)^{4/3}\left(\frac{m_e kT}{\hbar^2}\right)^2\right] \\
&= \frac{\alpha y^2}{2}\left(1 - \frac{\pi^2}{3\alpha^2 y^4}\right),
\end{aligned}
$$

(1.2.52)

gives the same thermodynamic functions as found by neglecting corrections for relativistic effects ($\sim 1/\alpha^2$) in (1.2.32). We can see from (1.2.48) that the adiabat of the non-relativistic electron gas is determined by $T\varrho^{-2/3} = $ const., $P\varrho^{-5/3} = $ const. regardless of the degree of degeneracy. Note that

$E_{e,\text{kin}} = \frac{3}{2}P_e$, where $E_{e,\text{kin}} = E_e - (m_e c^2/\mu_Z m_u)$. The same relation between ϱ, T and P along the adiabat is valid for any monatomic perfect nonrelativistic gas with constant μ and μ_Z.

1.2.5 Ultrarelativistic Gas

When the electron kinetic energy is well above its rest energy, the magnitude of α in the integrals (1.2.10) becomes negligible, and using (1.2.49) they may be rewritten in the form

$$I_{n\pm} = F_2(\pm\beta). \quad I_{P\pm} = I_{E\pm} = F_3(\pm\beta). \tag{1.2.53}$$

In an ultrarelativistic equilibrium gas the inequality $\beta \geq 0$ will always hold and degeneracy cannot be small because of the intense pair production.

Fermi integrals with an integer subscript have properties which make it possible to write the thermodynamic functions of ultrarelativistic gas as polynomials in T and β. Using (1.2.49) it is easy to show that

$$\frac{dF_\nu(x)}{dx} = \nu F_{\nu-1}(x), \quad F_0(x) = \int_0^\infty \frac{dy}{1 + e^{y-x}} = \ln(1 + e^x),$$
$$F_0(x) - F_0(-x) = x. \tag{1.2.54}$$

Integrating the first of equations (1.2.54) successively provides[4]

$$F_1(x) + F_1(-x) = \frac{x^2}{2} + 2F_1(0),$$

$$F_2(x) - F_2(-x) = \frac{x^3}{3} + 4F_1(0)x, \tag{1.2.55}$$

$$F_3(x) + F_3(-x) = \frac{x^4}{4} + 6F_1(0)x^2 + 2F_3(0).$$

The integrals $F_\nu(0)$ are given by (1.2.36), so we have $F_1(0) = \pi^2/12$, $F_2(0) = \frac{3}{2}\zeta(3) = 1.803$, and $F_3(0) = 7\pi^4/120$, so finally we can write the thermodynamic functions for e^+e^- pairs in the form

$$\frac{\varrho}{\mu_Z m_u} = \frac{1}{3\pi^2}\left(\frac{kT}{\hbar c}\right)^3 (\beta^3 + \pi^2\beta),$$

$$E_{e^-} + E_{e^+} = \frac{1}{4\pi^2\varrho}\left(\frac{kT}{\hbar c}\right)^3 \left(\beta^4 + 2\pi^2\beta^2 + \frac{7\pi^4}{15}\right) kT,$$

$$P_{e^-} + P_{e^+} = \frac{1}{3}\varrho\left(E_{e^-} + E_{e^+}\right), \tag{1.2.56}$$

$$S_{e^-} + S_{e^+} = \frac{k}{3\pi^2\varrho}\left(\frac{kT}{\hbar c}\right)^3 \left(\pi^2\beta^2 + \frac{7\pi^4}{15}\right).$$

---

[4] We can evaluate this integral by substituting $z = \exp[(x-y)/2]$.

In the limit of strong degeneracy $\beta \gg (1, \alpha)$ the contribution of positrons is negligible and the first of equations (1.2.56) and (1.2.29) gives

$$\beta = \left(\frac{3\pi^2 \varrho}{\mu_Z m_u}\right)^{1/3} \frac{\hbar c}{kT} \left[1 - \frac{\pi^2}{3}\left(\frac{\mu_Z m_u}{3\pi^2 \varrho}\right)^{2/3}\left(\frac{kT}{\hbar c}\right)^2\right]$$

$$= \alpha y \left(1 - \frac{\pi^2}{3\alpha^2 y^2}\right).$$

This provides thermodynamic functions (1.2.33) without terms $\sim y^{-2}$ corresponding to deviations from the ultrarelativistic limit. For an ultrarelativistic gas of low density, as $\beta \to 0$, we have

$$\beta = \frac{3\varrho}{\mu_Z m_u}\left(\frac{\hbar c}{kT}\right)^3\left[1 - \frac{1}{\pi^6}\left(\frac{3\pi^2 \varrho}{\mu_Z m_u}\right)^2\left(\frac{\hbar c}{kT}\right)^6\right]$$

$$= \frac{y^3 \alpha^3}{\pi^2}\left(1 - \frac{y^6 \alpha^6}{\pi^6}\right),$$

thereby recovering the thermodynamic functions yielded by (1.2.41) not including departures from ultrarelativism $\sim \alpha^{-2}$. We see from (1.2.56) that the relations $T\varrho^{-1/3} = \mathrm{const.}$, $P\varrho^{-4/3} = \mathrm{const.}$ are valid along an adiabat of an ultrarelativistic gas. The explicit expressions for the thermodynamic functions in terms of ϱ and T obtained above enable us to find explicit expressions for adiabatic indices and heat capacities in all the limiting cases using (1.1.11–17). Figure 1.2 presents the areas where the asymptotic expressions are valid with an accuracy $\sim 1\%$. Some asymptotic expressions involving more expansion terms are listed in [166]; tables and interpolation coefficients based on them can be found in [167] (see also [64a, 67a]).

1.2.6 Integral Representation [64b, 67a]

Applying the Mellin integral transformation to I_P from (1.2.10) we obtain

$$M(t, \alpha) = \int_0^\infty ds \, s^{t-1} I_{P-}(s, \alpha),$$

$$s = \exp(-\beta).$$

Changing the order of integration and using mathematical relations [1] for the gamma functions $\Gamma(t)$ and McDonald function $K_\nu(\alpha)$,

$$\int_0^\infty \frac{\sigma^{t-1} d\sigma}{1 + \sigma} = \Gamma(t)\Gamma(1 - t) = \frac{\pi}{\sin \pi t},$$

$$K_\nu(\alpha) = \frac{\sqrt{\pi}(\alpha/2)^\nu}{\Gamma[\nu + (1/2)]} \int_0^\infty \exp(-\alpha \cosh \theta) \sinh^{2\nu} \theta \, d\theta,$$

gives, upon substituting $x = \beta \sinh \theta$,

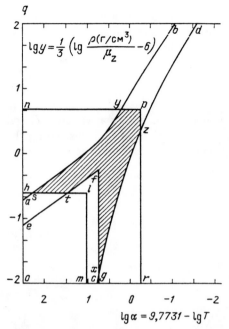

$$\lg \alpha = 9{,}7731 - \lg T$$

Fig. 1.2. Applicability regions of the asymptotic formulae on the $(\lg \alpha, \lg y)$ plane $\lg y = \frac{1}{3}(\lg(\rho/\mu z) - 5.989)$, $\lg \alpha = 9.7731 - \lg T$:

A) To the left of the line ayb the degenerated gas approximation (1.2.30) with corrections is valid.

B) To the right of the line czd the low density approximation (1.2.38) valid.

C) Within the region $oefg$ the almost non-degenerated, almost non-relativistic approximation (1.2.46) is valid.

D) $ohlm$ is the region of the non-relativistic gas approximation (1.2.48).

E) To the right and above the broken line npr the ultra-relativistic gas approximation (1.2.56) is valid.

The next regions allow different approximation:

1) $nqby$ the A and E approximations
2) To the right of the broken line rzd the B and E approximations
3) cxg the B and C approximations
4) $oetlm$ the C and D approximations
5) ahs the A and D approximations

In the shaded region it is necessary to calculate numerically the integrals in thermodynamic functions using, for example, the Gauss method

$$M(t, \alpha) = \frac{3\pi \alpha^2}{\sin \pi t} \frac{K_2(\alpha t)}{t^2}. \tag{1.2.57}$$

The reciprocal transformation of (1.2.57) leads to the integral representation of I_P, namely

$$I_{P-} = \frac{1}{2\pi i} \int_{C-i\infty}^{C+i\infty} M(t, \alpha) s^{-t} dt = \frac{3\alpha^2}{2i} \int_{C-i\infty}^{C+i\infty} \frac{e^{\beta t} K_2(\alpha t)}{t^2 \sin \pi t} dt,$$

$$0 < C < 1.$$

It is shown in [67a] that in the same way as for an ultrarelativistic gas the expression for the sum $(I_{P-} + I_{P+})$ may be transformed more simply than for I_{P-} above, so we take

$$I_{P-} + I_{P+} = -3\alpha^2 i \int_{C-i\infty}^{C+i\infty} \frac{\cosh(\beta t) K_2(\alpha t)}{t^2 \sin \pi t} dt. \tag{1.2.58}$$

The path of integration (1.2.58) goes along the imaginary axes and is closed by a semicircle of radius C around $t = 0$ to the right of this axis. Using the theory of residues [139], we finally obtain [67a] in the limit as $C \to 0$ the exact expression for the pressure (see (1.2.9))

$$P = P_{e-} + P_{e+}$$
$$= \frac{kT}{12\pi^2} \left(\frac{kT}{\hbar c}\right)^3 \left[\left(\frac{7\pi^4}{15} + 2\pi^2\beta^2 + \beta^4\right) - \alpha^2 \left(\pi^2 + 3\beta^2\right)\right.$$
$$\left. + \frac{3}{2}\alpha^4 \left(\ln \frac{4\pi}{\alpha} + \frac{3}{4}\right) + \frac{2}{\pi}\alpha^4 \int_{-1}^{1} dx \left(1 - x^2\right)^{3/2} \psi_F \left(\beta + \alpha x\right)\right], \tag{1.2.59}$$

where

$$\psi_F(x) \equiv \psi \left(\frac{1}{2} + \frac{ix}{2\pi}\right) + \psi \left(\frac{1}{2} - \frac{ix}{2\pi}\right), \quad \psi_F(-x) = \psi_F(x),$$

and

$$\psi(x) = \Gamma'(x)/\Gamma(x)$$

is the digamma function [1].

Using thermodynamic relations for the thermodynamic potential $\Omega = -PV$, we have

$$n_- - n_+ = \frac{1}{kT} \left(\frac{\partial P}{\partial \beta}\right)_T = \frac{1}{\pi^2} \left(\frac{kT}{\hbar c}\right)^3 \left[\frac{1}{3}(\beta^3 + \pi^2\beta) - \frac{\beta\alpha^2}{2}\right.$$
$$\left. + \frac{\alpha^3}{2\pi} \int_{-1}^{1} dx\, x \left(1 - x^2\right)^{1/2} \psi_F(\beta + \alpha x)\right] \tag{1.2.60}$$

and

$$\varrho E = \varrho(E_{e-} + E_{e+}) = -\alpha \frac{\partial P}{\partial \alpha} - P$$
$$= \frac{kT}{4\pi^2} \left(\frac{kT}{\hbar c}\right)^3 \left[\left(\frac{7\pi^4}{15} + 2\pi^2\beta^2 + \beta^4\right)\right.$$
$$- \frac{1}{3}\alpha^2 \left(\pi^2 + 3\beta^2\right) - \frac{1}{2}\alpha^4 \left(\ln \frac{4\pi}{\alpha} - \frac{1}{4}\right)$$
$$\left. - \frac{2}{\pi}\alpha^4 \int_{-1}^{1} dx \left(1 - x^2\right)^{1/2} x^2 \psi_F(\beta + \alpha x)\right]. \tag{1.2.61}$$

The expression for the potential $\Phi = (\mu_{te}/\varrho)(n_- - n_+) = E + (P/\varrho) - TS$ gives the general expression for the entropy per unit mass (cf. (1.2.9))

$$S = S_{e-} + S_{e+} = \frac{1}{T}\left[E + \frac{P}{\varrho} - \frac{\beta kT}{\varrho}(n_- - n_+)\right]. \tag{1.2.62}$$

The integrals in (1.2.59–61) may be calculated from the Chebyshev quadrature formula [1, 67a]:

$$\int_{-1}^{1} f(x)(1-x^2)^{-1/2}dx \approx \frac{\pi}{n}\sum_{k=1}^{n} f\left[\cos\frac{(2k-1)\pi}{n}\right]. \tag{1.2.63}$$

These formulae (1.2.59–61) are useful for calculations in the high-energy region, where the integral terms may be neglected. As shown in [67a] a relative accuracy of $\sim 10^{-3}$ is achieved in the absence of integral terms when $\varrho > 10^8 \mu_Z$ g cm^{-3} or $T \geq 10^{10}$ K. For lower values of $\varrho > 10^{5.8}\mu_Z$ g cm^{-3} or $T > 10^{9.5}$ K, five terms of the expansion (1.2.63) are needed to achieve the same accuracy. For still lower values of ϱ and T other relations are more useful (see Fig. 1.2).

The first terms in (1.2.59–61) coincide with corresponding quantities from (1.2.56), but relations (1.2.59–61) include additional analytic terms, increasing their range of application even without the integral terms.

Another integral presentation valid for the case of arbitrary relativism and weak degeneracy was suggested in [64b]. The integral I_{P-} from (1.2.10) is expanded in powers of $[1 + 2\exp(\alpha - \beta)]^{-1}$, so that

$$I_{P-} = \int_0^\infty \frac{x^4 dx}{(x^2+\alpha^2)^{1/2}} \frac{2\exp[x-(x^2+\alpha^2)^{1/2}]}{1+2\exp(\alpha-\beta)}$$

$$\times\left(1 - \frac{1-2\exp[\alpha-(x^2+\alpha^2)^{1/2}]}{1+2\exp(\alpha-\beta)}\right)^{-1}$$

$$= \sum_{n=0}^{N}[1+2\exp(\alpha-\beta)]^{-(n+1)}\beta_n(\alpha), \tag{1.2.64}$$

with

$$\beta_n(\alpha) = \int_0^\infty \frac{2x^4 dx}{(x^2+\alpha^2)^{1/2}}\exp[x-(x^2+\alpha^2)^{1/2}]$$

$$\times\{1-2\exp[\alpha-(x^2+\alpha^2)^{1/2}]\}^n dx. \tag{1.2.65}$$

The applicability of this expansion is wider than (1.2.44) and it gives a good precision even for moderate degeneracy in the ultrarelativistic case. Approximate spline fits for the integrals $\beta_n(\alpha)$ have been carried out in [64b] and below the dashed region in Fig. 1.2 the representation (1.2.64) gives relative precision better then 10^{-3} for $N = 5$. Other thermodynamical functions are calculated as in (1.2.60–62).

In the same region below the dashed region in Fig. 1.2 good precision is given by a simple approximation of the integrals in (1.2.10) by the trapezoidal rule [607b]. Presenting I_{P-} in the form

$$I_{P-} = 2 \int_0^\infty \frac{t^4(2\alpha + t^2)^{3/2} dt}{1 + \exp(t^2 + \alpha - \beta)} \tag{1.2.66}$$

by the substitution $x^2 = (t^2 + \alpha)^2 - \alpha^2 = (t^2 + 2\alpha)t^2$, we may use a formula

$$\int_0^\infty \exp(-t^2) f(t) dt \approx h \left(\frac{1}{2} f(0) + \sum_{k=1}^\infty \exp(-k^2 h^2) f(kh) \right). \tag{1.2.67}$$

This formula gives precision $\sim \exp(-k^2 h^2)$ for even functions $f(t)$, whose only singularities are branch points on the imaginary axis symmetrically placed with respect to the origin. The form (1.2.67) looks simpler than (1.2.64), but in fact it is less universal. In (1.2.64) the values $\beta_n(\alpha)$ are specified as functions of one parameter α, while in (1.2.67) the values $f(kh)$ depend on α and β, which needs a much larger number of calculations. When (1.2.60, 61) do not give good enough precision for derivatives, all integrals in (1.2.10) can be written in the same form (1.2.66) and calculated using (1.2.67).

1.2.7 The General Case

In the absence of small parameters the thermodynamic functions should be obtained by calculating integrals (1.2.10) numerically. There exist rather powerful methods analogous to the Gauss method, applied for this purpose in [46]. The integrals in (1.2.10) are written in the form $f(x)x^p e^{-x}$, where $f(x)$ is limited to a finite interval and is accurately approximated by a polynomial of order $\leq 2n - 1$ in the interval $(0, N)$, where N is sufficiently large. The calculations are based on the following quadrature formula:

$$\int_0^\infty f(x)x^p e^{-x} dx = \sum_{i=1}^n A_i f(x_i), \tag{1.2.68}$$

where x_i are the roots of the Laguerre polynomial $L_n^{(p)}$ and the coefficients A_i come from the set of linear equations

$$\sum_{i=1}^n A_i x_i^k = (k+p)!,$$

$$k = 0, 1, \ldots, n-1.$$

Formula (1.2.68) is exact as long as $f(x)$ is a polynomial of order $\leq 2n - 1$ because the Laguerre polynomials $L_n^{(p)}$ are orthogonal in the interval $(0, \infty)$ with the weight function $x^p e^{-x}$. The value $p = 2$ can be used for the calculation of I_E and I_P from (1.2.10). The values of x_i and A_i for the five-points scheme ($n = 5$) are listed in Table 1.4 [29] for $p = 0, 1, 2, 3, 4$.

Table 1.4. Roots and coefficients for calculating integrals (1.2.68) with $n = 5, p = 0, 1, 2, 3, 4$ (see for example [29])

Roots x_i and coefficients A_i	$p = 0$	$p = 1$	$p = 2$	$p = 3$	$p = 4$
x_1	0.26356	0.61703	1.0311	1.4906	1.9859
x_2	1.4134	2.1130	2.8372	3.5813	4.3417
x_3	3.5964	4.6108	5.6203	6.6270	7.6320
x_4	7.0858	8.3991	9.6829	10.944	12.188
x_5	12.641	14.260	15.828	17.357	18.852
A_1	0.52176	0.34801	0.52092	1.2510	4.1856
A_2	0.39867	0.50288	1.0667	3.2386	12.877
A_3	0.075942	0.14092	0.38355	1.3902	6.3260
A_4	3.6118(-3)	8.7199(-3)	0.028564	0.11904	0.60475
A_5	2.3370(-5)	6.8973(-5)	2.6271(-4)	1.2328(-3)	6.8976(-3)

Another type of approximate analytical representation of the electron thermodynamical functions for the general case, based on Chandrasekhar-type expansions [218], is given in [345a]. To obtain approximate formulae in the form where thermodynamical identities are fulfilled exactly, it was suggested to calculate only P from (1.2.9) by means of expansion, and to find ϱ and E from identities in (1.2.60–62). These relations can be used when P is found from the approximate formula (1.2.68).

The expressions for the adiabatic index γ_1 and for the heat capacities in the general case for a constant nuclear composition are given in [46]

$$\gamma_1 = \left(\kappa + \frac{N^2}{M}\right)$$

$$\times \left[1 + \frac{(\mu_N/\mu_Z)\pi^4}{45(I_{n-} - I_{n+})} + \frac{(\mu_N/\mu_Z)(I_{P-} - I_{P+})}{3(I_{n-} - I_{n+})}\right]^{-1}, \tag{1.2.69}$$

$$C_v = \frac{k}{\mu_N m_u} M, \quad C_p = \frac{k}{\mu_N m_u}\left(M + \frac{N^2}{\kappa}\right),$$

where

$$M = \frac{3}{2} + \frac{\mu_N/\mu_Z}{I_{n-} - I_{n+}}\left\{\frac{4\pi^4}{15} + \sum_{+,-}(3I_{E\pm} + I_{P\pm} + \alpha^2 I_{5\pm} + \alpha^2 I_{6\pm})\right.$$

$$\left. - \frac{[3(I_{n-} - I_{n+}) + \alpha^2(I_{4-} - I_{4+})]^2}{\sum_{+,-}(I_{5\pm} + I_{6\pm})}\right\},$$

$$N = 1 + \frac{\mu_N/\mu_Z}{3\left(I_{n-} - I_{n+}\right)} \left\{ \frac{4\pi^4}{15} + \sum_{+,-} \left(3I_{E\pm} + I_{P\pm}\right) \right.$$

$$\left. - \frac{3\left(I_{n-} - I_{n+}\right)\left[3\left(I_{n-} - I_{n+}\right) + \alpha^2\left(I_{4-} - I_{4+}\right)\right]}{\sum_{+,-}\left(I_{5\pm} + I_{6\pm}\right)} \right\},$$

$$\kappa = 1 + \frac{\mu_N}{\mu_Z} \frac{I_{n-} - I_{n+}}{\sum_{+,-}\left(I_{5\pm} + I_{6\pm}\right)},$$

$$I_{4\pm} = \int_0^\infty \frac{dx}{1 + \exp\left(\sqrt{x^2 + \alpha^2} \pm \beta\right)},$$

$$I_{5\pm} = \int_0^\infty \frac{\sqrt{x^2 + \alpha^2}\,dx}{1 + \exp\left(\sqrt{x^2 + \alpha^2} \pm \beta\right)},$$

$$I_{6\pm} = \int_0^\infty \frac{x^2\,dx}{\sqrt{x^2 + \alpha^2}\left[1 + \exp\left(\sqrt{x^2 + \alpha^2} \pm \beta\right)\right]}. \tag{1.2.70}$$

The dimensionless chemical potential β along an isentrope satisfies the equation

$$T\frac{d\beta}{dT} = \frac{M}{N} \frac{I_{n-} - I_{n+}}{\sum_{+,-}\left(I_{5\pm} + I_{6\pm}\right)} - \frac{3\left(I_{n-} - I_{n+}\right) + \alpha^2\left(I_{4-} - I_{4+}\right)}{\sum_{+,-}\left(I_{5\pm} + I_{6\pm}\right)}. \tag{1.2.71}$$

Figures 1.3–5 show the curves $\gamma_1(T)$, $C_v(T)$, $C_p/C_v(T)$ for pure iron ($A = 56$, $\mu_N/\mu_Z = Z = 26$) based on (1.2.69–71) and obtained in [46].

Problem. Find relativistic corrections to the adiabatic index γ_1 in the perfect gas.

Solution. $\gamma_1 = (5/3)\left[1 - (\mu/\mu_Z)(kT/m_e c^2)\right]$. Equations (1.1.11, 2.13, 15, 18, 47) have been used, all corrections for degeneracy and pair production $\sim \alpha^{3/2} y^3$ and $e^{-2\alpha}$ being omitted.

1.3 Nuclear Equilibrium Equation of State in the Presence of Weak Interaction Processes

When matter is heated to several billions of degrees the time scales for nuclear reactions t_n become shorter than all macroscopic time scales and equilibrium of nuclear composition occurs. Under these conditions the concentrations of different nuclei are given by the relation between the respective chemical potentials of nuclei $\mu_{A,Z}$, neutrons μ_n and protons μ_p:

$$\mu_{A,Z} = Z\mu_n + (A - Z)\mu_n, \tag{1.3.1}$$

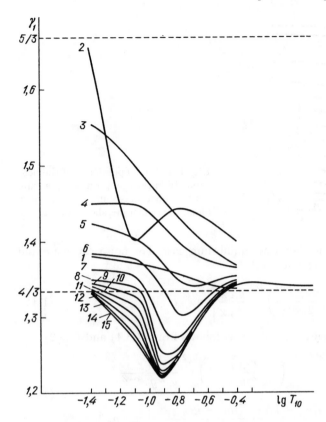

Fig. 1.3. The dependence of the adiabatic index γ_1 on temperature T for pure iron along the isentropes shown in Fig. 1.6.

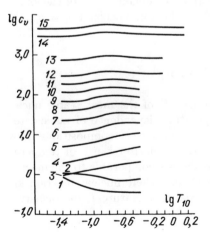

Fig. 1.4. The dependence of the heat capacity c_v on temperature T at a constant volume for pure iron along the isentropes shown in Fig. 1.6.

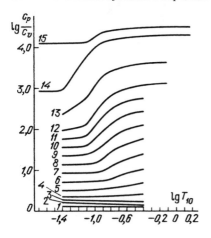

Fig. 1.5. The dependence of the heat capacity ratio c_p/c_v on temperature T at a constant volume for pure iron along the isentropes shown in Fig. 1.6.

in a similar way to the chemical equilibrium case. For non-relativistic and non-degenerate nuclei we have [145]

$$\mu_{A,Z} = -kT \ln \left[\left(\frac{m_{A,Z} kT}{2\pi \hbar^2} \right)^{3/2} \frac{g_{A,Z}}{n_{A,Z}} \right] m_{A,Z} c^2. \tag{1.3.2}$$

The equilibrium concentration of nuclei follows from (1.3.1) and (1.3.2)

$$n_{A,Z} = \left(\frac{2\pi \hbar^2}{kT} \right)^{3/2(A-1)} \left(\frac{m_{A,Z}}{m_p^Z m_n^{A-Z}} \right)^{3/2} \frac{g_{A,Z}}{g_p^Z g_n^{A-Z}}$$

$$\times \exp \left\{ \frac{[Zm_p + (A-Z)m_n - m_{A,Z}]c^2}{kT} \right\} n_p^Z n_n^{A-Z}. \tag{1.3.3}$$

Transforming the term in front of the exponent with the aid of the approximate equalities $m_n = m_p = m_u$, recalling that the denominator of the exponent is the nuclear binding energy $B_{A,Z}$, and taking into account $g_p = g_n = 2$ gives

$$n_{A,Z} = \left(\frac{2\pi \hbar^2}{m_p kT} \right)^{\frac{3(A-1)}{2}} \frac{A^{3/2}}{2^A} g_{A,Z} e^{\frac{B_{A,Z}}{kT}} n_p^Z n_n^{A-Z}.$$

Table 1.5 shows spins I and binding energies B of the most stable nuclei $g_{A,Z} = 2I_{A,Z} + 1$. The exponential temperature dependence of nuclear reaction rates (see Chap. 4) causes a transition in nuclear compositions from a frozen to a nuclear equilibrium state within a narrow range of temperatures in which the nuclear reaction time scales are comparable to the macroscopic (thermal, hydrodynamic) time scales and where nuclear reaction kinetics should be taken into account. At a fixed temperature T and density

$$\varrho = \sum_i n_{A_i Z_i} m_{A_i Z_i} + n_p m_p + n_n m_n$$

$$\approx \left(\sum A_i n_{A_i Z_i} + n_p + n_n \right) m_u \tag{1.3.4}$$

Table 1.5. Nuclear binding energy and spin of stable isotopes of the most abundant elements [135,180]

Atomic number	Element (isotope)	Binding energy, E_b, keV	Nuclear spin
1	^{1}H, ^{2}H	0, 2225	1/2, 1
2	^{3}He, ^{4}He	7718, 28297	1/2, 0
6	^{12}C, ^{13}C	92165, 97112	0, 1/2
7	^{14}N, ^{15}N	104663, 115496	1, 1/2
8	^{16}O, ^{16}O, ^{16}O	127624, 131766, 139813	0, 5/2, 0
10	^{20}Ne, ^{21}Ne, ^{22}Ne	160651, 167412, 177778	0, 3/2, 0
11	^{23}Na	186570	3/2
12	^{24}Mg, ^{25}Mg, ^{26}Mg	198262, 205594, 216688	0, 5/2, 0
13	^{27}Al	224959	5/2
14	^{28}Si, ^{29}Si, ^{30}Si	236544, 245018, 255627	0, 1/2, 0
15	^{31}P	262925	1/2
16	^{32}S, ^{33}S, ^{34}S	271789, 280432, 291847	0, 3/2, 0
17	^{35}Cl, ^{37}Cl	298220, 317112	3/2, 3/2
18	^{36}Ar, ^{38}Ar, ^{40}Ar	306727, 327354, 343822	0, 0, 0
20	^{40}Ca, ^{42}Ca, ^{43}Ca	342063, 361900, 369832	0, 0, 7/2
	^{44}Ca, ^{46}Ca, ^{48}Ca	380969, 398787, 416014	0, 0, 0
24	^{50}Cr, ^{52}Cr, ^{53}Cr	435061, 456364, 464304	0, 0, 3/2
	^{54}Cr	474024	0
25	^{55}Mn	482091	5/2
26	^{54}Fe, ^{56}Fe, ^{57}Fe	471779, 492280, 499926	0, 0, 1/2
	^{58}Fe	509969	0
28	^{58}Ni, ^{60}Ni, ^{61}Ni	506484, 526871, 534691	0, 0, 3/2
	^{62}Ni, ^{64}Ni	545288, 561788	0, 0

$B_{A,Z} = (Z m_p + (A - Z)m_n - m_{A,Z})c^2$

$m_n = m_p + m_e + 782.5$ keV

a relation between the concentrations n_n and n_p is necessary to find the nuclear composition.

The mutual transformations between protons and neutrons, both free and bound in nuclei, occur via the weak nuclear forces (see Chap. 5). At high temperatures the time scale for the weak processes t_β is much longer than the nuclear reaction time scale t_n and can reach the order of the macroscopic hydrodynamic or thermal time scales. Neutrinos produced by weak interactions leave the star easily. Under these conditions thermodynamic equilibrium amongst the weak interactions is not reached, with the exception of hot neutron stars, which are not transparent for neutrinos with energy $E_{\nu_e} \geq 1$ MeV. The thermodynamic functions of an equilibrium ν_e, $\tilde{\nu}_e$ gas of neutrinos with $kT \gg m_{\nu_e 0}c^2$ are[5] analogous to functions (1.2.56) of the gas of ultrarelativis-

---

[5] The estimates following from grand unification theories give $m_{\nu 0} \approx 10^{-1}$–$10^{-6}$ eV, but the possibility of $m_{\nu 0} = 0$ cannot be completely excluded; experiments of ^{3}H decay [127, 544a, 346a] give the limit $m_{\tilde{\nu}_e 0} < 5.1$ eV.

tic electrons, where $\beta = \mu_{\nu_e}/kT$, and $E_{\nu_e \tilde{\nu}_e}$, $P_{\nu_e \tilde{\nu}_e}$, $S_{\nu_e \tilde{\nu}_e}$ are half the $E_{e\pm}$, $P_{e\pm}$, $S_{e\pm}$ values, because of the difference in their statistical weights. If we want to find μ_{ν_e}, we have to replace $\varrho/\mu_Z m_u$ by a term involving the lepton charge Q_{ν_e}: $2(Q_{\nu_e} - n_{e-} + n_{e+}) = 2(n_{\nu_e} - n_{\tilde{\nu}_e})$ in the left part of the first relation in (1.2.56). After similar replacements, all the equations of Sect. 1.2 will apply to an equilibrium gas of neutrinos and the relation between n_p and n_n will be determined by the relations between chemical potentials

$$\mu_n = \mu_p + \mu_{te} + \mu_{\tilde{\nu}_e}, \quad \mu_{\tilde{\nu}_e} = -\mu_{\nu_e}. \tag{1.3.5}$$

The second of relations (1.3.5) accounts for the equilibrium of the reaction $\nu_e + \tilde{\nu}_e \to e^+ + e^-$ and for equality (1.2.11). The equilibrium of neutrinos ν_μ and ν_τ could be described in a similar way.

When neutrinos fly away freely, an accurate computation of the relation between n_p and n_n reduces to solving the kinetic equations for the β process

$$\frac{dN_n}{dt} = -\frac{dN_p}{dt} = \sum_i \left(W^+_{A_i Z_i} - W^-_{A_i Z_i} \right) n_{A_i Z_i},$$

$$N_n = \sum_i (A_i - Z_i) n_{A_i Z_i} + n_n, \tag{1.3.6}$$

$$N_p = \sum_i Z_i n_{A_i Z_i} + n_p.$$

for a given initial relation between N_n and N_p. In calculations of the nuclear equilibrium of iron group elements in [328] the parameter N_n/N_p is considered an independent quantity. Free protons and neutrons must be included in the sum in the first of equations (1.3.6). The rates of the β reactions $W^+_{A,Z} = W_{A,Z}(e^+ - \text{decay}) + W_{A,Z}(e^- - \text{capture})$ and $W^-_{A,Z} = W_{A,Z}(e^- - \text{decay}) + W_{A,Z}(e^+ - \text{capture})$ are considered in Sect. 5.2.

If changes in T and ϱ are not significant in a star during an interval $t \gg t_\beta$, a kinetic equilibrium with respect to β processes sets in with $dN_n/dt = 0$ in (1.3.6). In this case the composition of matter is entirely determined by (1.3.6) [117–119, 224]. If neutrinos fly away freely, the composition may sometimes be approximately computed using (1.3.5) with $\mu_{\nu_e} = 0$. The results of such calculations can be found in [114], where iron nuclei ^{56}Fe, including the first seven excited states, ^{4}He, n and p have been taken into account in nuclear equilibrium. Increasing the temperature first causes the iron nuclei to split into ^{4}He and nucleons and then leads to a pure nucleon composition. At high densities most of the free nucleons are represented by neutrons. Figure 1.6 from [46] shows matter isentropes on the (T, ϱ) plane and indicates the areas where $\gamma_1 < 4/3$ and which are necessary for stability analysis (see Chap. 12, Vol. 2). The results used in the nuclear equilibrium area were taken from [114].

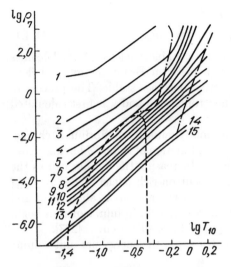

Fig. 1.6. Matter isentropes on the (T, ρ) plane. Isentropes from [114] are plotted for the equilibrium chemical composition and $10^9 < T < 10^{10}$ K, $10^5 < \rho < 10^{10}$ g/cm^3. Dashed line separating the regions where $\gamma_1 > 4/3$ and $\gamma_1 < 4/3$ and is plotted following calculation data [46]. Dot–dash line separating the region where $\gamma_1 < 4/3$ is based on the data from [114], iron disintegration taken into account. The isentropes are labeled as follows: $1 - S_{10} = 0.003981$, $2 - S_{10} = 0.01$, $3 - S_{10} = 0.01585$, $4 - S_{10} = 0.02512$, $5 - S_{10} = 0.03981$, $6 - S_{10} = 0.0631$, $7 - S_{10} = 0.1$, $8 - S_{10} = 0.1585$, $9 - S_{10} = 0.2512$, $10 - S_{10} = 0.3981$, $11 - S_{10} = 0.631$, $12 - S_{10} = 1.0$, $13 - S_{10} = 2.512$, $14 - S_{10} = 10$, $15 - S_{10} = 15.85$, $S_{10} = S/10^{10}$ erg·g^{-1}·K^{-1}

1.4 Matter at Very High Densities, Neutronization, Particle Interaction

At very high densities and for the case of strong electron and nucleon degeneracy, matter can be approximately considered as being cold with $T = S = 0$ [110].

1.4.1 Cold Neutronization
Along the Minimum Energy States (MES)

At densities $\varrho < 8.1 \times 10^6$ g cm^{-3} matter at MES consists of ^{56}Fe, which has the most stable nuclei.[6] When ε_{Fe} from (1.2.21) reaches the magnitude of ε_β at

---

[6] Binding energy per nucleon $B_n = B_{A,Z}/A$ determined according to (1.3.3, 1.4.3) is maximized for ^{62}Ni (see Table 1.5). However, if we want to compare the stability of different nuclei we have to determine their energy with respect to a certain fixed state, say the proton–electron state. According to these techniques, the formation of ^{56}Fe nuclei is accompanied by the maximum energy release.

$$\varepsilon_\beta(A, Z) = (m_{A,Z-1} - m_{A,Z})c^2 - m_e c^2$$
$$= B_{A,Z} - B_{A,Z-1} + (m_n - m_p)c^2 - m_e c^2, \tag{1.4.1}$$

then the process of electron capture onto a stable nucleus (A, Z) becomes energetically possible. At a high enough electron Fermi energy a nucleus $(A, Z - 1)$ that is normally β^- radioactive becomes stable. The process of electron capture onto nuclei is called neutronization and was first calculated in [216].

At zero temperature the state of thermodynamical equilibrium is characterized by a minimum total energy E_{tot} as a function of A and Z having a constant number of nucleons per unit volume. Increasing the density causes the equilibrium to shift towards increasingly neutron-rich nuclei. When $\varrho \geq \varrho_{nd}$ the binding energy of the last neutron Q_n in the nucleus is about zero and free neutrons start to form in equilibrium. Still in equilibrium, neutrons begin to separate from nuclei at the density $\varrho_{nd} = 4.3 \times 10^{11}$ g cm^{-3} [267].

In the absence of free neutrons and at $\varrho \leq \varrho_{nd}$ the energy per unit volume including the rest energy is [267]

$$E_{tot}(A, Z, n_b) = n_{A,Z}(m_{A,Z}c^2 + W_L) + E_e(n_e),$$
$$n_e = Z n_{A,Z}, \quad n_b = A n_{A,Z}. \tag{1.4.2}$$

Stable nuclear masses are determined by using the binding energy from Table 1.5. Experimental measurements of the masses $m_{A,Z}$ of nuclei far from the stability range have not yet been carried out. Instead, semi-empirical formulae (that is, theoretical but corrected with respect to available experimental data) have been applied. The simplest formula for the binding energy that involves all the general effects is that obtained by von Weizsäcker and has the form [23]

$$B_{A,Z} = 15.568 - 17.226 A^{2/3} - 0.698 \frac{Z^2}{A^{1/3}}$$
$$- 23.279 \frac{(A - 2Z)^2}{A} + \frac{34}{A^{3/4}} \delta \text{ MeV.} \tag{1.4.3}$$

Here, $\delta = 1$ for even A and Z, $\delta = 0$ for odd A and all Z, and $\delta = -1$ for even A and odd Z. Formula (1.4.3) accounts for the statistical and droplet properties of nuclei, the nuclear Coulomb and surface energies and nuclear interaction energy corrected for non-symmetry $(A \neq 2Z)$, and nucleon coupling effects. Von Weizsäcker's formula is used in [77] for calculations of the equilibrium composition in the case of cold neutronization. Similar calculations in [267] apply the more accurate but much more complex Meyer and Swiatecki formula [486]. Cold electron energy is given by (1.2.22–23), and W_L is the electrostatic interaction energy, which accounts for the presence of positive point charges within a homogeneous background of negative charges. The minimum of W_L corresponds to a body-centred cubic lattice and is given by [267]

$$W_{L,min} = -1.819620 Z^2 e^2 / b, \quad n_{A,Z} b^3 = 2. \tag{1.4.4}$$

Table 1.6. Equilibrium nuclear composition at densities below the neutron drip (from [267])

Nucleus	B_n, MeV	Z/A	ρ_{max}, g·cm^{-3}	μ_{te}, MeV	$\Delta\rho/\rho$, %
^{56}Fe	8.7905	0.4643	8.1(6)	0.95	2.9
^{62}Ni	8.7947	0.4516	2.7(8)	2.6	3.1
^{64}Ni	8.7777	0.4375	1.2(9)	4.2	7.9
^{84}Se	8.6797	0.4048	8.2(9)	7.7	3.5
^{82}Ge	8.5964	0.3902	2.2(10)	10.6	3.8
^{80}Zn	8.4675	0.3750	4.8(10)	13.6	4.1
^{78}Ni	8.2873	0.3590	1.6(11)	20.0	4.6
^{76}Fe	7.9967	0.3421	1.8(11)	20.2	2.2
^{124}Mo	7.8577	0.3387	1.9(11)	20.5	3.1
^{122}Zr	7.6705	0.3279	2.7(11)	22.9	3.3
^{120}Sr	7.4522	0.3166	3.7(11)	25.2	3.5
^{118}Kr	7.2002	0.3051	(4.3(11))	(26.2)	...

Here B_n is the binding energy per nucleon, ρ_{max} is the maximum density at which the nucleon still exists, μ_{te} is the chemical potential of electrons at this density, $\Delta\rho/\rho$ is the relative density increment between two successive nuclear compositions. $\rho_{max} = 4.3 \cdot 10^{11}$ g·cm^{-3} corresponds to the onset of neutron evaporation

The electrostatic interaction causes the energy and pressure of matter to decrease, since the mean distance between mutually repelling nuclei is larger than that between charges of different sign attracting each other [110]. Table 1.6 from [267] gives an equilibrium composition corresponding to the minimized energy E_{tot} from (1.4.2) at a fixed baryon density. Note that between densities ϱ_{max} and $\varrho_{max}(1 + (\Delta\varrho/\varrho))$ the pressure is nearly constant (or increases slowly because of nuclear pressure). The value of $\varrho = E_{tot}/c^2$ includes kinetic and interaction energies. The pressure P and relativistic adiabatic index Γ are given by [267]

$$P = n_b^2 \frac{\partial(E_{tot}/n_b)}{\partial n_b}\bigg|_S, \quad \Gamma = \frac{n_b}{P}\frac{\partial P}{\partial n_b}\bigg|_S = \frac{\varrho + P/c^2}{P}\left(\frac{\partial P}{\partial \varrho}\right)_S. \quad (1.4.5)$$

Table 1.7 represents calculated thermodynamic functions for the equilibrium neutronization case. The data from Table 1.7 for $\varrho < 10^4$ g cm^{-3}, when the interaction energy W_L becomes comparable to the kinetic energy of the electrons, are taken from [201].

Table 1.7. Thermodynamic functions in equlibrium corresponding to minimized total energy (from [267])

ρ, g·cm^{-3}	P, dyn·cm^{-2}	n_b, cm^{-3}	Z	A	Γ
7.86	1.01(9)	4.73(24)	26	56	$\cdots$
7.90	1.01(10)	4.76(24)	26	56	$\cdots$
8.15	1.01(11)	4.91(24)	26	56	$\cdots$
11.6	1.21(12)	6.99(24)	26	56	$\cdots$
16.4	1.40(13)	9.90(24)	26	56	$\cdots$
45.1	1.70(14)	2.72(25)	26	56	$\cdots$
212	5.82(15)	1.27(26)	26	56	$\cdots$
1150	1.90(17)	6.93(26)	26	56	$\cdots$
1.044(4)	9.744(18)	6.295(27)	26	56	1.796
2.622(4)	4.968(19)	1.581(28)	26	56	1.744
6.587(4)	2.431(20)	3.972(28)	26	56	1.706
1.654(5)	1.151(21)	9.976(28)	26	56	1.670
4.156(5)	5.266(21)	2.506(29)	26	56	1.631
1.044(6)	2.318(22)	6.294(29)	26	56	1.586
2.622(6)	9.755(22)	1.581(30)	26	56	1.534
6.588(6)	3.911(23)	3.972(30)	26	56	1.482
8.293(6)	5.259(23)	5.000(30)	28	62	1.471
1.655(7)	1.435(24)	9.976(30)	28	62	1.437
3.302(7)	3.833(24)	1.990(31)	28	62	1.408
6.589(7)	1.006(25)	3.972(31)	28	62	1.386
1.315(8)	2.604(25)	7.924(31)	28	62	1.369
2.624(8)	6.676(25)	1.581(32)	28	62	1.357
3.304(8)	8.738(25)	1.990(32)	28	64	1.355
5.237(8)	1.629(26)	3.155(32)	28	64	1.350
8.301(8)	3.029(26)	5.000(32)	28	64	1.346
1.045(9)	4.129(26)	6.294(32)	28	64	1.344
1.316(9)	5.036(26)	7.924(32)	34	84	1.343
1.657(9)	6.860(26)	9.976(32)	34	84	1.342
2.626(9)	1.272(27)	1.581(33)	34	84	1.340
4.164(9)	2.356(27)	2.506(33)	34	84	1.338
6.601(9)	4.362(27)	3.972(33)	34	84	1.337
8.312(9)	5.662(27)	5.000(33)	32	82	1.336
1.046(10)	7.702(27)	6.294(33)	32	82	1.336
1.318(10)	1.048(28)	7.924(33)	32	82	1.336
1.659(10)	1.425(28)	9.976(33)	32	82	1.335
2.090(10)	1.938(28)	1.256(34)	32	82	1.335
2.631(10)	2.503(28)	1.581(34)	30	80	1.335
3.313(10)	3.404(28)	1.990(34)	30	80	1.335
4.172(10)	4.628(28)	2.506(34)	30	80	1.334
5.254(10)	5.949(28)	3.155(34)	28	78	1.334
6.617(10)	8.089(28)	3.972(34)	28	78	1.334
8.332(10)	1.100(29)	5.000(34)	28	78	1.334
1.049(11)	1.495(29)	6.294(34)	28	78	1.334
1.322(11)	2.033(29)	7.924(34)	28	78	1.334
1.664(11)	2.597(29)	9.976(34)	26	76	1.334
1.844(11)	2.892(29)	1.105(35)	42	124	1.334
2.096(11)	3.290(29)	1.256(35)	40	122	1.334
2.640(11)	4.473(29)	1.581(35)	40	122	1.334
3.325(11)	5.816(29)	1.990(35)	38	120	1.334
4.188(11)	7.538(29)	2.506(35)	36	118	1.334
4.299(11)	7.805(29)	2.572(35)	36	118	1.334

1.4.2 Free Nucleon Formation.
MES at Subnuclear Densities and Nucleon Interaction

At $\varrho > \varrho_{nd}$ the neutron bound states in nuclei are occupied and any further density increase leads to free neutron formation. After several attempts to develop the equation of state for this range of densities (see overview [266]) a correct approach has now been developed [265]. The approach is based on the following principles:

1. The same description is applied to the nucleon interaction energy both inside and outside the nuclei.
2. The expression for the nuclear surface energy includes the presence of surrounding neutrons and becomes equal to zero if the inner and outer material is identical.
3. The Coulomb interaction energy between electrons and lattice nuclei and between protons inside the nucleus must be taken into account.

Let $n_{A,Z}$ and n_n be the nuclear concentration and the concentration of free nucleons outside the nuclei, respectively, $V_{A,Z}$ is the nucleus volume. The expression for the total energy E_{tot} per unit volume may then be written as

$$E_{tot}(A, Z, n_n, n_{A,Z}, V_{A,Z}) = n_{A,Z}(W_{A,Z} + W_L)$$
$$+ (1 - V_{A,Z}n_{A,Z})E_n(n_n) + E_e(n_e). \tag{1.4.6}$$

To obtain the equilibrium composition and the equation of state we must minimize E_{tot} with respect to its arguments at a constant nucleon concentration n_b:

$$n_b = An_{A,Z} + (1 - V_{A,Z}n_{A,Z})n_n. \tag{1.4.7}$$

The lattice Coulomb energy W_L and cold electron energy in (1.4.6) are determined by (1.4.4) and (1.2.22–23), respectively. The energy of a nucleus as represented by a liquid droplet may be written as

$$W_{A,Z}(A, Z, V_{A,Z}, n_n) = [(1 - x)m_nc^2 + xm_pc^2 + W(k, x)]A$$
$$+ W_{Coul}(A, Z, V_{A,Z}, n_n) + W_{surf}(A, Z, V_{A,Z}, n_n), \quad x = \frac{Z}{A}, \tag{1.4.8}$$

where $W(k, x)$ is the energy per baryon in homogeneous nuclear matter with concentration $n = k^3/1.5\pi^2 = A/V_{A,Z}$, W_{Coul} is the Coulomb interaction energy of protons in the nucleus, and W_{surf} is the nuclear surface energy. The expression for the energy E_n of the neutron gas can be obtained in a similar way to (1.4.8) at $x = 0$:

$$E_n(n_n) = n_nW(k_n, 0) + m_nc^2, \quad n_n = k_n^3/1.5\pi^2. \tag{1.4.9}$$

Both the functions $W(k, x)$ from (1.4.8, 9) and W_{Coul} and W_{surf} from (1.4.8) have been evaluated in [266]. A rather crude estimate of the surface energy has been made in [265] using a dimensional argument. More accurate expressions for W_{surf} have been obtained in [313] by the Thomas–Fermi method, in [256] by a variational method and in [504, 548] by the Hartree–Fock method. The sum $[W(k,x)A + W_{\text{Coul}} + W_{\text{surf}}]$ represents the binding energy of the nucleus (A, Z) involving the surrounding neutron density, approximated as for the case of normal nuclei in a vacuum by the von Weizsäcker formula (1.4.3). Minimization of E_{tot} with respect to its arguments at a constant n_b from (1.4.7) leads to four equations determining the equilibrium with respect to the following processes: (1) compression and expansion of the nucleus by pressure of the external neutrons, (2) neutron exchange between the nucleus and surrounding neutron gas, (3) proton–neutron transformations inside the nucleus, and (4) detachment of the nucleus, minimizing the total energy. The solution of these equations determines the properties of equilibrium material consisting of nuclei and free neutrons. Table 1.8 from [265] shows the results of these computations. As the density increases, nuclear masses and volumes grow until the nuclei touch each other at the density $\varrho_{\text{nn}} = 2.4 \times 10^{14}$ g cm^{-3}. In [265] a phase transition of the first kind to a homogeneous nuclear fluid is assumed to occur at the density ϱ_{nn}.

More accurate calculations of the surface energy have affected only the rate of increase of the nuclear charge $Z(\varrho)$, which has slowed down compared with the results of [265]. Figure 1.7 shows the results for $Z(\varrho)$ calculated by different authors. The decrease in the magnitude of Z at higher densities reduces the changes of Z during the phase transition to homogeneous nuclear matter. The discrepancy in the $Z(\varrho)$ dependence obtained in [256, 313, 504, 548] illustrates the insufficiency of our knowledge both of the physical laws governing nucleon interactions and of the computation techniques available.

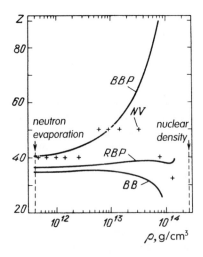

Fig. 1.7. The nucleus charge Z dependence in neutron star's crust according to the results of the following authors: BBP — [265], NV — [504], RBP — [548], BB — [31]

Table 1.8. Matter parameters at the minimized total energy and in the presence of free neutron (from [265])

ρ, g·cm^{-3}	P, dyn·cm^{-2}	n_b, cm^{-3}	Z	A	Γ
4.460(11)	7.890(29)	2.670(35)	40	126	0.40
5.228(11)	8.352(29)	3.126(35)	40	128	0.36
6.610(11)	9.098(29)	3.951(35)	40	130	0.40
7.964(11)	9.831(29)	4.759(35)	41	132	0.46
9.728(11)	1.083(30)	5.812(35)	41	135	0.54
1.196(12)	1.218(30)	7.143(35)	42	137	0.63
1.471(12)	1.399(30)	8.786(35)	42	140	0.73
1.805(12)	1.638(30)	1.077(36)	43	142	0.83
2.202(12)	1.950(30)	1.314(36)	43	146	0.93
2.930(12)	2.592(30)	1.748(36)	44	151	1.06
3.833(12)	3.506(30)	2.287(36)	45	156	1.17
4.933(12)	4.771(30)	2.942(36)	46	163	1.25
6.248(12)	6.481(30)	3.726(36)	48	170	1.31
7.801(12)	8.748(30)	4.650(36)	49	178	1.36
9.611(12)	1.170(31)	5.728(36)	50	186	1.39
1.246(13)	1.695(31)	7.424(36)	52	200	1.43
1.496(13)	2.209(31)	8.907(36)	54	211	1.44
1.778(13)	2.848(31)	1.059(37)	56	223	1.46
2.210(13)	3.931(31)	1.315(37)	58	241	1.47
2.988(13)	6.178(31)	1.777(37)	63	275	1.49
3.767(13)	8.774(31)	2.239(37)	67	311	1.51
5.081(13)	1.386(32)	3.017(37)	74	375	1.53
6.193(13)	1.882(32)	3.675(37)	79	435	1.54
7.732(13)	2.662(32)	4.585(37)	88	529	1.56
9.826(13)	3.897(32)	5.821(37)	100	683	1.60
1.262(14)	5.861(32)	7.468(37)	117	947	1.65
1.586(14)	8.595(32)	9.371(37)	143	1390	1.70
2.004(14)	1.286(33)	1.182(38)	201	2500	1.74
2.520(14)	1.900(33)	1.484(38)	...	...	1.81
2.761(14)	2.242(33)	1.625(38)	...	...	1.82
3.085(14)	2.751(33)	1.814(38)	...	...	1.87
3.433(14)	3.369(33)	2.017(38)	...	...	1.92
3.885(14)	4.286(33)	2.280(38)	...	...	1.97
4.636(14)	6.403(33)	2.715(38)	...	...	2.03
5.094(14)	7.391(33)	2.979(38)	...	...	2.05

1.4.3 The Density Above Its Nuclear Value
Inside Atomic Nuclei

The phase transition results in the formation of homogeneous nuclear matter (NM) incorporating neutrons and a small quantity of protons and electrons. As long as the NM density does not exceed $2\rho_0$ ($\rho_0 = 2.8 \times 10^{14}$ g cm^{-3} is the density of the free NM), the Brueckner–Bethe–Goldstone method based on perturbation theory can be applied to calculate the equation of state [22]. At

$\varrho > 2\varrho_0$ the variation principle developed by Pandharipande [537] is used for calculations. At high densities heavy hyperon production [9] and possible π^- meson production (or pion condensation) should also be taken into account. The equation of state for nuclear matter at $\varrho \geq \varrho_0$ is developed in [275, 479] for various assumptions. Table 1.9 from [479] presents a rather realistic set of matter parameters, with interaction potential parameters based on experimental data from high-energy physics and including hyperon production. The possibility of pion condensation scarcely affects this equation of state [275]. Note that, as the density increases, our knowledge of the physics of the strong interaction grows more and more uncertain and our computational techniques less and less accurate. This leads to an uncertainty of the order of 50% even in the realistic case given by Table 1.9.

As Zel'dovich first pointed out [103, 110], the principle of causality implies that the speed of sound v_s should not exceed that of light; this puts a limit on the equation of state, namely $P \leq \varepsilon = \varrho c^2$. Such a limit is particularly

Table 1.9 Realistic matter parameters at high density (from [479])

n_b, cm^{-3}	$\dfrac{E}{n_b} - m_n c^2$, MeV	ρ, g·cm^{-3}	P, dyn·cm^{-2}
1.0(38)	12.6	1.70(14)	1.19(33)
1.5(38)	16.6	2.55(14)	2.93(33)
2.0(38)	21.2	3.42(14)	6.00(33)
2.5(38)	26.0	4.31(14)	1.09(34)
3.0(38)	32.2	5.21(14)	1.83(34)
4.0(38)	46.9	7.04(14)	4.09(34)
5.0(38)	64.4	8.95(14)	7.61(34)
6.0(38)	83.7	1.09(15)	1.26(35)
7.0(38)	109	1.31(15)	1.99(35)
8.0(38)	134	1.54(15)	2.85(35)
9.0(38)	160	1.76(15)	3.71(35)
1.0(39)	189	2.01(15)	4.02(35)
1.1(39)	215	2.26(15)	5.02(35)
1.25(39)	254	2.66(15)	6.76(35)
1.4(39)	295	3.08(15)	8.81(35)
1.5(39)	324	3.37(15)	1.03(36)
1.7(39)	383	4.00(15)	1.38(36)
2.0(39)	475	5.04(15)	2.02(36)
2.5(39)	639	7.02(15)	3.40(36)
3.0(39)	814	9.36(15)	5.20(36)

At $n_b > 1.0 \cdot 10^{39}$ the following asymptotic formula apply

$$E = n_b \left(15.05 + 3.03 \left(\frac{n_b}{10^{39}} \right)^{1.33} \right) \cdot 10^{-4} \quad \text{erg} \cdot \text{cm}^{-3}$$

$$P = 403 \left(n_b/10^{39} \right)^{2.33} \cdot 10^{33} \quad \text{dyn} \cdot \text{cm}^{-3}$$

$$(\rho = E/c^2)$$

important because it is valid at the highest possible densities, where data on nuclear interactions are far from complete.

1.4.4 Calculation of Corrections for Finite Temperature

Temperature effects have been calculated for densities at which nuclear interactions become important. These calculations have been carried out by generalizing techniques applied earlier in cold matter studies [465] (see also [314]). Protons, neutrons, helium nuclei and a species of heavy nuclei have been studied in equilibrium and their total energy has been minimized. Since neutrinos escape freely and equilibrium with respect to β processes is not achieved, a relation $N_p/(N_n + N_p) = Y_e$ has been considered as given. The expressions from (1.4.6) and (1.4.8) for any kind of energy included in E_{tot} have been written with temperature corrections, and the energy of nuclear motion due to finite temperature together with the energy of the helium nuclei have been also added to E_{tot}. At a finite temperature the concentration of free protons n_p and helium nuclei n_α add to the independent variables, which are arguments of E_{tot} in (1.4.6). The quantity n_α is determined by the equilibrium condition with respect to the partition in protons and neutrons, as in (1.3.3), but with a finite volume (presumed to be equal to a certain value) of helium nuclei. To obtain n_p the equilibrium condition of proton exchange between protons in nuclei and a gas has been added.

Figure 1.8 from [456] illustrates the chemical properties at high temperatures and densities for $Y_e = 0.25$. It is curious that in the high-density regime $\varrho > 10^{14}$ g cm^{-3}, when nuclei begin to occupy more than half the volume, instead of nuclei immersed in a nucleon gas of lower density, globules of lower density (bubbles) start to form in the more dense nuclear matter. This analysis suggests that nuclei exist up to very high temperatures of ~ 20 MeV $\approx 2 \times 10^{11}$ K. This important conclusion follows from calculations of nuclear interactions and surrounding gas effects on nuclear properties. It has been pointed out in [465] that the diagram shown in Fig. 1.8 is relatively insensitive to changes in Y_e within the limits $0.2 \leq Y_e \leq 0.5$. These results are also weakly influenced by a variety of heavy nuclei species in equilibrium [346, 456]. Figures 1.9 and 1.10 show isentropes of matter and the dependence of the adiabatic index $\gamma_1(\varrho)$ for these isentropes. It is evident that the existence of nuclei causes γ_1 to decrease significantly in comparison with a nucleon gas with non-zero entropy. We can easily see from Fig. 1.8 that at densities above nuclear the matter is always homogeneous.

The Fermi energy of a perfect neutron gas determined in a similar way to (1.2.21) is

$$\varepsilon_{Fn} = m_n c^2 (\sqrt{1 + y_n^2} - 1), \quad y_n = \frac{p_{Fn}}{m_n c} = \left(\frac{\varrho_n}{6.2 \times 10^{15}}\right)^{1/3},$$

$$p_{Fn} = \left(\frac{3\pi^2 \varrho_n}{m_n}\right)^{1/3} \hbar. \tag{1.4.10}$$

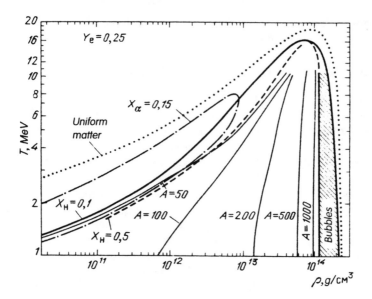

Fig. 1.8. Chemical properties of hot dense matter for $Y_e = 0.25$ from [456]. In the region confined by the solid curve the relative weight amount of nuclei $X_H > 0.1$, in the domain within the dashed curve $X_H > 0.5$. In the region inside the dashed-dot curve the relative weight amount of helium $X_\alpha > 0.15$. Thin solid curves define nuclear masses. The bubble region is hatched. The dotted curve shows the boundary of the two-phase region for homogeneous matter bulk equilibrium.

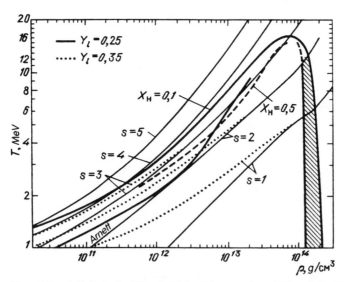

Fig. 1.9. Adiabats for $Y_l = 0.25$ (solid curves) and $Y_l = 0.35$ (dotted curves) with indications of dimensionless entropy s (in k/m_u) from [456]. The lines $X_H = 0.1$ and $X_H = 0.5$ have the same meaning as in Fig. 1.8. At $\rho > 10^{12}$ g·cm^{-3} the matter is not transparent for neutrinos and the lepton charge per baryon Y_l is a conserved quantity. Also shown is Arnett's trajectory of collapsing star center [255a].

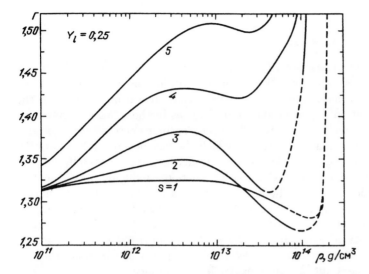

Fig. 1.10. Adiabatic index $\Gamma = \gamma_1 = (\partial \ln P/\partial \ln \rho)_s$ as a function of density ρ for $Y_l = 0.25$ and for the indicated s values (see Fig. 1.9) from [456]. The curves are smoothed in the dashed regions to remove some inessential details.

At $\varrho \approx \varrho_n \geq \varrho_0$ we have $y_n \geq 0.36$ and $\varepsilon_{Fn} \geq 61$ MeV. The highest possible temperature in gravitational collapse does not usually exceed ~ 20 MeV, so at $\varrho \geq \varrho_0$ temperature corrections to the equation of state are not important.

1.4.5 Non-equilibrium Neutronization with Increasing Density in Cold Matter

Equilibrium of the nuclear composition can be achieved at high temperatures when all reaction channels are open. As the material cools, most reaction channels close, so strictly speaking it is impossible to achieve the minimum energy states (MES) in cold matter. Matter at low temperatures is always out of equilibrium, but the difference between the equilibrium and the non-equilibrium states depends on the path by which the low-temperature state with given ϱ and T is achieved. The possible routes have been studied in [60, 61, 287]: cold matter contraction, and matter cooling at a fixed density accompanied by non-equilibrium changes in composition.

Now consider cold matter contraction, suppose that at low density it consists of the most stable element ^{56}Fe and contracts slowly at a temperature close to zero. When the density reaches 1.24×10^9 g cm^{-3} and the Fermi energy $\varepsilon_{Fe} = 3.81$ MeV, electron capture onto ^{56}Fe becomes energetically allowable[7] Since odd–odd nuclei are less stable, the first capture is immediately followed by a second, and a chain of successive reactions proceeds:

---

[7] This process is accompanied by electron capture onto the 109 keV excited state of the nucleus ^{56}Mn. The onset of capture onto the ground state of ^{56}Mn corresponds

$$^{56}\text{Fe} + e^- \to {}^{56}\text{Mn} + \nu_e,$$
$$^{56}\text{Mn} + e^- \to {}^{56}\text{Cr} + \nu_e. \tag{1.4.11}$$

In the second of reactions (1.4.11) the quantity $\varepsilon_\beta = 1.6$ MeV [99]. At $\varepsilon_{\text{Fe}} = 3.81$ MeV this reaction is not in equilibrium and is accompanied by heating ([56], see Chap. 5), which can be neglected, since it has no effect on the nuclear composition. Upon formation of ^{56}Cr, as further contraction occurs, the transformations $^{56}\text{Cr} \to {}^{56}\text{K} \to {}^{56}\text{Ar}$, etc. become energetically allowable and proceed until a nucleus forms in which the last neutron has separation energy close to zero: $Q_n \approx 0$. Any further density increase followed by electron capture leads to cold neutron evaporation, therefore reducing both A and Z. As the properties of even and odd nuclei differ, a fraction of the evaporating neutrons remove an energy of ~ 1 MeV/neutron, which heats the matter [78].

The number of nuclei per baryon does not change during the non-equilibrium neutronization. If A_0 is the atomic number of the initial nucleus, the concentrations of nuclei and electrons at a given density may then be written, taking account of charge neutrality, in the form

$$n_{A,Z} = \frac{\varrho}{A_0 m_u} \quad n_e = \frac{Z\varrho}{A_0 m_u}. \tag{1.4.12}$$

The nuclear mass is here taken to be approximately $A m_u$, the density $\varrho = N m_u$, and $N = N_n + N_p$ is the total concentration of baryons. Before the onset of neutron evaporation at $A = A_0$ the nuclear charge Z remains constant over the range of densities in which

$$\varepsilon_\beta(A_0, Z+2) < \varepsilon_{\text{Fe}}(n_e) < \varepsilon_\beta(A_0, Z). \tag{1.4.13}$$

When the left-hand part of inequality (1.4.13) turns into equality, the successive electron capture process starts at constant ε_{Fe} and P_e, covering the range of densities $\varrho_{Z+2} < \varrho < \varrho_Z$:

$$\varepsilon_{\text{Fe}}(n_e) = \varepsilon_\beta(A_0, Z+2) = \text{const.},$$
$$\frac{\varrho_Z}{\varrho_{Z+2}} = \frac{Z+2}{Z}. \tag{1.4.14}$$

Table 1.10, based on data from [379], shows the properties of material at cool neutronization before neutron evaporation takes place. After the onset of neutron evaporation, the neutron concentration is determined by

$$n_n = \frac{\varrho}{m_u}\left(1 - \frac{A}{A_0}\right), \quad m_n \approx m_p \approx m_u. \tag{1.4.15}$$

Neutron evaporation from the nucleus becomes possible when the evaporating neutron energy is above the neutron Fermi energy (1.4.10). The evaporation condition after the start of electron capture described by (1.4.14) is

---

to $\varrho = 1.15 \times 10^9$ g cm^{-3}, $\varepsilon_{\text{Fe}} = 3.7$ MeV, but this capture proceeds very slowly because of a strong inhibition (see Sect. 5.1 and (5.3.14)).

Table 1.10. Matter properties at a cold non-equilibrium neutronisation of iron nuclei

$\rho, 10^{11}$ g·cm^{-3}	$A; Z$	ε_{Fe}, MeV	ε_{Fn}, MeV	$P, 10^{30}$ dyn·cm^{-2}	$P_n, 10^{30}$ dyn·cm^{-2}
0.012	56; 24	3.81	0	5.5(-4)	0
0.14	56; 22	8.83	0	0.014	0
0.57	56; 20	14.0	0	0.080	0
1.6	56; 18	19.3	0	0.28	0
3.098	54; 18	23.84	0.069	0.569	3(-4) (6.5)
3.9	56; 16	25.2	0	0.73	0
5.01	56; 16	26.98	0	0.933	
6.17	54; 16	28.85	0.09	1.20	
6.233	48; 16	29.52	0.232	1.25	0.0059 (5.7)
7.24	46; 14	29.19	0.26	1.29	
7.664	42; 14	29.66	0.468	1.41	0.0344 (3.6)
9.689	36; 12	30.47	0.693	1.75	0.0921 (4.1)
10.1	56; 14	31.78	0	1.97	0
10.23	40; 12	31.00	0.44	1.66	
12.1	48; 12	32.06	0.43	2.07	0.029
12.88	35; 10	31.60	0.58	1.86	
14.13	32; 9	31.47	0.65	1.86	
14.88	30; 10	31.36	0.987	2.12	0.422 (4.5)
15.0	40; 10	32.35	0.79	2.25	0.13
16.8	36; 9	32.50	0.99	2.38	0.23
17.43	24; 8	32.47	1.41	2.48	0.537 (5.2)
19.2	32; 8	32.64	1.22	2.59	0.39
22.2	28; 7	32.78	1.49	2.88	0.65
25.65	18; 6	33.47	2.04	3.57	1.36 (6.0)
26.3	24; 6	32.92	1.82	3.35	1.07
44.12	12; 4	35.38	2.52	6.06	3.31 (7.8)

P is the total pressure, P_n is the neutron pressure, left hand values are from [78], right hand values from [560], where P_n is not given. In the middle of the columns there are computation results based on tables [379,135] before the onset of neutron evaporation, and equation (1.4.20). The last column shows in parenthesis temperature T (10^9 K) at which the spacified composition forms via non-equilibrium β-captures [78].

$$Q_n(A, Z) = B_{A,Z} - B_{A-1,Z} \le -\varepsilon_{Fn}(n_n). \qquad (1.4.16)$$

Since the function $Q_n(A, Z)$ oscillates according to the parity of A and Z, a chain of successive reactions proceeds

$$(A, Z) + e^- \to (A, Z - 1) + \nu_e \to (A - k_1, Z - 1) + k_1 n + \nu_e,$$
$$(A - k_1, Z - 1) + e^- \to (A - k_1, Z - 2) + \nu_e \qquad (1.4.17)$$
$$\to (A - k_1 - k_2, Z - 2) + k_2 n + \nu_e,$$

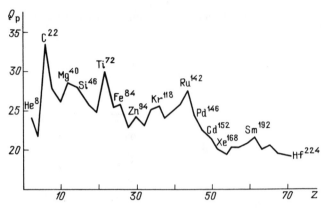

Fig. 1.11 The last proton separation energy Q_p as a function of Z for nuclei at the neutron-drip boundary $Q_n = 0$. The dependence is based on P. Nemirovsky's quantitative estimates.

where $k_1 \approx k_2 = 3$ or 4 for different nuclei [560]. The non-equilibrium neutronization involving neutron evaporation has been evaluated in [78] according to (1.4.12–16) with the binding energy determined by the von Weizsäcker formula, while in [560] the nucleon interaction in the nuclei has been calculated and the binding energy computed using techniques from [265] and also taking into account the effects of nuclear shells. The results of these calculations are shown in Table 1.10.

It has been pointed out in [560] that pycnonuclear reactions play an important role in the cold neutronization process because their rate increases rapidly with increasing density (see Chap. 4). At $\varrho = 1.4 \times 10^{12}$ g cm^{-3} the nuclei $(A, Z) = (32, 9)$ produced by ^{56}Fe coalesce and after having captured two electrons and evaporated eight neutrons form nuclei $(A, Z) = (56, 16)$. The thermal effect of this process and of further merger reactions have not been treated in [560]. Cold non-equilibrium neutronization has been studied up to the formation of nuclei with $(A, Z) = (99, 19)$, at $\varrho = 5.13 \times 10^{13}$ g cm^{-3} (cf. the nucleus (375, 74) from Table 1.8 or the nuclei with $Z = 30$–40 from Fig. 1.7 for equilibrium composition at the same density). Heat release in pycnonuclear reactions and further rapid non-equilibrium electron capture and neutron evaporation may increase the temperature sufficiently for the nuclear composition to evolve close to equilibrium.

The last proton separation energy

$$Q_p(A, Z) = B_{A,Z} - B_{A-1,Z-1}$$
$$= \varepsilon_\beta(A, Z) + Q_n(A, Z - 1) - (m_n - m_p)c^2 + m_e c^2 \qquad (1.4.18)$$

at the boundary $Q_n = 0$ has been studied by Nemirovsky (see [61]). Figure 1.11 shows the estimates he has obtained. The separation energy is roughly approximated by [61]

$$Q_p = \left(33 - \frac{Z}{7}\right) \text{MeV}, \quad A = 4Z \quad \text{for} \quad Q_n = 0, \quad Z \geq 6. \quad (1.4.19)$$

If we neglect the free neutron Fermi energy, assume neutronization with neutron evaporation to proceed along the line $Q_n = 0$ and also use (1.2.21, 4.12, 19), we then obtain from (1.4.14) the relation determining $Z(\varrho)$ and $A(\varrho)$:

$$\left[1 + \left(\frac{Z\varrho}{10^6 A_0}\right)^{2/3}\right]^{1/2} = 1.96\left(33 - \frac{Z}{7}\right) + 2.53; \quad A = 4Z. \quad (1.4.20)$$

Here we have used the equalities $m_e c^2 = 0.511$ MeV and $(m_n - m_p)c^2 = 1.293$ MeV. The results of calculations based on (1.4.20) are also shown in Table 1.10. We see that the difference between the $P(\varrho)$ values obtained by three different methods does not exceed 20%, being more than 50% larger than the equilibrium pressure at the same density (see Table 1.7).

1.4.6 Non-equilibrium Composition
During Cooling of Hot Dense Matter

When the temperature of cooling material falls below 4–5×10^9 K, the reactions between charged particles slow down considerably and the concentration of nuclei becomes frozen. Still possible under these conditions are reactions with neutrons, neutron photodetachment and capture, e^- decays at $\varepsilon_\beta > \varepsilon_{\text{Fe}}$ and e^- captures at $\varepsilon_\beta < \varepsilon_{\text{Fe}}$. In kinetic equilibrium with respect to β processes, when $dN/dt = 0$ in (1.3.6) there is a large excess of free neutrons in the matter [224]. Nuclei can join and detach neutrons at a finite temperature if

$$-\varepsilon_{\text{Fn}} < Q_n < Q_{nb} \approx 20kT - \varepsilon_{\text{Fn}}. \quad (1.4.21)$$

Figure 1.12 from [61, 287] shows how a non-equilibrium chemical composition is achieved during matter cooling under conditions of an excess of neutrons $x_n > 0.5$. The (A, Z) plane for nuclei is divided into three regions: region I with $Q_n > Q_{nb}$, region II with $-\varepsilon_{\text{Fn}} < Q_n < Q_{nb}$, $\varepsilon_\beta > \varepsilon_{\text{Fe}}$, region III with $-\varepsilon_{\text{Fn}} < Q_n < Q_{nb}$, $\varepsilon_\beta < \varepsilon_{\text{Fe}}$. At a high neutron concentration neutron capture causes the nuclei present in region I to pass rapidly into regions II and III, where equilibrium is achieved with respect to neutron capture and detachment. In region II β decays lead to an increase in Z and nuclei pass above the line cd. In region III β captures decrease Z and cause the nuclei to pass below the line ab. We see that at partial equilibrium at $\varepsilon_{\text{Fe}} \gg Q_{nb} \gg kT$, when the temperature has no effect on β processes, the nuclear composition is determined by the narrow domain $abcd$ on the (A, Z) plane. It is impossible for nuclei to pass beyond the boundary of this domain because β processes and neutron detachments are disallowed. At $T \leq 5 \times 10^8$ K only one nucleus remains at the boundary (1.4.16).

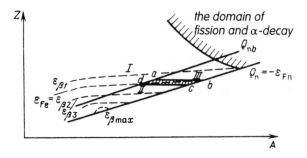

Fig. 1.12. Chemical composition formation at cooling corresponding to the limited equilibrium stage. The line $Q_n = -\epsilon_{Fn}$ confines the domain where nuclei exist. The line Q_{nb} shows the boundary between region 1 where neutron photodetachment is impossible, and regions 2 and 3. The dashed lines correspond to $\epsilon_\beta = \text{const}$, $\epsilon_{\beta 1} < \epsilon_{\beta 2} < \ldots < \epsilon_{\beta,max}$. The inequality $Q_n > Q_{nb}$ is valid in region 1, $Q_n < Q_{nb}$, $\epsilon_{Fe} < \epsilon_\beta$ in region 2, $Q_n < Q_{nb}$, $\epsilon_{Fe} > \epsilon_\beta$ in region 3. The solid line with a hatched band to the right confines the domain of fission and α-decay. The hatched region marked *abcd* corresponds to nuclei (A, Z) in limited equilibrium at given $Q_{nb}(T)$ and $\epsilon_{Fe}(\rho)$.

Neglecting ε_{Fn} and using (1.4.18) and (1.4.19) gives for the nuclear composition a set of approximate relations similar to (1.4.20):

$$A = 4Z, \quad Z = 7 \left\{ 33 - 0.511 \left[\left(\frac{\varrho}{\mu_Z 10^6} \right)^{2/3} + 1 \right]^{1/2} + 1.293 \right\}. \quad (1.4.22)$$

Figure 1.13 shows $Z(\varrho)$ as a function of ϱ for $x_n = 1/2$. At $\varrho = \varrho_2 = 2.24 \times 10^{12}$ g cm^{-3} a nucleus with $Z = 6$ and $\varepsilon_{\beta,max}$ appears; at $\varrho = \varrho_1 = 1.2 \times 10^{11}$ g cm^{-3} a nucleus with $Z = 150$ is formed, characterized by $Z^2/A = Z/4 = 37.5$ and by a fission time $\tau_f = 3 \times 10^7$ yr [164], which depends on Z^2/A. At $\varrho < \varrho_1$ the quantity Z^2/A increases, while the fission time decreases. Nuclear fission together with α decay leads to an increasing number of seed nuclei. At $\varrho > \varrho_2$ or $\varrho < \varrho_1$ the chemical composition may thus tend towards the equilibrium in MES. An essential non-equilibrium on cooling is only possible if $\varrho_1 < \varrho < \varrho_2$.

The non-equilibrium that forms on cooling differs mostly at low densities from the non-equilibrium produced by cold neutronization when in the first case Z values are very high (see Table 1.10 and Fig. 1.13). On the other hand, the two cases differ significantly from MES, where $Z(\varrho)$ increases with increasing density, while both the non-equilibrium compositions have $Z(\varrho)$ decreasing rapidly in response to a density increase.

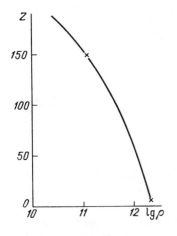

Fig. 1.13. $Z(\rho)$ dependence from (1.4.22) for non-equilibrium composition which forms at $x_n = 1/2$ upon dense matter cooling. The crosses denote approximate boundaries of non-equilibrium.

1.4.7 Thermodynamic Properties of Matter and Coulomb Interactions

Coulomb interactions in normal matter determine the state of aggregation, the degree of ionization and corrections to the thermodynamic functions. In rarefied gases the interactions are, for the most part, localized within atoms and ions that are not totally stripped. If the degree of ionization is evaluated from the Saha equation (1.1.8), these interactions will be automatically included in the result in an approximate way.

The principal parameter that determines the degree of ionization of rarefied gases is the quantity $\lambda_{ij} = I_{ij}/kT$ (see Sect. 1.1). When the temperature decreases and λ_{ij} increases, the gas becomes neutral, whereupon after the thermal energy falls below the atomic interaction energy it turns into liquid and eventually into a solid crystal with an ordered structure at minimized energy[8]. When the material contracts, the mean interatomic separation $l = n_A^{-1/3} = Z^{1/3}n_e^{-1/3}$ decreases[9] and at $\varrho = \varrho_{i1}$ becomes of the order of the atomic size $a_Z = Z^{-1/3}a_0$ in the Thomas–Fermi (TF) model [555], where $a_0 = \hbar^2/m_e e^2$ is the Bohr radius. The onset of ionization by pressure is at $\varrho = \varrho_{i1}$. The matter becomes totally ionized at $\varrho = \varrho_{i2}$, when the interelectron separation becomes of the order of the radius of the electron orbit closest to the nucleus $a_{Z0} = a_0 Z^{-1}$ [71]. Using (1.2.21) we can find the density ϱ_{i1} from the relation

$$\frac{l}{a_Z} = \frac{Z^{2/3}}{n_e^{1/3}a_0} = (3\pi^2)^{1/3}\frac{Z^{2/3}\alpha}{y} = 1, \quad n_e = \frac{y^3}{3\pi^2}\frac{m_e^3 c^3}{\hbar^3},$$

$$\varrho_{i1} = \mu_Z m_u \frac{m_e^3 c^3}{\hbar^3} Z^2 \alpha^3 = 3\pi^2 \times 10^6 \mu_Z Z^2 \alpha^3 = 11.4 \mu_Z Z^2,$$

$$(1.4.23)$$

---

[8] With the exception of helium-3 and helium-4, which remain liquid at normal pressure and zero temperature because of quantum effects.

[9] Here n_e also includes bound electrons.

where $\alpha = e^2/\hbar c = 1/137$ is the fine structure constant. Similarly, we have for the density ϱ_{i2}

$$\frac{Z}{n_e^{1/3}a_0} = (3\pi^2)^{1/3}\frac{Z\alpha}{y} = \theta_{i2}, \quad \theta_{i2} \geq 1;$$

$$\rho_{i2} = \frac{\mu_Z m_u}{\theta_{i2}^3}\frac{m_e^3 c^3}{\hbar^3}Z^3\alpha^3 = \frac{3\pi^2}{\theta_{i2}^3}\cdot 10^6 \mu_Z Z^3 \alpha^3 \tag{1.4.24}$$

$$= \frac{11.4}{\theta_{i2}^3}\mu_Z Z^3.$$

Note that if $Z < [\theta_{i2}/\alpha(3\pi^2)^{1/3}] \approx 44\theta_{i2}$, total ionization by pressure occurs when the electrons are non-relativistic. For iron, ^{56}Fe, we have $\varrho_{i1} = 1.7 \times 10^4$ g cm^{-3} and $\varrho_{i2} = 4.4 \times 10^5 \theta_{i2}^{-3}$ g cm^{-3}.

Consider phase transformations and Coulomb corrections to the thermodynamic functions in a dense, totally ionized material. As the parameter $\beta_Z = Z^{2/3}/n_e^{1/3}a_0 \ (\approx E_c/Z\varepsilon_{Fe})$ is small, corrections at $T = 0$ can be found by an iterative method [3, 555]. The principal corrections to the electron kinetic energy arise from electrostatic interactions between ions in the lattice and free electrons, and from electron exchange interactions. To a first approximation the electron distribution may be considered to be uniform. The simplest way to evaluate the electrostatic interaction energy E_c is to apply the Wigner–Seitz approximation (WS)¡, which describes electron and ion interactions as occurring only inside a spherical cell of radius l_{WS}:

$$\frac{4\pi}{3}l_{WS}^3 = n_{A,Z}^{-1}, \quad l_{WS} = \left(\frac{3}{4\pi}\right)^{1/3}l. \tag{1.4.25}$$

The energy per nucleus in this approximation is [3, 555]:

$$E_c = -\int_0^{l_{ws}}\frac{Ze}{r}dq_e + \int_0^{l_{ws}}\frac{q_e dq_e}{r} = -\frac{9}{10}\frac{(Ze)^2}{l_{WS}} =$$

$$= -\frac{9}{10}\left(\frac{4}{9\pi}\right)^{1/3}Z^{5/3}\alpha m_e c^2 y, \tag{1.4.26}$$

where[10] we have used the equality

$$l_{WS} = \left(\frac{9\pi}{4}\right)^{1/3}\frac{\hbar}{m_e c}\frac{Z^{1/3}}{y}$$

and where $q_e = (4\pi/3)en_e r^3$ is the electric charge of the electrons inside radius r. A second-order correction to the TF model arises from non-uniformity of the electron density inside the cell and is evaluated in [555]. This correction

---

[10] The exact value of electrostatic energy per nucleus W_L from (1.4.4) is 1.00454 less than the value E_c from (1.4.26).

includes the correction E_{TF} to the Fermi energy and the correlation correction E_{cor} to the electrostatic energy:

$$E_{\mathrm{TF}} = -\frac{162}{175}\left(\frac{4}{9\pi}\right)^{2/3}\sqrt{1+y^2}Z^{7/3}\alpha^2 m_e c^2,$$

$$E_{\mathrm{cor}} = \left[0.031\ln\left(\frac{e^2 m_e}{\hbar^2}l_{WS}\right) - 0.048\right]Z\alpha^2 m_e c^2. \tag{1.4.27}$$

The last term becomes important with decreasing density. The exchange interaction energy is given generally by the expression [555]

$$E_{\mathrm{ex}} = -\frac{3Z}{4\pi}\alpha m_e c^2 y\varphi(y),$$

$$\varphi(y) = \frac{1}{4y^4}\left[\frac{9}{4} + 3\left(\beta^2 - \frac{1}{\beta^2}\right)\ln\beta - 6(\ln\beta)^2 - \left(\beta^2 + \frac{1}{\beta^2}\right)\right.$$

$$\left. - \frac{1}{8}\left(\beta^4 + \frac{1}{\beta^4}\right)\right] \quad, \beta = y + \sqrt{1+y^2}; \tag{1.4.28}$$

$$\varphi(y) = \begin{cases} 1 & \text{for } y \ll 1, \\ -1/2 & \text{for } y \gg 1. \end{cases}$$

In the non-relativistic limit E_{ex} is given in [3]. These equations together with (1.2.19) thus give the principal corrections to the cold gas kinetic energy (1.2.22) per unit mass arising from the Coulomb interaction:

$$E_q = \frac{1}{Am_u}(E_c + E_{\mathrm{TF}} + E_{\mathrm{cor}} + E_{\mathrm{ex}})$$

$$= \left\{-\frac{3}{10\pi^2}\left(\frac{4}{9\pi}\right)^{1/3}\alpha Z^{2/3}y\right.$$

$$-\frac{54}{175\pi^2}\left(\frac{4}{9\pi}\right)^{2/3}\alpha^2 Z^{4/3}\sqrt{1+y^2}$$

$$+\frac{0.031\alpha^2}{3\pi^2}\ln\left[\left(\frac{9\pi}{4}\right)^{1/3}\frac{\alpha Z^{1/3}}{y}\right]$$

$$\left. -\frac{0.016}{\pi^2}\alpha^2 - \frac{\alpha}{4\pi^3}y\varphi(y)\right\}\frac{m_e^4 c^5}{\hbar^3}\frac{y^3}{\varrho}. \tag{1.4.29}$$

From the thermodynamic relation $P = \varrho^2(dE/d\varrho)$, $(dy/d\varrho) = (1/3)(y/\varrho)$ we obtain the correction to the pressure

$$P_q = P_c + P_{TF} + P_{cor} + P_{ex}$$

$$= -\frac{m_e^4 c^5}{\hbar^3} y^3 \left[\frac{1}{10\pi^2} \left(\frac{4}{9\pi} \right)^{1/3} \alpha Z^{2/3} y \right.$$

$$+ \frac{18}{175\pi^2} \left(\frac{4}{9\pi} \right)^{2/3} \alpha^2 Z^{4/3} \frac{y^2}{\sqrt{1+y^2}}$$

$$\left. - \frac{0.031}{9\pi^2} \alpha^2 + \frac{\alpha}{4\pi^3} \frac{\chi(y)}{y^3} \right], \tag{1.4.30}$$

where

$$\chi(y) = \frac{y^4}{3} \frac{d}{dy} [y\varphi(y)]$$

$$= \frac{1}{32} \left(\beta^4 + \frac{1}{\beta^4} \right) + \frac{1}{4} \left(\beta^2 + \frac{1}{\beta^2} \right) - \frac{9}{16} - \frac{3}{4} \left(\beta^2 - \frac{1}{\beta^2} \right) \ln \beta$$

$$+ \frac{3}{2} (\ln \beta)^2 - \frac{y}{3} \left(1 + \frac{y}{\sqrt{1+y^2}} \right) \left[\frac{1}{8} \left(\beta^3 - \frac{1}{\beta^5} \right) \right.$$

$$\left. - \frac{1}{4} \left(\beta - \frac{1}{\beta^3} \right) - \frac{3}{2} \left(\beta + \frac{1}{\beta^3} \right) \ln \beta + \frac{3}{\beta} \ln \beta \right],$$

$$\chi(y) = \begin{cases} y^4/3 & \text{for } y \ll 1, \\ -y^4/6 & \text{for } y \gg 1. \end{cases}$$

Using this relation and recalling (1.2.32) and (1.2.33) gives, in the limiting cases,

$$\frac{P_{qr}}{P_e} = -\frac{6}{5} \left(\frac{4}{9\pi} \right)^{1/3} \alpha Z^{2/3} - \frac{216}{175} \left(\frac{4}{9\pi} \right)^{2/3} \alpha^2 Z^{4/3}$$

$$- \frac{0.124 \, \alpha^2}{3} \frac{1}{y} + \alpha/2\pi \tag{1.4.31}$$

$$= -4.56 \times 10^{-3} Z^{2/3} - 1.78 \times 10^{-5} Z^{4/3}$$

$$- \frac{2.2 \times 10^{-6}}{y} + 1.16 \times 10^{-3}$$

for $y \gg 1$ [555], and

$$\frac{P_{qnr}}{P_e} = -\frac{3}{2}\left(\frac{4}{9\pi}\right)^{1/3}\alpha\frac{Z^{2/3}}{y} - \frac{54}{35}\left(\frac{4}{9\pi}\right)^{2/3}\alpha^2 Z^{4/3}$$

$$-\frac{0.155\alpha^2}{3y^2} - \frac{5\alpha}{4\pi y}$$

$$= -5.7 \times 10^{-3}\frac{Z^2}{y} - 2.23 \times 10^{-5}Z^{4/3}$$

$$-\frac{2.75 \times 10^{-6}}{y^2} - \frac{2.9 \times 10^{-3}}{y}$$

$$(1.4.32)$$

for $y \ll 1$. Equations (1.4.30, 32) apply only to y, corresponding to $P_q/P_e \ll 1$.

The Coulomb interaction leads with decreasing density to quite another form of the equation of state. The approximate analytical relation is [647]

$$\varrho = \varrho_0(\zeta + \varphi)^3, \quad \zeta^5 \equiv P/P_0, \tag{1.4.33}$$

where

$$\varrho_0 = \frac{32}{3}\pi^{-3}AZm_u a_0^{-3} = 3.88ZA \text{ g cm}^{-3},$$

$$P_0 = Z^{10/3}\frac{2^8(2\pi)^{1/3}}{15\pi^4}\left(\frac{e^2}{\hbar c}\right)^2\frac{m_e c^2}{a_0^3}$$

$$= 9.52 \times 10^{13} \text{ dyn cm}^{-2},$$

$$\varphi = \frac{1}{20}3^{1/3} + \frac{1}{8}\left(\frac{3}{4}\pi^{-2}Z^{-2}\right)^{1/3}, \quad a_0 = \hbar^2/m_e e^2.$$

$$(1.4.34)$$

In the limit $\varrho \gg \varrho_0$, $\zeta \gg \varphi$ the equation of state (1.4.33) reduces to the first term in (1.2.32).

The quantum properties of matter cause ions to oscillate in the lattice at zero absolute temperature with a frequency determined by their interaction with electrons. According to the WS approximation the ions that are disturbed from equilibrium come under the influence of a restoring force created by electrons in the cell of magnitude

$$F(r) = -\frac{Zeq_e}{r^2} = -\frac{4}{3}Ze^2 n_e r.$$

The frequency of harmonic oscillation is therefore [125, 555]

$$\omega_i^2 = -\frac{F(r)}{Am_u r} = \frac{4\pi}{3}\frac{Ze^2 n_e}{Am_u} = \frac{4}{9\pi}\frac{Zy^3}{Am_u}\alpha\frac{m_e^3 c^4}{\hbar^2}. \tag{1.4.35}$$

The three-dimensional oscillator has a zero-point energy $E_{zp} = (3/2)\hbar\omega_i$, which should be compared with E_c from (1.4.26). We have

$$f_{pc} = \frac{E_{zp}}{E_c} = \frac{5}{3}\left(\frac{4}{9\pi}\right)^{1/6}\left(\frac{m_e}{m_u}\right)^{1/2}\left(\frac{y}{\alpha A Z^{7/3}}\right)^{1/2}$$

$$= \left(0.11\frac{y}{AZ^{7/3}}\right)^{1/2}.$$

(1.4.36)

When $f_{pc} \geq 1$, quantum oscillations cause the crystal to be destroyed even at zero temperature. Nevertheless, zero-point oscillations in cold stars do not destroy the crystal structure because of the too-high densities $\varrho > 7.5 \times 10^8 \ \mu_Z A^3 Z^7$ g cm^{-3} required for this.

Crystal destruction by ion thermal motion is more realistic [485, 620]. Crystal melting occurs at $T = T_m$, when the kinetic energy kT of oscillations is about $1/150$ the Coulomb energy of the WS cell [544, 245, 578][11]. Using (1.4.26) we have

$$kT_m = 0.003\alpha m_e c^2 Z^{5/3} y,$$

$$T_m = 1.3 \times 10^5 Z^{5/3}\left(\frac{\varrho}{\mu_Z 10^6}\right)^{1/3} \text{K}.$$

(1.4.37)

Combined account has been taken of quantum and thermal effects on the crystal melting point parameters in [321b].

Thermodynamic properties of crystals have been well studied [145]. At low temperatures only degrees of freedom having long wavelengths and low frequencies are excited. These modes of vibration (phonons) have properties analogous to those of a photon gas. On the other hand, at high temperatures all possible modes of vibration may be excited, the crystal vibration energy being twice the kinetic energy of matter in the gaseous state at the same temperature. The thermal energy of an ionic lattice per unit mass is usually given by the Debye interpolation formula [145]

$$E_{iT} = \frac{3kT}{Am_u}\mathcal{D}\left(\frac{\theta}{T}\right),$$

$$\mathcal{D}(x) = \frac{3}{x^3}\int_0^x \frac{z^3 dz}{e^z - 1} \quad \text{(the Debye function)},$$

(1.4.38)

$$\theta = 0.775\frac{\hbar\omega_i}{k} = \frac{3.5 \times 10^3 \sqrt{\varrho}}{\mu_Z} \text{K};$$

θ is the Coulomb lattice Debye temperature [145, 620]. Here ω_i is the ion vibration frequency in the crystal (1.4.35). The limiting cases of the function $\mathcal{D}(x)$ are[12]

---

[11] In [578] computations using a Monte Carlo method have given $\Gamma = Z^2 e^2/kT l_{WS}$ $= 178 = \Gamma_m$ at the melting point of the crystal; l_{WS} is given in (1.4.25).

[12] In [227] in the limit $\theta \gg T$ the energy has been taken to be $4/3$ the energy in (1.4.38, 39).

$$
\mathcal{D}(x) = \begin{cases} \dfrac{\pi^4}{5x^3} - 3e^{-x} & \text{for } x \gg 1, \\[2mm] 1 - \dfrac{3x}{8} + \dfrac{x^2}{20} & \text{for } x \ll 1. \end{cases} \tag{1.4.39}
$$

The entropy per unit mass and the pressure in the ionic crystal are [145]

$$
S_{\mathrm{iT}} = \frac{k}{Am_{\mathrm{u}}} \left[-3\ln(1 - e^{-\theta/T}) + 4\mathcal{D}\left(\frac{\theta}{T}\right) \right],
$$

$$
P_{\mathrm{iT}} = \frac{3}{2}\frac{\varrho kT}{Am_{\mathrm{u}}} \mathcal{D}\left(\frac{\theta}{T}\right) = \frac{\varrho E_{\mathrm{iT}}}{2}. \tag{1.4.40}
$$

In the limiting cases,

$$
S_{\mathrm{iT}} = \begin{cases} \dfrac{3k}{Am_{\mathrm{u}}}\left(\ln\dfrac{\theta}{T} + \dfrac{4}{3}\right) & \text{for } T \gg \theta, \\[3mm] \dfrac{4\pi^4 kT^3}{5Am_{\mathrm{u}}\theta^3} & \text{for } T \ll \theta. \end{cases} \tag{1.4.41}
$$

Direct calculations based on the simplified Debye–Einstein model give [321b] the following formulae for E_{N}, P_{N} and free energy F_{N} of nuclei in the semi-degenerate crystal (in the notation of (1.4.38))

$$
E_{\mathrm{N}} = \frac{kT}{Am_{\mathrm{u}}} \left[1.713\frac{\theta}{T} + 2\mathcal{D}\left(0.892\frac{\theta}{T}\right) + \frac{2.009\theta/T}{e^{2.009\theta/T} - 1} \right]
$$

$$
P_{\mathrm{N}} = \frac{\varrho E_{\mathrm{N}}}{2}, \qquad TS_{\mathrm{N}} = E_{\mathrm{N}} - F_{\mathrm{N}}
$$

$$
F_{\mathrm{N}} = \frac{kT}{Am_{\mathrm{u}}} \left[1.713\frac{\theta}{T} - \frac{2}{3}\mathcal{D}\left(0.892\frac{\theta}{T}\right) \right. \tag{1.4.40a}
$$

$$
\left. + 2\ln(1 - e^{-0.892\theta/T}) + \ln(1 - e^{-2.009\theta/T}) \right].
$$

The following states of ionized matter are possible depending on the relations between the temperatures T, T_{m} and θ:

$T < \theta < T_m$	$\theta < T < T_m$	$\theta < T_m < T$ $T_m < \theta < T$	$T < T_m < \theta$ $T_m < T < \theta$
Quantum (degenerate) crystal	Classical ionic crystal	Classical liquid (non-ideal plasma)	Quantum liquid

The melting of an ionic crystal at $T \sim T_{\mathrm{m}} > \theta$ is accompanied by the release of heat, and on a further increase in temperature there is another transition to a perfect gas with $E_{\mathrm{iT}} = (3/2)kT/Am_{\mathrm{u}}$. According to [544, 245, 578] the perfect gas approximation applies from

$$
T_{\mathrm{g}} \approx 150T_{\mathrm{m}} \approx 2 \times 10^7 Z^{5/3}[\varrho/(\mu_Z 10^6)]^{1/3} \text{ K.} \tag{1.4.42}
$$

In the range $T_{\mathrm{m}} < T < 150T_{\mathrm{m}}$ interpolation can be used. The interpolation formulae for nuclear thermodynamic functions in fully ionized plasma were obtained in [384a]. From model computations we have for the "average nuclei" approximation

$$E_{\mathrm{N}} = \frac{kT}{\mu_{\mathrm{N}} m_{\mathrm{u}}} \left(\frac{3}{2} + f \right),$$

$$P_{\mathrm{N}} = \frac{\varrho kT}{\mu_{\mathrm{N}} m_{\mathrm{u}}} \left(1 + \frac{f}{3} \right), \tag{1.4.43}$$

$$f = -\Gamma^{3/2} \left[\frac{0.899962}{(0.702482 + \Gamma)^{1/2}} - \frac{0.274105}{1.319505 + \Gamma} \right],$$

with the corresponding relation for "average nuclei"

$$\Gamma = \frac{\langle Z \rangle^2 \langle n_{AZ} \rangle^{1/3} e^2}{kT} \left(\frac{4\pi}{3} \right)^{1/3} = \frac{2.27 \times 10^5}{T} \left(\frac{\mu_{\mathrm{N}}}{\mu_Z} \right)^{5/3} \left(\frac{\varrho}{\mu_Z} \right)^{1/3},$$

$$\langle n_{AZ} \rangle^{1/3} = \left(\frac{\varrho}{\mu_{\mathrm{N}} m_{\mathrm{u}}} \right)^{1/3}, \tag{1.4.43a}$$

$$\langle Z \rangle^2 = \left(\frac{\mu_{\mathrm{N}}}{\mu_Z} \right)^2,$$

where μ_{N} and μ_Z are determined in (1.2.16, 17), and for Γ see Footnote 11 on p. 54. Equations (1.4.43) are valid for the liquid state when $\Gamma \leq 80$. Another approximate expression for f obtained in [384a] using Monte Carlo computations is valid for $1 < \Gamma \leq 160$

$$f = -0.896434\Gamma + 0.86185\Gamma^{1/4} - 0.5551. \tag{1.4.44}$$

The negative values of E_{N} and P_{N} in (1.4.43–44) at large Γ are but small corrections to the electron energy and pressure under conditions of strong degeneracy.

The problem of latent heat is not quite understood. Mestel and Ruderman [485] give arguments for a very small value of latent heat at constant T and ϱ. Van Horn [620] proceeds from other considerations and assumes a phase transition of the first kind and the lattice latent heat at constant T and ϱ

$$\delta U_{\mathrm{Coul}} \approx -\frac{3}{4} kT_{\mathrm{m}}. \tag{1.4.45}$$

This problem has not yet been accurately solved.

The density range $\varrho_{i1} < \varrho < \varrho_{i2}$ hardly lends itself to a quantitative description, for the degree of ionization is controlled by both thermal energy and pressure. The cold matter ionization process is developed in the TF approximation in [126]. At the density

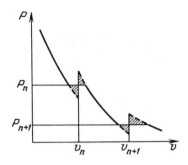

Fig. 1.14. Pressure plotted schematically as a function of specific volume, shell effects are taken into account. Lines of constant pressure P may be found by equating hatched areas.

$$\varrho_n = \left(\frac{\kappa}{\pi n}\right)^6 \mu_Z m_{\mathrm{u}} \frac{m_{\mathrm{e}}^3 c^3}{\hbar^3} Z^4 \alpha^3, \quad n = 1, 2 \ldots,$$

$$\kappa = 3 \left(\frac{3}{4\pi}\right)^{1/6} \ln \frac{1 + \sqrt{3}}{\sqrt{2}} \approx 1.56$$

(1.4.46)

shell corrections lead to a pressure jump at each subsequent ionization. The dependence of the non-monotonic correction to pressure on density is given in [126]

$$\delta P = -12 \left(\frac{9\pi}{4}\right)^{1/6} \kappa \frac{Z^{5/3} v^{4/3}}{\ln(Z^3 v)} [S_0] \frac{m_{\mathrm{e}}^4 c^5}{\hbar^3} \alpha^5,$$

(1.4.47)

where

$$v = \frac{A m_{\mathrm{u}}}{\varrho} \alpha^3 \frac{m_{\mathrm{e}}^3 c^4}{\hbar^3}, \quad S_0 = \kappa Z^{1/2} v^{1/6} - \pi/2,$$

(1.4.48)

and $[f(S_0)] = f(S_0)$ at $-\pi/2 < S_0 < \pi/2$, and is periodically continued beyond these limits. Pressure jumps (1.4.47) at $\varrho = \varrho_n$ and $v = v_n = (1/Z^3)(\pi n/\kappa)^6$ correspond to phase transitions of the first kind. The electron pressure is constant around $\varrho = \varrho_n$, and the quantity P_n can be found by using the Maxwell rule of equal areas (see Fig. 1.14). The shell effects vanish, according to (1.4.46), at

$$\varrho > \varrho_1 = 0.015 \mu_Z m_{\mathrm{u}} \frac{m_{\mathrm{e}}^3 c^3}{\hbar^3} \alpha^3 Z^4,$$

(1.4.49)

when all the energy levels merge into a continuous spectrum. Physically $\varrho_1 \le \varrho_{i2}$ from (1.4.24). The assumption that the electrons become relativistic with $y = 1$ and $\varrho_1 = \varrho_{i2}$ leads to $\theta_{i2} = 1.11$ and to the inequality $Z \le 49$ which allows the non-relativistic TF treatment to be applied to the ionization process [126].

The thermodynamic properties of a hydrogen–helium gaseous mixture at finite temperature have been studied in [403], where their mutual interaction and ionization by pressure have also been calculated. H_2, H, H^+, H^-, He, He^+, and He^{++} have all been included. The free energy F in a volume V at a given temperature T for a total number of ions N_{i} is evaluated in the form

$$F = F_0 + F_C, \tag{1.4.50}$$

where F_0 is the perfect gas free energy and F_C corresponds to the Coulomb interaction. The pressure P, entropy S and energy E of the system are given by

$$P = -\left(\frac{\partial F}{\partial V}\right)_{T,N_i},$$
$$S = -\left(\frac{\partial F}{\partial T}\right)_{V,N_i}, \tag{1.4.51}$$
$$E = F + TS.$$

To find specific values per unit mass one must divide these F, E and S by ϱV. The Coulomb interaction is calculated here in the form of corrections to a perfect gas

$$P = P_0 + P_C, \quad S = S_0 + S_C, \quad E = E_0 + E_C. \tag{1.4.52}$$

On the basis of the TF theory with the Debye–Hückel potential around the nucleus (TFDH) we have [403]

$$F_C = -\frac{kTV}{12\pi}\kappa_{\mathcal{D}}^3,$$
$$\kappa_{\mathcal{D}}^2 = \frac{4\pi e^2}{kT}(n_i + n_e\theta_e), \quad n_i = \frac{N_i}{V} \quad n_e = \frac{N_e}{V}, \tag{1.4.53}$$
$$\theta_e = F'_{1/2}(\eta)F_{1/2}(\eta),$$

where $F_\alpha(\eta)$ is the Fermi function from (1.2.49), ηkT is the chemical potential of the free electrons, $\kappa_{\mathcal{D}}^{-1}$ is the Debye radius (see also (2.4.47)), and all ions are assumed to have the same mean charge $Z = 1$. When the mixture with $x_H = 0.739$, $x_{He} = 0.261$ considered above changes its composition from perfectly neutral to fully ionized, the quantity Z changes from unity to 1.0811. Using (1.4.51) we find the corrections to the thermodynamic functions

$$P_C = -F_C\left(\frac{1}{V} + \frac{3}{\kappa_{\mathcal{D}}}\frac{\partial\kappa_{\mathcal{D}}}{\partial V}\right),$$
$$S_C = -F_C\left(\frac{1}{T} + \frac{3}{\kappa_{\mathcal{D}}}\frac{\partial\kappa_{\mathcal{D}}}{\partial T}\right), \tag{1.4.54}$$
$$E_C = \frac{kT^2V}{4\pi}\kappa_{\mathcal{D}}^2\frac{\partial\kappa_{\mathcal{D}}}{\partial T}.$$

The results of the TFDH treatment are valid when

$$\Gamma_e = \frac{e^2}{kT}\left(\frac{4\pi}{3}n_e\right)^{1/3} \ll 1. \tag{1.4.55}$$

The TFDH theory is applied in [403], when $\Gamma_e < 0.1$; while for $\Gamma_e > 1$ calculations based on the Monte-Carlo method are used, and when $0.1 < \Gamma_e < 1$, the results of both methods are interpolated.

The Saha equation is used to determine the degree of ionization. This equation involves, in addition to (1.1.8), a finite free-electron degeneracy, which affects the electron's chemical potential. We have then, instead of (1.1.8),

$$\frac{y_{i,j-1}}{y_{ij}} = \frac{g_{i,j-1}}{g_{ij}} e^{\beta + I_{ij}/kT}. \tag{1.4.56}$$

The screening DH potential causes the ionic and molecular levels to shift, so the number of bound states decreases and such states become shallower. Including them in the Saha equation accounts for ionization by pressure. Calculation of this ionization represents a complicated self-consistent problem, in which the concentrations of free electrons n_e and ions n_i determine the DH radius and potential, and these in turn determine the concentrations via the shift of levels determined by the Saha equation. The problem is further complicated because the Schrödinger equation with the DH potential has no analytical solution [240], and numerical calculations are required. The results of such calculations from [403] are shown in Figs. 1.15 and 1.16. Tables for the equation of state with a Coulomb interaction computed in the same way can be found in [358]. The phase transitions in hydrogen have been studied in [550a].

A phenomenological treatment of the pressure ionization was carried out in [345a], where the influence of Coulomb interaction on the ionization was taken into account by applying a correction to the chemical potential of the ideal electrons, so that the exponent in (1.4.56) was $\beta + \Delta\beta + I_{ij}/kT$. To find $\Delta\beta$, the Coulomb correction to the free energy is postulated as

$$F_C = \frac{\Omega(T)}{V}(N_{e0}^2 - N_e^2). \tag{1.4.57}$$

Here, N_{e0} is a constant, the total number on electrons (bound or free). Then

$$\Delta\beta = -\frac{1}{kT}\left(\frac{\partial F_C}{\partial N_e}\right)_{V,T} = 2\frac{\Omega}{kT}n_e, \tag{1.4.58}$$

and we also have

$$P_C = \Omega(n_{e0}^2 - n_e^2), \quad S_C = -V\frac{d\Omega}{dT}(n_{e0}^2 - n_e^2), \tag{1.4.59}$$

where $n_{e0} = N_{e0}/V$. The function $\Omega(T)$ is chosen in [345a] as

$$2\Omega = a_0^3(kT + 20\chi_0), \tag{1.4.60}$$

where

$$a_0 = 5.23 \times 10^{-9}/\langle Z \rangle \text{ cm}, \quad \chi_0 = 2.16 \times 10^{-11}\langle Z \rangle^2 \text{ ergs}, \tag{1.4.61}$$

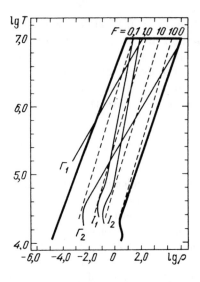

Fig. 1.15. Region in the $(\lg \rho, \lg T)$ – plane within which the thermodynamic functions have been computed [403]. To the left of the line $\Gamma_e = \Gamma_1 = 0.1$ calculations are based on the TFDH theory, to the right of the line $\Gamma_e = \Gamma_2 = 1.0$ on the Monte–Carlo method (see (1.4.55)). To the left of the line I_1 hydrogen and helium are partially ionized, between the lines I_1 and I_2 hydrogen is totally ionized, helium is partially ionized, to the right of the line I_2 both gases are totally ionized. The dashed lines represent constant values $F = 4F_{1/2}(\eta)/\sqrt{\pi}$, characterizing the degeneracy, in the non-degenerated plasma $F \ll 1$.

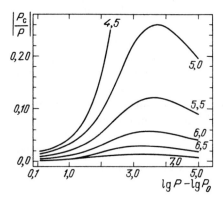

Fig. 1.16. Ratio of the pressure correction term P_c to the total pressure P at $T = \text{const}$. Each curve is labeled by the corresponding value $\lg T$ satisfying $\lg P_0 = 2.74118 \lg T - 4.58825$.

where an average nuclear charge $\langle Z \rangle$ is defined in (1.4.43a). This is loosely equivalent to saying that the statistical weight g of the less ionized state is decreasing relative to that of the more ionized state. Alternatively, the correction, or part of it, can be considered as a reduction in the ionization potential χ. The numerical constants in (1.4.61) are chosen to ensure that when the mean interelectron distance is of the order of a_0 and the density is of the order of ϱ_{i2} from (1.4.24), the gas becomes fully pressure ionized. Another choice of F_c with several free parameters, varied experimentally to give a good fit, was used in [544*].

The use of a static screened Coulomb potential to calculate level shifts and to estimate the cut-off of the internal partition function is, rigorously speaking, invalid. The atom is situated in a fluctuating time-dependent screening potential, where the very existence of stationary atom states cannot be stated. A consistent phenomenological approach accounting for these effects in the problem of the equation of state in the partially ionized non-ideal plasma was developed in [403b, 487a, 335a] (the MHD method).

The influence of fluctuating screening is described by use of the *weighted* internal partition functions Z_{int}^w, where energy level shifts are not assumed.

$$Z_{int}^w = \sum_i w_i e^{-(E_i - E_1)/kT}. \tag{1.4.62}$$

The total partition function is the product of all possible inputs $Z = \prod_q Z_q$, and the free energy is obtained as $F = -kT \ln Z$. The formulae for occupation probabilities w_i, connected with perturbations of charged and neutral particles, are given in [403b] in a way that provides convergence (soft cut-off) for large principal "n" quantum numbers ($w_i \sim n^{-6}$). The parameters of these formulae were chosen to fit a quantum-mechanical Stark ionization theory. The weighting coefficients are intended to describe the influence of the fluctuating potential, which implies wandering of an atom between some static levels, which are supposed to be unperturbed in the MHD method.

While the results for the equation of state obtained by this method do not differ strongly from the previous ones, on the basis of a static DH potential, they are claimed to have better thermodynamical and statistical mechanical consistency. In addition, this approach is supported by spectroscopic data in precision plasma experiments. The same appoach is applied for the calculation of opacity tables in the region where radiative transitions between levels are important (see Sect. 2.7). The results of the calculations are presented in the form of tables [407a] and programs exist for their creation for a given composition in the given range of ϱ and T (see [327a]).

In the situation where the fluctuating potential has a non-zero average value, it seems to be more consistent to take into account the shift of the levels connected with the static screening together with inclusion of the weighting coefficients like (1.4.62) responsible for the pure fluctuating (with zero average) part of the screening potential. This, however, would lead to serve complications and lack of analytical solutions, which may be considered an advantage of the MHD method.

The description of many different approaches to the equation of state problem in a non-ideal plasma can be found in the book [321c].

Problem. Find a smooth interpolation of the neutron star cold matter equation of state.

Solution. An interpolation smooth to the first derivative of the equation of state $P(\varrho)$ from Table 1.9 [479] is given by the following formulae (ϱ is the total density of the mass-energy with interactions taken into account):

$$P = \begin{cases} P^{(1)} = b_1 \varrho^{5/3}/(1 + c_1 \varrho^{1/3}), & \varrho \leq \varrho_1 \\ P^{(k)} = a \times 10^{b_k (\lg \varrho - 8.419)^{c_k}} & \varrho_{k-1} \leq \varrho \leq \varrho_k, \end{cases} \tag{1}$$
$$k = 2, 3, \ldots, 6;$$

$$a = 10^{26.1673},$$
$$b_1 = 10^{12.40483},$$
$$b_2 = 1,$$
$$b_3 = 2.5032,$$
$$b_4 = 0.70401515,$$
$$b_5 = 0.16445926,$$
$$b_6 = 0.86746415.$$

$$c_1 = 10^{-2.257},$$
$$c_2 = 1.1598,$$
$$c_3 = 0.356293,$$
$$c_4 = 1.2972138,$$
$$c_5 = 2.117802,$$
$$c_6 = 1.237985,$$

$$\varrho_1 = 10^{9.419},$$
$$\varrho_2 = 10^{11.5519},$$
$$\varrho_3 = 10^{12.26939},$$
$$\varrho_4 = 10^{14.302},$$
$$\varrho_5 = 10^{15.0388},$$
$$\varrho_6 \gg \varrho_5,$$

Continuity of the derivatives $dP/d\varrho$ in the points $\varrho = \varrho_k$ from (1) is achieved by smoothing in the form

$$P(\varrho) = \begin{cases} P^{(k)}, & \varrho \in [\varrho_{k-1} + \xi_{k-1}, \varrho_k - \xi_k], \quad k = 1, 2, \ldots, 6, \\ \theta_k P^{(k)} + (1 - \theta_k) P^{(k+1)}, & \varrho \in [\varrho_k - \xi_k, \varrho_k + \xi_k], \\ \varrho_0 + \xi_0 = 0, \end{cases} \tag{2}$$

where

$$\theta_k = \theta_k(\varrho) = \frac{1}{2} - \frac{1}{2} \sin \left(\frac{\pi}{2\xi_k} (\varrho - \varrho_k) \right),$$

$$\xi_k = 0.01 \varrho_k.$$

2 Radiative Energy Transfer.
Heat Conduction

The energy produced in the central regions by nuclear reactions and gravitational contraction is radiated at the stellar surface. Photons usually have a much longer mean free path than particles, and in the absence of convection radiative transfer by photons is the principal energy transfer mechanism. In very dense stars (such as white dwarfs and neutron stars), in which matter is degenerate, the radiative energy density is relatively small whereas the mean free path of the degenerate gas particles is large. Heat conduction by electrons is therefore dominant in white dwarfs, whereas in neutron stars neutron heat conduction is also important[1].

Loss of energy in a star occurs because photons emerge through transparent layers and escape freely to infinity. The radiative spectrum forms in an intermediate region with optical depth $\tau = \int_{r_f}^{\infty} \sigma_{fm} n dr \leq 1$ (where σ_{fm} is the cross-section of interaction between a photon and a matter particle, n is the number density of particles, and r_f is the radius of the photosphere). The most general description of radiation in stellar interiors, both for transparent and opaque regions, is provided by the radiative transfer equation, which is the kinetic equation for photons.

2.1 Radiative Transfer Equation

All photons travel at the speed of light c and are not influenced by any external force. (We put aside strong gravitational fields which lead, according to general relativity, to the gravitational redshift and curvature of the path.) The photon distribution function $f_\nu(\mathrm{cm}^{-3}\,\mathrm{Hz}^{-1}\,\mathrm{ster}^{-1})$ depends on time t, coordinates x_i, photon energy $h\nu$, and the direction of its movement defined by a unit vector l_i [159, 215]. The number of photons dn_ν in a volume element dV in the frequency range $d\nu$ moving within a solid angle $d\Omega$ is given by

$$dn_\nu = f_\nu(t, x_i, \nu, l_i) dV d\nu d\Omega. \qquad (2.1.1)$$

---

[1] Neutrino energy losses, which predominate at very high temperatures, are discussed in Sect. 5.2.

As free photons travel in straight lines with the same velocity c, the amount of energy dE_ν passing in time dt in the range $d\nu d\Omega$ through area $d\sigma$ equals the photon energy within the volume

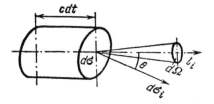

Fig. 2.1. Geometry for deriving the energy transfer equation.

$$dV = (l_i d\sigma_i)cdt, \tag{2.1.2}$$

where $d\sigma_i$ is normal to this area and of length $d\sigma$ (Fig. 2.1). The amount of energy is

$$dE_\nu = h\nu dn_\nu = f_\nu d\nu d\Omega h\nu dV = f_\nu ch\nu d\nu d\Omega dt(l_i d\sigma_i). \tag{2.1.3}$$

In the theory of radiative transfer the function f_ν, which determines the concentration of quanta according to (2.1.1), is usually replaced by the spectral intensity $I_\nu(\text{erg cm}^{-2} \text{ s}^{-1} \text{ ster}^{-1} \text{ Hz}^{-1})$, where

$$I_\nu = ch\nu f_\nu, \tag{2.1.4}$$

is related to the energy flux (2.1.3). For this reason, it is natural to use the radiation intensity as a variable for continuously moving photons, rather than their concentration.

Any loss in the intensity I_ν of a beam as it passes through a region of matter of thickness ds is caused by absorption, and may be written as

$$dI_\nu \mid_a = -\alpha_\nu \rho I_\nu ds. \tag{2.1.5}$$

The corresponding gain in intensity due to proper emission of the medium is

$$dI_\nu \mid_e = j_\nu \rho ds. \tag{2.1.6}$$

Here, $\alpha_\nu(\text{cm}^2 \text{ g}^{-1})$ is the monochromatic mass absorption coefficient, and $j_\nu(\text{erg g}^{-1} \text{ s}^{-1} \text{ ster}^{-1} \text{ Hz}^{-1})$ is the mass emission coefficient. Part of the radiation lost by the incident beam is reemitted in other directions. If the frequency of the radiation is conserved this process is called coherent scattering. The change in intensity caused by scattering is given by

$$dI_\nu \mid_s = \left[-\sigma_\nu I_\nu + \int_{\Omega'} \sigma_{\nu l} \left(l_i' \cdot l_i \right) I_\nu \left(l_i' \right) \frac{d\Omega'}{4\pi} \right] \rho ds. \tag{2.1.7}$$

The first term in the right hand side involves a decrease in the intensity in the direction l_i, whereas the second term involves an increase in the radiation intensity in the direction l_i caused by the gain in photons scattered out of directions l_i'. The cross-section depends on the angle between l_i and l_i', i.e. on $\cos\tilde{\theta} = l_i \cdot l_i'$. The scattering coefficient $\sigma_\nu(\text{cm}^2\,\text{g}^{-1})$ averaged over all directions is given by

$$\sigma_\nu = \int_0^\pi \sigma_{\nu l}(\cos\tilde{\theta})\frac{\sin\tilde{\theta}d\tilde{\theta}}{2}\left(=\int\sigma_{\nu l}\frac{d\Omega'}{4\pi}=\int\sigma_{\nu l}\frac{d\Omega}{4\pi}\right), \qquad (2.1.8)$$

and normalization of $\sigma_{\nu l}$ is chosen so that $\sigma_{\nu l} = \sigma_\nu$ for isotropic scattering. Equating the convective part of the change in intensity

$$\left(\frac{1}{c}\frac{\partial I_\nu}{\partial t}+\frac{\partial I}{\partial s}\right)ds$$

to the sum (2.1.5–7), we obtain the transfer equation [219]

$$\frac{1}{c}\frac{\partial I_\nu}{\partial t}+\frac{\partial I_\nu}{\partial s}=-\alpha_\nu\rho I_\nu+j_\nu\rho-\sigma_\nu\rho I_\nu+\rho\int_{\Omega'}\sigma_{\nu l}(\cos\tilde{\theta})I_\nu(l_i')\frac{d\Omega'}{4\pi}. \quad (2.1.9)$$

Here, $\partial I_\nu/\partial s$ is the derivative along the beam direction. In Cartesian coordinates

$$\frac{\partial I_\nu}{\partial s}=l_i\frac{\partial I_\nu}{\partial x_i}. \qquad (2.1.10)$$

In the case of plane symmetry $I_\nu(z,\theta,\nu,t)$, $\frac{\partial I_\nu}{\partial x}=\frac{\partial I_\nu}{\partial y}=0$, $l_z=\cos\theta$, where θ measures the direction of the beam with respect to the z axis. The left-hand side of (2.1.9) then becomes

$$\frac{1}{c}\frac{\partial I_\nu}{\partial t}+\cos\theta\frac{\partial I}{\partial z}. \qquad (2.1.11)$$

In the case of spherical symmetry $I_\nu = I(r,\theta,\nu,t)$, where θ is the angle between the direction of the beam and the radius vector from the origin $\frac{\partial r}{\partial s}=\cos\theta$, $\frac{\partial\theta}{\partial s}=-\sin\theta/r$ (Fig. 2.2). In this case, the left-hand side of (2.1.9) takes the form

$$\frac{1}{c}\frac{\partial I_\nu}{\partial t}+\cos\theta\frac{\partial I_\nu}{\partial r}-\frac{\sin\theta}{r}\frac{\partial I_\nu}{\partial\theta}. \qquad (2.1.12)$$

If the medium is in local thermodynamic equilibrium (LTE) j_ν and α_ν are related by Kirchhoff's law[2]

---

[2] To account for stimulated emission the quantity α_ν in (2.1.9) has to be replaced by $\alpha_\nu^* = \alpha_\nu(1-e^{-h\nu/kT})$ (2.1.13a).

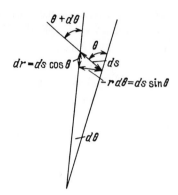

Fig. 2.2. Radiation transfer in spherical geometry.

$$j_\nu = \alpha_\nu B_\nu(T), \tag{2.1.13}$$

where

$$B_\nu(T) = \frac{2h\nu^3}{c^2} \frac{1}{e^{\frac{h\nu}{kT}} - 1} \quad (\text{erg cm}^{-2}\ \text{s}^{-1}\ \text{ster}^{-1}\ \text{Hz}^{-1}) \tag{2.1.14}$$

is the equilibrium radiation intensity, also called the Planck function [145].

The angle $\tilde\theta$ measures the direction of the scattered ray with respect to the direction of the incident ray, $l_i l_i' = \cos\tilde\theta$, and serves to express the angular dependence of the cross-section. It may be written in terms of the angles of incident (θ, ϕ) and scattered (θ', ϕ') rays as follows:

$$\cos^2\tilde\theta = \cos^2\theta\cos^2\theta' + \sin^2\theta\sin^2\theta'\cos^2(\phi - \phi')$$
$$+ 2\cos\theta\cos\theta'\sin\theta\sin\theta'\cos(\phi - \phi'). \tag{2.1.15}$$

If the frequency changes upon scattering, the scattering is called incoherent. The last two terms in (2.1.9) must then be replaced by

$$\rho \int \left[\frac{\nu}{\nu'} K(\tilde\theta, \nu' \to \nu) I_{\nu'}' \left(1 + \frac{c^2}{2h\nu^3} I_\nu \right) - K(\tilde\theta, \nu \to \nu') \right.$$
$$\left. \times I_\nu \left(1 + \frac{c^2}{2h\nu'^3} I_{\nu'}' \right) \right] d\nu' \frac{d\Omega'}{4\pi}, \tag{2.1.16}$$

$$\int K(\tilde\theta, \nu \to \nu') d\nu' \frac{d\Omega'}{4\pi} = \sigma_\nu,$$

involving the stimulated scattering processes (see Sect. 5.4) which are not present in coherent scattering. Instead of using the real angular dependence for scattering, the isotropic scattering approximation is often used, in which the last term in (2.1.9), upon integrating over $d\phi'$, takes the form

$$\frac{1}{2}\rho\sigma_\nu \int_0^\pi I_\nu(\theta')\sin\theta' d\theta'. \tag{2.1.17}$$

2.2 The Eddington Approximation. Radiative Heat Transfer

Photons in stellar interiors usually have mean free paths much smaller than the stellar size; the radiative heat transfer approximation is then quite appropriate for treatment of the radiative transfer. The parameters of the photosphere with $\tau \leq 1$ represent boundary conditions for the equations of stellar structure (see Chap. 6). This treatment only uses averaged properties of the radiation field. The Eddington approximation allows us to obtain a unique description of both the transparent and opaque regions, and is therefore often used to evaluate the boundary conditions [197]. Exact solutions of the transfer equation are treated by stellar atmosphere theory, which provides theoretical spectra and compares them with observations [159, 197, 219].

2.2.1 Moments of the Transfer Equation

The approximate description of the radiation field is based on the treatment of angle- and frequency-averaged quantities. In order to obtain these, we consider moments of the transfer equation (2.1.9). Integrating (2.1.9) over $d\Omega d\nu$ and using (2.1.10) written in Cartesian coordinates for the case of LTE, (2.1.13–14) gives

$$\frac{\partial S}{\partial t} + \frac{\partial F_i}{\partial x_i} = -\rho c[\int_0^\infty \alpha_\nu S_\nu d\nu - \alpha_P B(T)]. \tag{2.2.1}$$

Integrating analogously with the weight function l_i yields

$$\frac{1}{c}\frac{\partial F_i}{\partial t} + c\frac{\partial P_{ik}}{\partial x_k} = -\kappa\rho F_i. \tag{2.2.2}$$

Here, S_ν (erg cm^{-3} Hz^{-1}) and S(erg cm^{-3}) are the monochromatic and total radiative energy density:

$$S_\nu = \frac{1}{c}\int_\Omega I_\nu d\Omega, \quad S = \int_0^\infty S_\nu d\nu; \tag{2.2.3}$$

F_i is the radiative energy flux density (erg cm^{-2} s^{-1}):

$$F_i = \int_{\Omega,\nu} I_\nu l_i d\Omega d\nu; \tag{2.2.4}$$

and $B(T)$ is the density of equilibrium radiation energy (erg cm^{-3}):

$$B(T) = \frac{1}{c}\int_{\Omega,\nu} B_\nu(T)d\Omega d\nu = \frac{8\pi(kT)^4}{c^3h^3}\int_0^\infty \frac{x^3dx}{e^x - 1} = aT^4,$$

$$\int_0^\infty \frac{x^3dx}{e^x - 1} = \frac{\pi^4}{15} \quad [145], \quad a = \frac{8\pi^5k^4}{15c^3h^3} \quad \text{(see Sect. 1.1).} \tag{2.2.5}$$

The right-hand terms of (2.1.9) involving scattering cancel after integrating (2.2.1) by (2.1.8) which yields the equality

$$\sigma_\nu \int_\Omega I_\nu(l_i)d\Omega = \int_\Omega d\Omega \int_{\Omega'} \frac{d\Omega'}{4\pi}\sigma(\cos\tilde{\theta})I_\nu(l_i'). \tag{2.2.6}$$

The quantity P_{ik} represents the radiation field pressure tensor (erg cm^{-3}) defined by

$$P_{ik} = \frac{1}{c}\int_{\Omega,\nu} I_\nu l_i l_k d\Omega d\nu. \tag{2.2.7}$$

The term in (2.2.2) involving radiation cancels by $\int_\Omega \cos\theta d\Omega = 0$ and the last term in (2.1.9) becomes zero only for even functions $\sigma_{\nu l}(x)$ satisfying the relation $\int \sigma_{\nu l}(\cos\theta)\cos\theta d\Omega = 0$. If the electrons are relativistic [21] this term remains after integrating and should be added to (2.2.2).

As the radiation flux is dominated by the frequencies at which absorption and scattering are minimized, the Rosseland average of $(\alpha_\nu + \sigma_\nu)$ with respect to frequency

$$\frac{1}{\kappa} = \frac{\int_0^\infty \frac{1}{\kappa_\nu}\frac{dB_\nu(T)}{dT}d\nu}{\int_0^\infty \frac{dB_\nu(T)}{dT}d\nu}, \tag{2.2.8}$$

is used in (2.2.2) [229] where $B_\nu(T)$ is given by (2.1.14) and $\kappa_\nu = \alpha_\nu + \sigma_\nu$. In (2.2.1) using the Planck average of $\langle\alpha_\nu\rangle = \alpha_P$

$$\alpha_P = \frac{4\pi\int_0^\infty \alpha_\nu B_\nu(T)d\nu}{cB(T)}, \tag{2.2.9}$$

is more correct [197].

2.2.2 Plane Atmosphere

In the case of plane symmetry, only the following F_i and P_{ik} components remain in (2.2.1) and (2.2.2):

$$F = F_z = \int_{\Omega,\nu} I_\nu \cos\theta d\Omega d\nu, \tag{2.2.10}$$

$$P = P_{zz} = \frac{1}{c}\int_{\Omega,\nu} I_\nu \cos^2\theta d\Omega d\nu. \tag{2.2.11}$$

We then have for the stationary case

$$\frac{dF}{dz} = -\rho c\left[\int_0^\infty \alpha_\nu S_\nu d\nu - \alpha_P B(T)\right], \tag{2.2.12}$$

$$c\frac{dP}{dz} = -\kappa\rho F. \tag{2.2.13}$$

In a stellar atmosphere the condition of radiative equilibrium is taken to correspond to that of constant flux F in (2.2.12), that is $F = \text{const.} = L/4\pi R^2$, where L is the luminosity and R is the stellar radius, so

$$\int_0^\infty \alpha_\nu S_\nu d\nu = \alpha_P B(T). \tag{2.2.14}$$

In the Eddington approximation it is assumed that

$$P = \frac{1}{3}S, \tag{2.2.15}$$

just as in the case of equilibrium radiation. This follows by removing the average $\langle \cos^2 \theta \rangle = \frac{1}{4\pi} \int_\Omega \cos^2 \theta d\Omega = 1/3$ from the integral in (2.2.11). We then have for a grey atmosphere, when $\alpha_\nu = \alpha_P = \kappa = \text{const.}$,

$$S = B(T) = \alpha T^4. \tag{2.2.16}$$

Defining the optical depth

$$\tau = \int_z^\infty \kappa \rho dz \tag{2.2.17}$$

we therefore solve (2.2.13) in the form

$$\frac{dS}{d\tau} = 3\frac{F}{c}, \quad S = 3\frac{F}{c}\tau + \text{const.} \tag{2.2.18}$$

In order to find the constant in (2.2.18) it should be remembered that only photons moving in the directions $0 \le \theta \le \pi/2$ remain as $\tau \to 0$. The average $\langle \cos \theta \rangle = 1/2$ for this interval, and Eddington assumed that

$$F = \int_0^\infty d\nu \int_0^{\pi/2} I_\nu \cos \theta d\Omega$$
$$\approx \frac{1}{2} \int_0^\infty d\nu \int_0^{\pi/2} I_\nu d\Omega = \frac{cS(0)}{2}. \tag{2.2.19}$$

We then find $S(0) = \text{const.} = 2\frac{F}{c}$ in (2.2.18) and hence

$$S = 2\frac{F}{c}\left(1 + \frac{3}{2}\tau\right). \tag{2.2.20}$$

Taking into account (2.2.16) and setting $\frac{acT_{\text{ef}}^4}{4} = F$ this can be transformed into

$$T^4 = \frac{1}{2}T_{\text{ef}}^4\left(1 + \frac{3}{2}\tau\right). \tag{2.2.21}$$

The boundary of the photosphere is determined by $T = T_{ef}$, and consequently by $\tau = 2/3 = 0.667$; the outer boundary ($\tau = 0$) has temperature $T(0) = \frac{1}{2^{1/4}} T_{ef} = 0.841 T_{ef}$.

It has been shown by Chandrasekhar [219] that a more rigorous approximation is to use the boundary condition

$$F = \frac{cS(0)}{\sqrt{3}}. \tag{2.2.22}$$

This leads to

$$\text{const.} = \frac{\sqrt{3}F}{c} \qquad S = 3\frac{F}{c}\left(\frac{1}{\sqrt{3}} + \tau\right),$$

$$T^4 = \frac{3}{4}T_{ef}^4\left(\frac{1}{\sqrt{3}} + \tau\right), \tag{2.2.23}$$

with the photosphere boundary given by $\tau = 4/3 - \frac{1}{\sqrt{3}} = 0.756$, and

$$T(0) = \left(\frac{\sqrt{3}}{4}\right)^{1/4} T_{ef} = 0.811 T_{ef}.$$

2.2.3 Spherical Symmetry

In this case, we have non-zero terms

$$F = F_r = \int_{\Omega,\nu} I_\nu \cos\theta d\Omega d\nu, \tag{2.2.24}$$

$$P = P_{rr} = \frac{1}{c}\int_{\Omega,\nu} I_\nu \cos^2\theta d\Omega d\nu, \tag{2.2.25}$$

but, contrary to (2.2.10–11), θ is measured from the radius vector at a given point and changes with non-radial movement. Using the left-hand side of the transfer equation in the form (2.1.12) gives the following for the moments in the stationary case:

$$\frac{dF}{dr} + 2\frac{F}{r} = -\rho c\left[\int_0^\infty \alpha_\nu S_\nu d\nu - \alpha_P B(T)\right], \tag{2.2.26}$$

$$\frac{dP}{dr} - \frac{S - 3P}{r} = -\kappa\rho\frac{F}{c}. \tag{2.2.27}$$

In the case of spherical symmetry $F \sim 1/r^2$, $S \sim 1/r^2$ as $\tau \to 0$, therefore instead of (2.2.15) we now use the Eddington variable factor [159]:

$$P = fS, \qquad f = \begin{cases} 1/3 & \text{for } \tau \to \infty, \\ 1 & \text{for } \tau \to 0. \end{cases} \tag{2.2.28}$$

In radiative equilibrium

$$F = \frac{L}{4\pi r^2}. \tag{2.2.29}$$

As $\tau \to 0$ and large radii the radiation is considerably diluted, $\cos\theta \approx 1$ and instead of (2.2.19) we have

$$F = \frac{L}{4\pi r^2} = cS(\tau \to 0). \tag{2.2.30}$$

From (2.2.28–30) the solution of (2.2.27) as $\tau \to 0$ has the form

$$S = \frac{L}{4\pi r^2 c}(\tau + 1). \tag{2.2.31}$$

At large τ the radiation in stellar interiors is always close to equilibrium and relations (2.2.15), (2.2.16) are valid. From (2.2.27) we then have the radiation heat conduction approximation

$$L(r) = -\frac{4acT^3}{3\kappa\rho}\frac{dT}{dr}4\pi r^2. \tag{2.2.32}$$

Substituting (2.2.16) into (2.2.31), solving via T and differentiating gives, as $\tau \to 0$,

$$\frac{dT}{dr} = -\frac{1}{2}T_0 R_0^{1/2} r^{-3/2}, \quad T_0^4 = \frac{L}{4\pi acR_0^2}. \tag{2.2.33}$$

Interpolating (2.2.32) and (2.2.33) then yields an approximate equation for determining the temperature dependence $T(r)$ at any optical depth τ [520]:

$$\frac{dT}{dr} = -\frac{3\kappa\rho L}{16\pi acT^3 r^2} - \frac{f}{2}T_0 R_0^{1/2} r^{-3/2},$$

$$f = \begin{cases} 1 - \frac{3\tau}{2} & \text{for } \tau < 2/3 \\ 0 & \text{for } \tau \geq 2/3. \end{cases} \tag{2.2.34}$$

According to calculations [520] the temperature distribution in stellar atmospheres does not sensitively depend on the values of R_0 and T_0 corresponding to $\tau = 0$.

Instead of applying the variable Eddington factor (2.2.28) we may make use of two free parameters [18]

$$P = \mu_0^2 S, \quad 3P - S = 2\frac{F}{c}(\Lambda - \mu_0) \tag{2.2.35}$$

for connecting the three moments S, F and P in the spherical case. We then have, instead of (2.2.27),

$$\mu_0^2 \frac{dS}{dr} + 2(\Lambda - \mu_0)\frac{F}{cr} + \kappa\rho\frac{F}{c} = 0. \tag{2.2.36}$$

In [18] it has been assumed, as in the Eddington approximation, that $\mu_0 = \frac{1}{\sqrt{3}}$, and two possibilities for Λ have been considered:

$$\Lambda = 2\mu_0 \quad \text{and} \quad \Lambda = \mu_0 + \mu_0^2. \tag{2.2.37}$$

In both these cases at small τ the solution is found to be more accurate than in the Eddington approximation $\Lambda = \mu_0$. At large τ, relations (2.2.37) give a significantly greater error than $\Lambda = \mu_0$, which suggests that we should consider the interpolation

$$\Lambda = \mu_0 + \frac{\mu_0^2}{1 + (\kappa r \rho)^2}.$$

Another form of the Eddington approximation for the case of spherical symmetry is given in [614, 615].

2.3 Opacity: Absorption, Scattering and Electron Heat Conduction. The Rosseland Mean

As a photon with energy $h\nu \ll m_e c^2$ travels through matter, it can undergo scattering, or absorption and transfer of energy to the electron in three ways: (1) Increase of velocity of the free electron; (2) Transition of the electron from a bound state to a free state; and (3) Upward transition of the electron from a lower bound state to a higher bound state.

Scattering and electron heat conduction can play an important role in heat transfer. In order to obtain a rough description of the absorption processes one can treat them approximately but the quantitative results will then be too poor. An accurate treatment of these processes, especially of bound–bound transitions, involves many quantum-mechanical calculations related to the different states of ionization, excitation states of various elements and transitions from one state to another, as well as very time-consuming computations. Let us consider an approximate theory of these processes.

2.3.1 Free–Free Transitions

The absorption coefficient can be calculated more easily if we make use of the principle of detailed balance.

If the concentration of electrons and hydrogen-like atoms with charge Z is unity, the relation between the bremsstrahlung cross-section σ_{eff} and the cross-section σ_{aff} for free–free absorption of a photon with energy $h\nu$ is given by [215]

$$\sigma_{eff}/\sigma_{aff} = 8\pi\nu^2/hc^2 v, \tag{2.3.1}$$

where v is the electron velocity before absorption or after emission of a photon. For non-relativistic electrons the bremsstrahlung cross-section of a photon with frequency $\nu \gg mv^3/2\pi Ze^2$ is given by the classical approximation [215, 143]

$$\sigma_{eff} = \frac{32\pi^2}{3\sqrt{3}} \frac{Z^2 e^6}{m_e^2 c^3 h^2 \nu v^2}. \tag{2.3.2}$$

Introducing for the free–free transitions a Gaunt factor g_{ff} of order unity which involves quantum corrections to the classical formula gives, for σ_{aff},

$$\sigma_{aff} = \frac{4\pi}{3\sqrt{3}} \frac{Z^2 e^6}{m_e^2 ch v v^3} g_{ff}. \tag{2.3.3}$$

In addition to the spontaneous emission determined by (2.1.13) there exists stimulated emission which, being coherent under LTE conditions, is equivalent to reducing the absorption cross-section by a factor $(1 - e^{-h\nu/kT})$, so that [229]

$$\sigma_{aff}^* = \sigma_{aff}(1 - e^{-h\nu/kT}). \tag{2.3.4}$$

Integrating over electron velocities with the aid of the Fermi function (1.2.2) and (1.2.8) and summing over different ions yields the absorption coefficient for free–free transitions

$$\alpha_\nu^{ff} = \sum_{A,Z} \frac{X_{A,Z}}{Am_u} \left[\frac{8\pi m_e^3}{h^3} \int_0^\infty \frac{\sigma_{aff}^* v^2 dv}{1 + \exp\left(\frac{mv^2}{2kT} + \alpha - \beta\right)} \right], \tag{2.3.5}$$

α_ν^{ff} has the same meaning here as α_ν^* in (2.1.13a). $X_{A,Z}$ is the concentration of ions with A baryons in the nucleus and of charge Ze. Using (1.2.48–50) we have, for non-degenerate electrons with concentration n_e,

$$\beta - \alpha = \ln\left[\frac{n_e}{2} \left(\frac{2\pi\hbar^2}{m_e kT} \right)^{3/2} \right]. \tag{2.3.6}$$

Averaging within the brackets (2.3.5) then reduces to multiplying σ_{aff}^* by

$$vn_e\left(\frac{\overline{1}}{v}\right) = vn_e\sqrt{\frac{2m_e}{\pi kT}}, \tag{2.3.7}$$

and the Gaunt factor should be Maxwell averaged [128]. The relation between β and n_e is generally determined by (1.2.48), and replacing g_{ff} by its average reduces the integral (2.3.5) to (1.2.49).

2.3.2 Bound–Free Transitions

The nature of these transitions is substantially quantum. Krammers (1923) however, obtained a semi-classical formula using the correspondence principle, which states that increasing the quantum number n causes a discrete spectrum to transform gradually into a continuous one. The energy difference between two neighboring levels of the discrete spectrum of a hydrogen-like ion is [215, 144]

$$\Delta E = \frac{2\pi^2 m_e Z^2 e^4}{h^2} \left[\frac{1}{n^2} - \frac{1}{(n+1)^2} \right] \approx \frac{4\pi^2 m_e Z^2 e^4}{n^3 h^2}. \tag{2.3.8}$$

Recalling that the state with main quantum number n and a given spin orientation has statistical weight n^2, we may rewrite the correspondence principle for the emission cross-section of a photon with energy $h\nu$ in free–bound and free–free transitions σ_{efb} and σ_{eff} in the form [215]

$$\frac{n^2}{\Delta E} \sigma_{efb} = \sigma_{eff}. \tag{2.3.9}$$

We find from (2.3.2) that the cross-section for recombination onto the level corresponding to n is

$$\sigma_{efb} = \frac{128\pi^4}{3\sqrt{3}} \frac{Z^4 e^{10}}{n^5 m_e c^3 h^4 \nu \nu^2}. \tag{2.3.10}$$

Applying the principle of detailed balance to the recombination cross-section σ_{efb} and the cross-section σ_{abf} for photo-emission at a given polarization yields [215]

$$\frac{\sigma_{efb}}{\sigma_{abf}} = \frac{2h^2 \nu^2}{m_e^2 v^2 c^2}. \tag{2.3.11}$$

The final form of the cross-section per electron for the bound–free absorption from a given level n is

$$\sigma_{abf} = \frac{64\pi^4}{3\sqrt{3}} \frac{Z^4 e^{10} m_e}{n^5 c h^6 \nu^3} g_{bf},$$

$$\sigma_{abf}^* = \sigma_{abf}(1 - e^{-h\nu/kT}), \tag{2.3.12}$$

$$h\nu > E_b = \frac{2\pi^2 m_e Z^2 e^4}{h^2 n^2},$$

where g_{bf} is the Gaunt factor for the bound–free transitions, E_b is the ionization energy, and σ_{abf}^* involves the stimulated emission effect. In order to calculate the electron degeneracy we have to multiply the cross-section in (2.3.12) by the factor

$$q_{bf} = 1 - f_e(E_f) = [1 + \exp(\beta - \alpha - E_f/kT)]^{-1}, \tag{2.3.13}$$

where $E_f = h\nu - E_b$ is the energy of the escaping electron, and f_e is defined by (1.2.2). The factor q_{bf} determines the fraction of non-filled states in the continuum for escaping electrons.

The absorption coefficient α_ν^{bf} is found by summing (2.3.12) over electrons on different levels and over different ions:

$$\alpha_\nu^{fb} = \sum_{A,Z,n} \sigma_{abf}^* q_{bf} \frac{X_{A,Z}}{Am_u} N_{A,Z,n}. \tag{2.3.14}$$

Here, $N_{A,Z,n}$ is the number of electrons on level n of an ion with mass number A and charge Ze.

2.3.3 Krammers' Formula

Consider the Rosseland mean opacities (2.2.8) for free–free and bound–free transitions separately. We neglect the contribution of heavy elements to free–free transitions as they are small compared to their contribution to bound–free transitions, and we treat hydrogen and helium as totally ionized. Substituting α_ν^{ff} from (2.3.5) for κ_ν in (2.2.8) yields, for the non-degenerate electrons,

$$\kappa_{ff} = \frac{4}{3}\sqrt{\frac{2\pi}{3}} \frac{e^6 h^2 (x_H + x_{He})}{m_e^{3/2} cm_u (kT)^{7/2}} \bar{g}_{ff} \frac{n_e}{196.5}$$

$$= 3.68 \times 10^{22} \frac{\rho}{T^{7/2}} (x_H + x_{He})(1 + x_H)\bar{g}_{ff}. \tag{2.3.15}$$

where we have used the equalities $Z_H^2/A_H = Z_{He}^2/A_{He} = 1$, $n_e \approx \frac{\rho}{2m_u}(1+x_H)$, and $\bar{\nu}^3 = 196.5(kT/h)^3$ obtained from (2.2.8).

Applying the Saha formula (1.1.8) to each level for the case of high ionization gives for $y_{i,j} \approx 1$, $g_{i,j-1}/g_{i,j} = n^2$ [229]

$$N_{A,Z,n} = n_e n^2 \frac{h^3}{2(2\pi m_e kT)^{3/2}} e^{E_b/kT}. \tag{2.3.16}$$

Substituting (2.3.12), (2.3.16) in (2.3.14) then gives approximately

$$\alpha_\nu^{bf} = \frac{4}{3}\sqrt{\frac{2\pi}{3}} \frac{e^6 h^2}{m_e^{3/2} cm_u (kT)^{7/2}} \frac{Z^2}{A} n_e g_{bf}$$

$$\times \left[\frac{1}{n} \frac{E_b}{kT} e^{E_b/kT} \left(\frac{kT}{h\nu}\right)^3 \right] x_Z. \tag{2.3.17}$$

Only heavy element contributions are taken into account here, E_b is defined in (2.3.12) and x_Z is the mass concentration of heavy elements with the ratio

Z^2/A adopted as equal for all elements. Setting $Z^2/A = 6$, $n_e = \frac{\rho}{2m_u}(1+x_H)$ and replacing κ_ν in (2.2.8) by (2.3.17) gives, upon averaging,

$$\kappa_{bf} = 4.34 \times 10^{25} \frac{\rho}{T^{7/2}} x_Z(1 + x_H)\frac{\bar{g}_{bf}}{t}. \tag{2.3.18}$$

The factor $1/t$ arises here from averaging the quantity in the brackets of (2.3.17). The range of $t/\bar{g}_{bf}$ is approximately from unity to ten. At low temperatures, with decreasing degree of ionization, relation (2.3.16) no longer holds and the value yielded by (2.3.18) becomes too high. This may be compensated by increasing t. Relations (2.3.15) and (2.3.18) are called the Krammers formulae.

2.3.4 Bound–Bound Transitions

Line absorption may play a significant role if a large portion of the emitted spectrum occurs in the absorption region. It is therefore necessary to know not only the number of lines but their line widths as well.

The intrinsic line width γ_n at a frequency ν_0 is determined by interaction with zero vacuum oscillations whereas the absorption cross-section is determined by the classical oscillator formula for resonant absorption around ν_0 [128]. This formula follows from (2.3.23) with $\nu \approx \nu_{ij} = \nu_0$:

$$\sigma_R = \frac{e^2}{m_e c} f_{ij} \frac{\gamma_n/4\pi}{(\nu - \nu_0)^2 + (\gamma_n/4\pi)^2}, \tag{2.3.19}$$

where f_{ij} is the oscillator strength representing a correction for quantum effects. Doppler broadening is due to atomic motion and the absorption cross-section has the form [128]

$$\sigma_D = \frac{\pi e^2}{m_e c} f_{ij} \left(\frac{Am_u c^2}{2\pi kT}\right)^{1/2} \frac{1}{\nu_0} \exp\left[-\frac{Am_u c^2}{2kT}\left(\frac{\nu - \nu_0}{\nu_0}\right)^2\right]. \tag{2.3.20}$$

Collision broadening arises from the finite lifetime of excited atoms as a result of encounters with surrounding particles. The cross-section is then given by (2.3.19) with γ_c instead of γ_n [128]

$$\gamma_c = \frac{4\pi k}{h} 5.3 \times 10^8 \frac{\rho y}{T^{1/2}} \frac{n^4 + n'^4}{Z^{*2} \sum_j x_j \mu_j}. \tag{2.3.21}$$

Here n and n' are the initial and final main quantum numbers, respectively, Z^* is the effective nuclear charge, y is the mean number of free electrons per atom, μ_j is the molecular mass of j-th component, x_j is its molar fraction in the mixture and T is measured in kelvin.

The Stark effect, or static broadening, is related to the change in energy levels caused by a space-variable electric field, and to atomic motion in this field. The stimulated line emission is evaluated similarly to (2.3.4). The line absorption has been taken into account in the numerical calculations of the opacity tables. The tables [128, 129, 334] have been corrected by account of larger number of lines, leading to considerable increase of opacity in some regions of parameters where line excitation is important [428a,b,c; 551a].

2.3.5 Scattering from Electrons

The Thomson coefficient for scattering by non-relativistic free electrons is frequency independent and has the form

$$\sigma_{Tl} = \frac{2\pi}{\mu_i m_u} \left(\frac{e^2}{m_e c^2}\right)^2 (1 + \cos^2 \tilde{\theta}),$$

$$\sigma_T = \int_{\tilde{\Omega}} \sigma_{Tl} \frac{d\tilde{\Omega}}{4\pi} = \frac{8\pi}{3} \left(\frac{e^2}{m_e c^2}\right)^2 \frac{1}{\mu_i m_u} (cm^2\ g^{-1}), \qquad (2.3.22)$$

$$\int_{\tilde{\Omega}} (1 + \cos^2 \tilde{\theta}) \frac{d\tilde{\Omega}}{4\pi} = 4/3,$$

where $\tilde{\theta}$ is the scattering angle, and μ_i ($= \mu_Z$ from (1.2.17) for a totally ionized plasma) is the number of nucleons per free electron. The Rayleigh coefficient for scattering by bound electrons yields the same angular distribution (2.3.22). The scattering coefficient[3] for an arbitrary oscillating electron is

$$\sigma_{se\nu} = \frac{2}{\mu_a m_u} \frac{e^2}{m_e c^2} \frac{f_{ij} \nu^2 \left(\frac{\nu}{\nu_{ij}}\right)^2 \frac{\gamma}{2\pi}}{(\nu^2 - \nu_{ij}^2)^2 + \left(\frac{\gamma}{2\pi}\right)^2 \nu^2}, \qquad (2.3.23)$$

$$\sigma_{se\nu l} = \frac{3}{4} \sigma_{se\nu} (1 + \cos^2 \tilde{\theta}),$$

where ν_{ij} is the frequency of transition between levels (for a bound electron) with oscillator strength f_{ij} and lifetime $1/\gamma$, and μ_a is the number of nucleons per scattering particle. When $\nu, \gamma \ll \nu_{ij}$ we have Rayleigh scattering with $\sigma_R \sim \nu^4$. The decay of free electron oscillations of frequency ν_0 is determined by radiation reaction [143] and by $\gamma = 8\pi^2 e^2 \nu_0^2 / 3 m_e c^2$. The coefficient for scattering of a photon with frequency ν by this oscillator is given by (2.3.23) with $\nu_{ij} = \nu_0$, $f_{ij} = 1$, $\mu_a = \mu_i$. When $\nu \gg \nu_0$ this coefficient reduces to the Thomson scattering coefficient. The quantity

$$\sigma_e = \frac{8\pi}{3} \left(\frac{e^2}{m_e c^2}\right)^2 = 6.65 \times 10^{-25}\ cm^2, \qquad (2.3.24)$$

entering in (2.3.22), is called the Thomson scattering cross-section.

---

[3] Scattering coefficient equals the cross-section multiplied by $(\mu_a m_u)^{-1}$.

The total cross-section for relativistic scattering of a photon by an electron is [21]

$$\sigma_{er} = 2\pi \left(\frac{e^2}{m_e c^2}\right)^2 \frac{1}{x} \left[\left(1 - \frac{4}{x} - \frac{8}{x^2}\right) \ln(1+x)\right.$$

$$\left. + \frac{1}{2} + \frac{8}{x} - \frac{1}{2(1+x)^2}\right]$$

$$= 2\pi \left(\frac{e^2}{m_e c^2}\right)^2 \times \begin{cases} \frac{4}{3}(1-x) & \text{at } x \ll 1, \\ \frac{1}{x}[\ln(x+1) + 1/2] & \text{at } x \gg 1, \end{cases} \qquad (2.3.25)$$

where, in an arbitrary frame of reference

$$x = \frac{(\bar{p} + \bar{k})^2}{m_e^2 c^2} - 1, \qquad (2.3.26)$$

and $\bar{p}$ and $\bar{k}$ are the electron and photon 4-momenta before the encounter. In the centre-of-mass frame

$$x = 2t_\nu(t_\nu + \sqrt{t_\nu^2 + 1}), \quad t_\nu = \frac{h\nu}{m_e c^2}. \qquad (2.3.27)$$

The scattering angle is related to the scattered photon energy so that

$$\bar{p}\bar{k} - \bar{p}\bar{k}' - \bar{k}\bar{k}' = 0, \qquad (2.3.28)$$

where $\bar{k}'$ is the 4-momentum of the scattered photon. In the laboratory frame of reference (static electron)

$$m(\nu - \nu') - \nu\nu'(1 - \cos\theta) = 0,$$

and in the centre-of-inertia frame of reference scattering is coherent with $\nu = \nu'$. Other characteristics of the angular scattering by relativistic electrons are given in [21]. The coefficients σ_ν and $\sigma_{\nu\nu'l}$ entering into (2.1.9) and (2.1.16) depend on electron momenta and should be averaged over the electron distribution.

For the case of weak degeneracy the opacity owing to relativistic scattering may be approximately obtained by setting $t_\nu = t_0 = \beta_0 \frac{kT}{m_e c^2}$ in (2.3.25) and (2.3.27). This yields

$$\kappa_{se}^r \approx \sigma_{ser}(x_0) = \frac{8\pi}{3} \left(\frac{e^2}{m_e c^2}\right)^2 \frac{f(x_0)}{m_u \mu_i} = \frac{0.40}{\mu_i} f(x_0),$$

$$f(x_0)= \frac{3}{4}\left[\left(1 - \frac{4}{x_0} - \frac{8}{x^2}\right)\ln(1+x_0) + \frac{1}{2} + \frac{8}{x_0} - \frac{1}{2(1+x_0)^2}\right]\frac{1}{x_0}$$

$$= \begin{cases} 1 - x_0 & \text{at } x_0 \ll 1, \\ \frac{3}{4x_0}\left[\ln(1+x_0) + \frac{1}{2}\right] & \text{at } x_0 \gg 1. \end{cases} \qquad (2.3.29)$$

$$x_0 = 2\beta_0\frac{kT}{m_e c^2}\left(\beta_0\frac{kT}{m_e c^2} + \sqrt{\left(\beta_0\frac{kT}{m_e c^2}\right)^2 + 1}\right).$$

In the absence of positrons and at full ionization we have $\mu_i = \mu_Z = \frac{2}{1+x_H}$. When pairs are produced positrons scatter photons just as electron do, therefore in this case

$$\mu_i = \frac{2}{1 + x_H + \frac{4n_+ m_u}{\rho}}, \qquad (2.3.30)$$

where, in equilibrium, n_+ follows from (1.2.9) and (1.2.17). In the non-relativistic case of coherent scattering there is no correction for stimulated scattering analogous to (2.3.4) (see (2.1.16)) [129, 215]. Note that weak degeneracy is only possible for $kT < m_e c^2$ (see Sect. 1.2.3).

The opacity due to photons scattering by electrons is calculated in [315], degeneracy and stimulated processes being taken into account, for $kT < m_e c^2/2$. The results are approximated [532] with an accuracy better than 10% by

$$\kappa^r_{se} \approx \sigma_T (1 + 2.7 \times 10^{11}\rho T^{-2})^{-1}\left[1 + \left(\frac{T}{4.5 \times 10^8}\right)^{0.86}\right]^{-1}, \qquad (2.3.31)$$

$$kT < \frac{1}{2}m_e c^2, \quad -10 \le \beta - \alpha \le 4,$$

where σ_T is given in (2.3.22) and $\beta - \alpha$ in (1.2.8), (1.2.17).

At $\beta_0 = 4$ the accuracy of (2.3.29) is no worse than 20% compared to the results of exact calculations [559], carried out for $kT < 0.24 m_e c^2$.

2.3.6 The Electron Heat Conduction

Heat transfer by electrons becomes important at sufficiently low temperatures, when electron degeneracy is achieved and the electron mean free path increases rapidly. The heat conduction coefficient λ_e is calculated for degenerate electrons in Sects. 2.4.3 and 2.4.4 on the basis of solving the Boltzmann equation. The corresponding opacity owing to λ_e is

$$\kappa_e = \frac{4acT^3}{3\lambda_e\rho},$$
(2.3.32)

whereas the total opacity is

$$\frac{1}{\kappa_{tot}} = \frac{1}{\kappa} + \frac{1}{\kappa_e},$$
(2.3.33)

The opacity tables from [128, 129, 428a, b; 551b] include the λ_e contribution into opacity.

2.3.7 Other Sources of Opacity

At low temperatures absorption by negative ions and molecules becomes important in the outer layers of cold stars. The most significant absorber is the hydrogen ion H^- with binding energy 0.75 eV corresponding to the wavelength 16 500 Å. Most negative ions have ionization potential ~ 1 eV. The calculations of opacity tables [250] take into account molecules containing H, C, N, O and other abundant elements. In addition, dust formation is sometimes of great importance in determining the opacity at low temperatures. In Sect. 7.2 (Vol. 2) this problem is considered in more detail.

At the very high temperatures achieved during stellar collapse, $T \geq 10^9$ K, electron–positron pairs are produced and the nuclei undergo photo-excitation and photo-fission. The pair production cross-section owing to collision between a photon with energy $h\nu$ and an ion at rest is given, to an order of magnitude, by [21]

$$\sigma_{\gamma Z\pm} \sim \left(\frac{e^2}{m_e c^2}\right)^2 \frac{Z^2}{137} \left(\ln \frac{h\nu}{2m_e c^2}\right)^{1+\frac{4m_e c^2}{h\nu}}.$$
(2.3.34)

This cross-section is small compared with the scattering cross-section (2.3.25) up to high energies $\sim 10 m_e c^2$. The cross-section for two-quanta pair production, for photons moving in opposite directions

$$\sigma_{\gamma\gamma\pm} \sim \left(\frac{e^2}{m_e c^2}\right)^2 \frac{(m_e c^2)^2}{h\nu_1 h\nu_2} \ln \frac{h\nu_1 h\nu_2}{(m_e c^2)^2}$$
(2.3.35)

can exceed $\sigma_{\gamma Z\pm}$, but as shown in [559] the scattering from pairs always dominates over both (2.3.35) and photon–photon scattering because of pair production. The cross-sections for interaction with nuclei are always less than $\sigma_{e\pm}$, and the contribution of all these processes to the opacity at high temperatures is therefore not significant under astrophysical conditions.

2.3.8 Opacity Tables

The Krammers formulae (2.3.15) and (2.3.18) are too rough, therefore necessitating the use of opacity tables to calculate stellar evolution processes. In these tables κ is obtained by numerical calculation with the aid of (2.2.8). These computations include the following processes for various element mixtures: free–free transitions (2.3.5); bound–free transitions (2.3.14); bound–bound transitions (Sect. 2.3.4); and scattering from electrons with relativistic corrections [559], Rayleigh scattering from molecular hydrogen H_2, with negative ions H^- and He^-, and molecules H_2, H_2^+ taken into account.

The concentration of electrons and concentrations of ions involving degeneracy and pressure ionization have also been computed numerically. The total opacity κ is the sum of the opacities for all different mechanisms of interaction with radiation, including scattering:

$$\kappa = \sum_i \kappa_i. \tag{2.3.36}$$

The contribution of electron heat conduction has been included by applying (2.3.33).

The opacity values from [428a] are listed in Table 2.1 for a composition close to the solar: $x_H = 0.7$, $x_{He} = 0.28$. Previous opacity tables [128,129,334] which have been used for many computations of stellar evolution (see also [319]), appear to be incomplete in accounting for bound–bound opacity in some regions of ρ and T. The corrected opacities from [428a] exceed the previous ones by several times in some regions of parameters. The opacity tables for low temperatures (values from [250*], see also [250], in Table 2.1) are given for the same composition, and include in addition a large number of molecules and silicate grains. Water vapour, titanium oxide, and dust are the most significant low-temperature contributors to the opacity. Large discrepancies existing between different authors calculating low temperature opacity tables have been demonstrated in [504a].

Table 2.1. Opacities $\log_{10} \kappa$ (cm^2 g^{-1}) for the chemical composition $x_H = 0.7$, $x_{He} = 0.28$, $x_Z = 0.02$ for $\log_{10} T(K) = 3 \div 9.3$, in steps of 0.05 and $\log_{10} \rho$(g/cm^3) = -12, -11.5, -11, -10.5, -10, -9.5, -9, -8.5, -8, -7.5, -7, -6.5, -6, -5.5, -5, -4.5, -4, -3.5, -3, -2.5, -2, -1.5, -1, -0.5, 0, 0.5, 1, 1.5, 2, 3, 4, 5, 6, 7, 8, 9, 10.

$\log_{10} T(K)$	$\log_{10} \rho$(g/cm^3)																		
	-12	-11.5	-11	-10.5	-10	-9.5	-9	-8.5	-8	-7.5	-7	-6.5	-6	-5.5	-5	-4.5	-4	-3.5	-3
(3.00)	0.579	0.581	0.582	0.581	0.582	0.582	0.582	0.582	0.582	0.582	0.582	0.582	0.582	0.583	0.584	0.586	0.587	0.588	0.5
(3.05)	0.325	0.438	0.538	0.584	0.626	0.636	0.638	0.638	0.638	0.638	0.638	0.638	0.638	0.643	0.649	0.654	0.659	0.665	0.6
(3.10)	0.071	0.155	-0.107	-0.023	0.220	0.454	0.581	0.640	0.681	0.697	0.698	0.699	0.699	0.718	0.737	0.756	0.775	0.794	0.8
(3.15)	-2.397	-2.397	-2.396	-2.398	-1.109	-0.914	-0.737	-0.585	0.057	0.483	0.622	0.688	0.733	0.770	0.807	0.844	0.881	0.919	0.9
(3.20)	-2.254	-2.252	-2.251	-2.251	-2.250	-2.252	-2.252	-2.251	-2.248	-0.901	-0.790	-0.682	-0.580	-0.536	-0.492	-0.448	-0.404	-0.360	-0.3
(3.25)	-2.089	-2.034	-2.016	-2.007	-2.003	-2.001	-2.000	-1.999	-1.998	-1.995	-1.994	-1.988	-1.978	-1.945	-1.912	-1.879	-1.847	-1.814	-1.7
(3.30)	-3.013	-2.414	-2.041	-1.882	-1.825	-1.805	-1.796	-1.791	-1.788	-1.785	-1.782	-1.779	-1.774	-1.756	-1.738	-1.719	-1.701	-1.683	-1.6
(3.35)	-4.496	-3.911	-3.197	-2.539	-2.050	-1.780	-1.667	-1.623	-1.604	-1.594	-1.587	-1.581	-1.574	-1.557	-1.540	-1.523	-1.506	-1.489	-1.4
(3.40)	-4.800	-4.733	-4.463	-3.925	-3.224	-2.540	-2.001	-1.667	-1.505	-1.434	-1.401	-1.381	-1.365	-1.327	-1.289	-1.251	-1.213	-1.175	-1.1
(3.45)	-4.429	-4.481	-4.487	-4.404	-4.146	-3.660	-3.025	-2.370	-1.825	-1.471	-1.283	-1.189	-1.136	-1.049	-0.963	-0.876	-0.790	-0.703	-0.6
(3.50)	-4.131	-4.124	-4.087	-4.009	-3.876	-3.672	-3.376	-2.978	-2.480	-1.939	-1.462	-1.136	-0.948	-0.792	-0.636	-0.480	-0.323	-0.167	-0.0
(3.55)	-4.056	-4.000	-3.895	-3.731	-3.511	-3.250	-2.967	-2.678	-2.385	-2.083	-1.775	-1.476	-1.201	-0.977	-0.752	-0.528	-0.304	-0.079	0.1
(3.60)	-3.910	-3.934	-3.874	-3.704	-3.426	-3.073	-2.696	-2.328	-1.984	-1.666	-1.363	-1.070	-0.789	-0.513	-0.237	0.039	0.314	0.590	0.8
(3.65)	-3.338	-3.474	-3.546	-3.516	-3.354	-3.059	-2.665	-2.229	-1.800	-1.399	-1.034	-0.699	-0.391	-0.076	0.238	0.553	0.868	1.182	1.4
(3.70)	-2.605	-2.781	-2.903	-2.935	-2.858	-2.681	-2.428	-2.111	-1.736	-1.320	-0.895	-0.489	-0.114	0.228	0.570	0.912	1.255	1.597	1.9
(3.75)	-1.853	-2.040	-2.183	-2.245	-2.199	-2.060	-1.859	-1.622	-1.360	-1.070	-0.744	-0.376	0.023	0.367	0.711	1.055	1.399	1.742	2.0
(3.80)	-1.145	-1.304	-1.432	-1.499	-1.484	-1.383	-1.223	-1.024	-0.807	-0.574	-0.324	-0.049	0.256	0.570	0.883	1.197	1.510	1.824	2.1
(3.85)	-0.587	-0.649	-0.707	-0.738	-0.724	-0.657	-0.545	-0.396	-0.220	-0.028	0.180	0.409	0.662	0.798	1.066	1.335	1.603	1.871	2.1
(3.90)	-0.362	-0.264	-0.162	-0.074	0.002	0.079	0.167	0.271	0.396	0.541	0.708	0.893	1.101	1.238	1.462	1.689	1.916	2.143	2.3
(3.95)	-0.363	-0.205	0.018	0.272	0.507	0.684	0.815	0.920	1.020	1.125	1.244	1.382	1.544	1.700	1.893	2.088	2.284	2.479	2.6
(4.00)	-0.416	-0.270	-0.035	0.285	0.647	0.983	1.250	1.440	1.575	1.680	1.779	1.883	2.004	2.150	2.312	2.482	2.652	2.823	2.9
(4.05)	-0.444	-0.338	-0.143	0.156	0.541	0.964	1.366	1.702	1.948	2.120	2.244	2.348	2.451	2.577	2.708	2.857	3.007	3.157	3.3
(4.10)	-0.436	-0.362	-0.216	0.027	0.368	0.785	1.241	1.686	2.069	2.365	2.578	2.730	2.850	2.962	3.078	3.209	3.345	3.482	3.6
(4.15)	-0.424	-0.369	-0.260	-0.071	0.212	0.581	1.014	1.488	1.959	2.379	2.716	2.964	3.143	3.281	3.404	3.527	3.660	3.792	3.9
(4.20)	-0.420	-0.371	-0.280	-0.127	0.107	0.426	0.817	1.269	1.756	2.242	2.686	3.054	3.328	3.525	3.678	3.811	3.950	4.088	4.2
(4.25)	-0.424	-0.384	-0.308	-0.178	0.022	0.302	0.658	1.081	1.553	2.048	2.544	3.006	3.390	3.681	3.893	4.060	4.211	4.363	4.5
(4.30)	-0.425	-0.398	-0.337	-0.230	-0.058	0.191	0.521	0.925	1.383	1.872	2.377	2.882	3.350	3.741	4.039	4.264	4.441	4.614	4.7
(4.35)	-0.416	-0.393	-0.347	-0.263	-0.121	0.098	0.404	0.790	1.240	1.728	2.236	2.752	3.258	3.723	4.111	4.410	4.634	4.831	5.0
(4.40)	-0.402	-0.382	-0.336	-0.263	-0.147	0.035	0.307	0.675	1.118	1.607	2.120	2.645	3.168	3.671	4.123	4.497	4.781	4.998	5.1
(4.45)	-0.395	-0.375	-0.329	-0.253	-0.143	0.016	0.249	0.583	1.008	1.495	2.011	2.544	3.084	3.612	4.103	4.530	4.876	5.133	5.3
(4.50)	-0.394	-0.372	-0.337	-0.258	-0.139	0.024	0.240	0.536	0.927	1.397	1.911	2.445	2.996	3.543	4.065	4.527	4.917	5.210	5.4
(4.55)	-0.398	-0.379	-0.351	-0.284	-0.164	0.012	0.244	0.535	0.899	1.339	1.835	2.360	2.904	3.456	3.996	4.493	4.911	5.239	5.4
(4.60)	-0.404	-0.387	-0.368	-0.318	-0.213	-0.043	0.199	0.512	0.886	1.314	1.787	2.290	2.814	3.353	3.898	4.414	4.855	5.208	5.4
(4.65)	-0.412	-0.398	-0.381	-0.349	-0.268	-0.121	0.104	0.419	0.812	1.260	1.734	2.223	2.724	3.242	3.765	4.275	4.741	5.092	5.6
(4.70)	-0.416	-0.404	-0.390	-0.371	-0.315	-0.199	-0.010	0.275	0.655	1.110	1.604	2.107	2.607	3.110	3.612	4.089	4.541	4.904	5.2
(4.75)	-0.421	-0.411	-0.397	-0.381	-0.344	-0.260	-0.109	0.132	0.471	0.898	1.387	1.906	2.427	2.939	3.435	3.893	4.309	4.728	5.1
(4.80)	-0.425	-0.415	-0.401	-0.385	-0.359	-0.299	-0.184	0.010	0.303	0.686	1.146	1.654	2.185	2.716	3.232	3.702	4.091	4.482	4.8
(4.85)	-0.427	-0.416	-0.403	-0.387	-0.365	-0.319	-0.230	-0.076	0.169	0.506	0.931	1.416	1.938	2.473	3.008	3.506	3.918	4.299	4.6
(4.90)	-0.425	-0.415	-0.403	-0.389	-0.370	-0.332	-0.259	-0.134	0.068	0.363	0.749	1.208	1.715	2.244	2.785	3.307	3.756	4.128	4.5
(4.95)	-0.415	-0.405	-0.393	-0.380	-0.364	-0.334	-0.273	-0.170	0.000	0.258	0.607	1.035	1.520	2.041	2.579	3.113	3.601	4.017	4.4
(5.00)	-0.387	-0.376	-0.363	-0.348	-0.332	-0.313	-0.269	-0.186	-0.045	0.182	0.499	0.900	1.362	1.867	2.394	2.928	3.441	3.900	4.3
(5.05)	-0.339	-0.324	-0.308	-0.291	-0.271	-0.255	-0.227	-0.166	-0.055	0.136	0.421	0.796	1.239	1.725	2.236	2.759	3.276	3.758	4.1
(5.10)	-0.277	-0.257	-0.235	-0.211	-0.186	-0.168	-0.147	-0.100	-0.013	0.137	0.377	0.714	1.135	1.606	2.100	2.607	3.112	3.583	3.9
(5.15)	-0.252	-0.225	-0.196	-0.164	-0.131	-0.094	-0.063	-0.016	0.061	0.183	0.378	0.664	1.045	1.493	1.976	2.467	2.950	3.402	3.7
(5.20)	-0.286	-0.255	-0.220	-0.180	-0.136	-0.088	-0.037	0.024	0.110	0.230	0.406	0.651	0.979	1.388	1.848	2.325	2.787	3.214	3.5

$\log_{10} T(K)$	$\log_{10} \rho(g/cm^3)$																	
	-2.5	-2	-1.5	-1	-0.5	0	0.5	1	1.5	2	3	4	5	6	7	8	9	10
00)	0.590	0.589	0.562	0.341	-0.386	-1.357	-2.353	-3.353	-4.342	-5.304	-7.264	-9.243	-11.215	-13.126	-14.860	-16.366	-17.748	-19.092
05)	0.675	0.678	0.656	0.443	-0.277	-1.257	-2.253	-3.253	-4.242	-5.204	-7.164	-9.143	-11.115	-13.026	-14.760	-16.266	-17.648	-18.992
10)	0.831	0.847	0.834	0.600	-0.157	-1.156	-2.153	-3.153	-4.142	-5.104	-7.064	-9.043	-11.015	-12.926	-14.660	-16.166	-17.548	-18.892
15)	0.992	1.026	1.024	0.760	-0.038	-1.045	-2.053	-3.053	-4.042	-5.004	-6.964	-8.943	-10.915	-12.826	-14.560	-16.066	-17.448	-18.792
20)	-0.272	-0.229	-0.187	-0.163	-0.310	-0.985	-1.957	-2.953	-3.942	-4.904	-6.864	-8.843	-10.815	-12.726	-14.460	-15.966	-17.348	-18.692
25)	-1.748	-1.715	-1.682	-1.650	-1.623	-1.653	-2.028	-2.872	-3.844	-4.804	-6.764	-8.743	-10.715	-12.626	-14.360	-15.866	-17.248	-18.592
30)	-1.647	-1.629	-1.610	-1.593	-1.579	-1.614	-1.955	-2.777	-3.745	-4.704	-6.664	-8.643	-10.615	-12.526	-14.260	-15.766	-17.148	-18.492
35)	-1.456	-1.439	-1.422	-1.406	-1.395	-1.439	-1.818	-2.673	-3.644	-4.604	-6.564	-8.543	-10.515	-12.426	-14.160	-15.666	-17.048	-18.392
40)	-1.099	-1.061	-1.023	-0.986	-0.961	-1.050	-1.612	-2.561	-3.543	-4.504	-6.464	-8.443	-10.415	-12.326	-14.060	-15.566	-16.948	-18.292
45)	-0.530	-0.443	-0.357	-0.275	-0.244	-0.564	-1.431	-2.450	-3.442	-4.404	-6.364	-8.343	-10.315	-12.226	-13.960	-15.466	-16.848	-18.192
50)	0.145	0.301	0.455	0.584	0.457	-0.301	-1.307	-2.338	-3.342	-4.304	-6.264	-8.243	-10.215	-12.126	-13.860	-15.366	-16.748	-18.092
55)	0.370	0.594	0.815	0.983	0.712	-0.178	-1.196	-2.228	-3.242	-4.204	-6.164	-8.143	-10.115	-12.026	-13.760	-15.266	-16.648	-17.992
60)	1.142	1.417	1.674	1.676	0.939	-0.063	-1.087	-2.118	-3.142	-4.104	-6.064	-8.043	-10.015	-11.926	-13.660	-15.166	-16.548	-17.892
65)	1.811	2.121	2.359	2.007	1.064	0.045	-0.978	-2.008	-3.032	-4.004	-5.964	-7.943	-9.915	-11.826	-13.560	-15.066	-16.448	-17.792
70)	2.280	2.608	2.776	2.167	1.174	0.152	-0.870	-1.899	-2.921	-3.904	-5.864	-7.843	-9.815	-11.726	-13.460	-14.966	-16.348	-17.692
75)	2.429	2.756	2.908	2.278	1.281	0.259	-0.764	-1.790	-2.811	-3.804	-5.764	-7.743	-9.715	-11.626	-13.360	-14.866	-16.248	-17.592
80)	2.450	2.749	2.914	2.372	1.388	0.366	-0.657	-1.681	-2.702	-3.697	-5.664	-7.643	-9.615	-11.526	-13.260	-14.766	-16.148	-17.492
85)	2.406	2.664	2.836	2.446	1.493	0.472	-0.550	-1.573	-2.592	-3.587	-5.564	-7.543	-9.515	-11.426	-13.160	-14.666	-16.048	-17.392
90)	2.595	2.811	2.937	2.550	1.599	0.580	-0.444	-1.466	-2.483	-3.477	-5.464	-7.443	-9.415	-11.326	-13.060	-14.566	-15.948	-17.292
95)	2.867	3.046	3.113	2.675	1.708	0.687	-0.337	-1.359	-2.374	-3.368	-5.364	-7.343	-9.315	-11.226	-12.960	-14.466	-15.848	-17.192
00)	3.159	3.303	3.301	2.801	1.817	0.794	-0.230	-1.253	-2.265	-3.258	-5.264	-7.243	-9.215	-11.126	-12.860	-14.366	-15.748	-17.092
05)	3.449	3.558	3.478	2.921	1.926	0.902	-0.123	-1.146	-2.159	-3.149	-5.164	-7.143	-9.115	-11.026	-12.760	-14.266	-15.648	-16.992
10)	3.742	3.812	3.643	3.032	2.035	1.010	-0.015	-1.039	-2.052	-3.041	-5.064	-7.043	-9.015	-10.926	-12.660	-14.166	-15.548	-16.892
15)	4.035	4.061	3.792	3.133	2.146	1.118	0.092	-0.932	-1.945	-2.932	-4.954	-6.943	-8.915	-10.826	-12.560	-14.066	-15.448	-16.792
20)	4.325	4.293	3.919	3.225	2.261	1.226	0.200	-0.825	-1.838	-2.824	-4.844	-6.843	-8.815	-10.726	-12.460	-13.966	-15.348	-16.692
25)	4.601	4.492	4.024	3.310	2.377	1.334	0.307	-0.718	-1.732	-2.718	-4.735	-6.743	-8.715	-10.626	-12.360	-13.866	-15.248	-16.592
30)	4.848	4.650	4.111	3.390	2.493	1.443	0.415	-0.610	-1.625	-2.611	-4.626	-6.643	-8.615	-10.526	-12.260	-13.766	-15.148	-16.492
35)	5.055	4.768	4.185	3.467	2.604	1.551	0.523	-0.503	-1.518	-2.504	-4.516	-6.543	-8.515	-10.426	-12.160	-13.666	-15.048	-16.392
40)	5.211	4.858	4.252	3.540	2.709	1.660	0.632	-0.395	-1.410	-2.397	-4.408	-6.443	-8.415	-10.326	-12.060	-13.566	-14.948	-16.292
45)	5.332	4.931	4.315	3.610	2.806	1.769	0.740	-0.287	-1.303	-2.290	-4.299	-6.340	-8.315	-10.226	-11.960	-13.466	-14.848	-16.192
50)	5.411	4.992	4.374	3.678	2.895	1.882	0.848	-0.179	-1.196	-2.183	-4.190	-6.230	-8.215	-10.126	-11.860	-13.366	-14.748	-16.092
55)	5.457	5.044	4.431	3.745	2.977	1.999	0.957	-0.071	-1.088	-2.075	-4.083	-6.120	-8.115	-10.026	-11.760	-13.266	-14.648	-15.992
60)	5.475	5.090	4.486	3.809	3.056	2.118	1.066	0.037	-0.980	-1.968	-3.976	-6.011	-8.015	-9.926	-11.660	-13.166	-14.548	-15.892
65)	5.654	5.176	4.546	3.872	3.133	2.235	1.174	0.146	-0.872	-1.860	-3.869	-5.902	-7.915	-9.826	-11.560	-13.066	-14.448	-15.792
70)	5.586	5.211	4.599	3.933	3.207	2.347	1.283	0.254	-0.764	-1.752	-3.761	-5.793	-7.815	-9.726	-11.460	-12.966	-14.348	-15.692
75)	5.412	5.216	4.648	3.993	3.278	2.451	1.392	0.363	-0.656	-1.644	-3.654	-5.684	-7.715	-9.626	-11.360	-12.866	-14.248	-15.592
80)	5.191	5.176	4.690	4.051	3.347	2.548	1.502	0.472	-0.548	-1.536	-3.546	-5.575	-7.608	-9.526	-11.260	-12.766	-14.148	-15.492
85)	5.018	5.120	4.726	4.107	3.414	2.636	1.618	0.580	-0.439	-1.428	-3.438	-5.467	-7.499	-9.426	-11.160	-12.666	-14.048	-15.392
90)	4.881	5.066	4.760	4.161	3.479	2.717	1.738	0.689	-0.331	-1.320	-3.330	-5.359	-7.389	-9.326	-11.060	-12.566	-13.948	-15.292
95)	4.815	5.053	4.802	4.216	3.542	2.796	1.860	0.798	-0.222	-1.211	-3.222	-5.252	-7.280	-9.226	-10.960	-12.466	-13.848	-15.192
00)	4.734	5.023	4.837	4.269	3.604	2.871	1.978	0.907	-0.113	-1.103	-3.114	-5.144	-7.170	-9.126	-10.860	-12.366	-13.748	-15.092
05)	4.604	4.934	4.850	4.319	3.665	2.944	2.090	1.016	-0.004	-0.994	-3.005	-5.037	-7.062	-9.026	-10.760	-12.266	-13.648	-14.992
10)	4.398	4.751	4.808	4.360	3.723	3.014	2.195	1.125	0.104	-0.885	-2.897	-4.929	-6.953	-8.926	-10.660	-12.166	-13.548	-14.892
15)	4.159	4.509	4.685	4.380	3.778	3.081	2.290	1.238	0.213	-0.775	-2.788	-4.821	-6.844	-8.816	-10.560	-12.066	-13.448	-14.792
20)	3.919	4.245	4.484	4.357	3.825	3.146	2.376	1.357	0.322	-0.666	-2.679	-4.712	-6.736	-8.706	-10.460	-11.966	-13.348	-14.692

Table 2.1. Continued

$\log_{10} T(K)$	$\log_{10} \rho(\mathrm{g/cm^3})$																		
	-12	-11.5	-11	-10.5	-10	-9.5	-9	-8.5	-8	-7.5	-7	-6.5	-6	-5.5	-5	-4.5	-4	-3.5	-3
(5.25)	-0.363	-0.335	-0.300	-0.260	-0.211	-0.155	-0.093	-0.017	0.086	0.224	0.407	0.641	0.933	1.294	1.717	2.172	2.618	3.022	3
(5.30)	-0.425	-0.408	-0.385	-0.354	-0.313	-0.260	-0.194	-0.115	-0.006	0.142	0.335	0.575	0.861	1.194	1.581	2.008	2.439	2.827	3
(5.35)	-0.455	-0.448	-0.437	-0.420	-0.394	-0.355	-0.300	-0.227	-0.127	0.009	0.193	0.433	0.726	1.061	1.434	1.838	2.252	2.632	2.
(5.40)	-0.465	-0.463	-0.459	-0.452	-0.440	-0.419	-0.385	-0.329	-0.247	-0.131	0.029	0.251	0.542	0.890	1.270	1.666	2.065	2.438	2
(5.45)	-0.468	-0.467	-0.466	-0.464	-0.460	-0.452	-0.435	-0.405	-0.347	-0.255	-0.119	0.076	0.352	0.701	1.093	1.494	1.885	2.251	2
(5.50)	-0.469	-0.468	-0.468	-0.467	-0.466	-0.463	-0.457	-0.444	-0.415	-0.352	-0.245	-0.080	0.171	0.511	0.909	1.319	1.712	2.075	2
(5.55)	-0.469	-0.469	-0.468	-0.468	-0.468	-0.467	-0.464	-0.459	-0.447	-0.415	-0.344	-0.219	-0.004	0.315	0.713	1.137	1.541	1.906	2
(5.60)	-0.469	-0.469	-0.469	-0.469	-0.468	-0.468	-0.467	-0.464	-0.457	-0.445	-0.408	-0.327	-0.165	0.113	0.499	0.939	1.366	1.743	2
(5.65)	-0.469	-0.469	-0.469	-0.469	-0.469	-0.468	-0.468	-0.467	-0.462	-0.455	-0.438	-0.393	-0.290	-0.079	0.268	0.716	1.175	1.579	1
(5.70)	-0.469	-0.469	-0.469	-0.469	-0.469	-0.469	-0.468	-0.468	-0.466	-0.458	-0.450	-0.424	-0.365	-0.227	0.045	0.464	0.951	1.399	1.
(5.75)	-0.469	-0.469	-0.469	-0.469	-0.469	-0.468	-0.468	-0.467	-0.464	-0.458	-0.454	-0.439	-0.400	-0.315	-0.128	0.216	0.693	1.187	1.
(5.80)	-0.470	-0.470	-0.470	-0.470	-0.470	-0.469	-0.469	-0.468	-0.465	-0.459	-0.456	-0.448	-0.419	-0.360	-0.233	0.019	0.436	0.942	1.
(5.85)	-0.470	-0.470	-0.470	-0.470	-0.470	-0.469	-0.469	-0.469	-0.467	-0.464	-0.458	-0.454	-0.432	-0.388	-0.294	-0.112	0.223	0.693	1.
(5.90)	-0.470	-0.470	-0.470	-0.470	-0.470	-0.469	-0.469	-0.469	-0.468	-0.466	-0.460	-0.457	-0.443	-0.408	-0.336	-0.196	0.063	0.468	0.
(5.95)	-0.470	-0.470	-0.470	-0.470	-0.469	-0.469	-0.469	-0.468	-0.467	-0.465	-0.460	-0.458	-0.449	-0.423	-0.368	-0.259	-0.057	0.281	0.
(6.00)	-0.471	-0.470	-0.470	-0.470	-0.470	-0.469	-0.468	-0.467	-0.465	-0.462	-0.457	-0.455	-0.449	-0.429	-0.391	-0.306	-0.147	0.132	0.
(6.05)	-0.470	-0.470	-0.470	-0.470	-0.469	-0.468	-0.467	-0.466	-0.464	-0.461	-0.457	-0.452	-0.445	-0.429	-0.400	-0.337	-0.213	0.015	0.
(6.10)	-0.470	-0.470	-0.470	-0.470	-0.469	-0.468	-0.468	-0.466	-0.465	-0.462	-0.459	-0.453	-0.443	-0.426	-0.400	-0.353	-0.257	-0.076	0.
(6.15)	-0.470	-0.470	-0.470	-0.469	-0.469	-0.468	-0.467	-0.466	-0.464	-0.462	-0.459	-0.456	-0.447	-0.430	-0.401	-0.359	-0.283	-0.143	0.
(6.20)	-0.470	-0.470	-0.469	-0.469	-0.468	-0.468	-0.467	-0.465	-0.464	-0.462	-0.460	-0.457	-0.454	-0.442	-0.414	-0.370	-0.301	-0.186	0.
(6.25)	-0.469	-0.469	-0.469	-0.468	-0.468	-0.467	-0.466	-0.465	-0.465	-0.463	-0.462	-0.461	-0.461	-0.457	-0.436	-0.394	-0.325	-0.218	-0.
(6.30)	-0.470	-0.470	-0.470	-0.469	-0.469	-0.469	-0.469	-0.468	-0.468	-0.467	-0.467	-0.466	-0.465	-0.467	-0.456	-0.424	-0.362	-0.260	-0.
(6.35)	-0.471	-0.471	-0.470	-0.470	-0.470	-0.470	-0.470	-0.469	-0.469	-0.469	-0.469	-0.468	-0.468	-0.473	-0.467	-0.446	-0.402	-0.316	-0.
(6.40)	-0.473	-0.473	-0.473	-0.473	-0.473	-0.473	-0.473	-0.473	-0.473	-0.473	-0.473	-0.473	-0.473	-0.476	-0.472	-0.458	-0.431	-0.372	-0.
(6.45)	-0.472	-0.472	-0.472	-0.472	-0.472	-0.471	-0.471	-0.471	-0.471	-0.471	-0.471	-0.470	-0.470	-0.475	-0.474	-0.465	-0.447	-0.411	-0.
(6.50)	-0.473	-0.473	-0.473	-0.473	-0.473	-0.473	-0.473	-0.473	-0.473	-0.473	-0.473	-0.473	-0.473	-0.472	-0.474	-0.469	-0.458	-0.434	-0.
(6.55)	-0.473	-0.473	-0.473	-0.473	-0.473	-0.473	-0.473	-0.473	-0.473	-0.473	-0.473	-0.473	-0.473	-0.473	-0.473	-0.471	-0.465	-0.450	-0.
(6.60)	-0.474	-0.474	-0.474	-0.474	-0.474	-0.474	-0.474	-0.474	-0.474	-0.474	-0.474	-0.474	-0.474	-0.474	-0.473	-0.472	-0.469	-0.460	-0.
(6.65)	-0.474	-0.474	-0.474	-0.474	-0.474	-0.474	-0.474	-0.474	-0.474	-0.474	-0.474	-0.474	-0.474	-0.474	-0.475	-0.474	-0.471	-0.466	-0.
(6.70)	-0.475	-0.475	-0.475	-0.475	-0.475	-0.475	-0.475	-0.475	-0.475	-0.475	-0.475	-0.475	-0.475	-0.475	-0.475	-0.475	-0.475	-0.474	-0.470
(6.75)	-0.475	-0.475	-0.475	-0.475	-0.475	-0.475	-0.475	-0.475	-0.475	-0.475	-0.474	-0.474	-0.474	-0.474	-0.474	-0.474	-0.475	-0.473	-0.
(6.80)	-0.475	-0.475	-0.475	-0.475	-0.475	-0.475	-0.475	-0.475	-0.475	-0.475	-0.476	-0.475	-0.474	-0.474	-0.474	-0.474	-0.475	-0.474	-0.
(6.85)	-0.476	-0.476	-0.476	-0.476	-0.476	-0.476	-0.476	-0.476	-0.476	-0.476	-0.476	-0.476	-0.476	-0.476	-0.475	-0.475	-0.476	-0.476	-0.
(6.90)	-0.478	-0.478	-0.478	-0.478	-0.478	-0.478	-0.478	-0.478	-0.478	-0.478	-0.478	-0.478	-0.478	-0.478	-0.478	-0.478	-0.478	-0.477	-0.
(6.95)	-0.479	-0.479	-0.479	-0.479	-0.479	-0.479	-0.479	-0.479	-0.479	-0.479	-0.479	-0.479	-0.479	-0.479	-0.479	-0.479	-0.479	-0.478	-0.
(7.00)	-0.480	-0.480	-0.480	-0.480	-0.480	-0.480	-0.480	-0.480	-0.480	-0.480	-0.480	-0.480	-0.480	-0.480	-0.480	-0.480	-0.480	-0.480	-0.
(7.05)	-0.481	-0.481	-0.481	-0.481	-0.481	-0.481	-0.481	-0.481	-0.481	-0.481	-0.481	-0.481	-0.481	-0.481	-0.481	-0.481	-0.481	-0.481	-0.
(7.10)	-0.482	-0.482	-0.482	-0.482	-0.482	-0.482	-0.482	-0.482	-0.482	-0.482	-0.482	-0.482	-0.482	-0.482	-0.482	-0.482	-0.482	-0.482	-0.
(7.15)	-0.484	-0.484	-0.484	-0.484	-0.484	-0.484	-0.484	-0.484	-0.484	-0.484	-0.484	-0.484	-0.484	-0.484	-0.484	-0.484	-0.484	-0.484	-0.
(7.20)	-0.486	-0.486	-0.486	-0.486	-0.486	-0.486	-0.486	-0.486	-0.486	-0.486	-0.486	-0.486	-0.486	-0.486	-0.486	-0.486	-0.486	-0.486	-0.
(7.25)	-0.487	-0.487	-0.487	-0.487	-0.487	-0.487	-0.487	-0.487	-0.487	-0.487	-0.487	-0.487	-0.487	-0.487	-0.487	-0.487	-0.487	-0.487	-0.
(7.30)	-0.489	-0.489	-0.489	-0.489	-0.489	-0.489	-0.489	-0.489	-0.489	-0.489	-0.489	-0.489	-0.489	-0.489	-0.489	-0.489	-0.489	-0.489	-0.
(7.35)	-0.492	-0.492	-0.492	-0.492	-0.492	-0.492	-0.492	-0.492	-0.492	-0.492	-0.492	-0.492	-0.492	-0.492	-0.492	-0.492	-0.492	-0.492	-0.
(7.40)	-0.495	-0.495	-0.495	-0.495	-0.495	-0.495	-0.495	-0.495	-0.495	-0.495	-0.495	-0.495	-0.495	-0.495	-0.495	-0.495	-0.495	-0.495	-0.
(7.45)	-0.498	-0.498	-0.498	-0.498	-0.498	-0.498	-0.498	-0.498	-0.498	-0.498	-0.498	-0.498	-0.498	-0.498	-0.498	-0.498	-0.498	-0.498	-0.
(7.50)	-0.501	-0.501	-0.501	-0.501	-0.501	-0.501	-0.501	-0.501	-0.501	-0.501	-0.501	-0.501	-0.501	-0.501	-0.501	-0.501	-0.501	-0.501	-0.
(7.55)	-0.504	-0.504	-0.504	-0.504	-0.504	-0.504	-0.504	-0.504	-0.504	-0.504	-0.504	-0.504	-0.504	-0.504	-0.504	-0.504	-0.504	-0.504	-0.
(7.60)	-0.508	-0.508	-0.508	-0.508	-0.508	-0.508	-0.508	-0.508	-0.508	-0.508	-0.508	-0.508	-0.508	-0.508	-0.508	-0.508	-0.508	-0.508	-0.
(7.65)	-0.512	-0.512	-0.512	-0.512	-0.512	-0.512	-0.512	-0.512	-0.512	-0.512	-0.512	-0.512	-0.512	-0.512	-0.512	-0.512	-0.512	-0.512	-0.

$_{10}T(K)$	$\log_{10}\rho(\text{g/cm}^3)$																	
	-2.5	-2	-1.5	-1	-0.5	0	0.5	1	1.5	2	3	4	5	6	7	8	9	10
5)	3.666	3.961	4.217	4.254	3.852	3.208	2.456	1.480	0.431	-0.557	-2.569	-4.604	-6.628	-8.597	-10.360	-11.866	-13.248	-14.592
0)	3.423	3.688	3.935	4.070	3.841	3.261	2.532	1.604	0.541	-0.447	-2.460	-4.495	-6.520	-8.488	-10.260	-11.766	-13.148	-14.492
5)	3.214	3.457	3.691	3.867	3.787	3.303	2.604	1.723	0.650	-0.338	-2.350	-4.386	-6.412	-8.379	-10.160	-11.666	-13.048	-14.392
0)	3.017	3.241	3.466	3.660	3.689	3.328	2.671	1.834	0.759	-0.228	-2.240	-4.277	-6.304	-8.269	-10.060	-11.566	-12.948	-14.292
5)	2.828	3.051	3.268	3.469	3.567	3.331	2.732	1.936	0.868	-0.119	-2.130	-4.168	-6.195	-8.160	-9.958	-11.466	-12.848	-14.192
0)	2.650	2.870	3.081	3.283	3.423	3.305	2.784	2.027	0.984	-0.009	-2.020	-4.058	-6.086	-8.051	-9.848	-11.366	-12.748	-14.092
5)	2.482	2.706	2.911	3.113	3.277	3.253	2.826	2.109	1.106	0.101	-1.909	-3.948	-5.978	-7.943	-9.739	-11.266	-12.648	-13.992
0)	2.327	2.553	2.760	2.963	3.141	3.186	2.856	2.184	1.230	0.211	-1.798	-3.838	-5.868	-7.834	-9.630	-11.166	-12.548	-13.892
5)	2.185	2.412	2.620	2.823	3.011	3.150	2.872	2.253	1.353	0.320	-1.688	-3.728	-5.759	-7.724	-9.520	-11.066	-12.448	-13.792
0)	2.051	2.284	2.492	2.697	2.891	3.019	2.875	2.316	1.469	0.430	-1.577	-3.618	-5.650	-7.615	-9.411	-10.966	-12.348	-13.692
5)	1.916	2.164	2.376	2.581	2.781	2.934	2.865	2.373	1.577	0.540	-1.465	-3.507	-5.540	-7.505	-9.302	-10.866	-12.248	-13.592
0)	1.770	2.041	2.263	2.473	2.679	2.851	2.845	2.424	1.675	0.651	-1.354	-3.396	-5.430	-7.395	-9.192	-10.761	-12.148	-13.492
5)	1.603	1.910	2.149	2.365	2.576	2.763	2.812	2.467	1.762	0.767	-1.243	-3.285	-5.319	-7.285	-9.082	-10.652	-12.048	-13.392
0)	1.414	1.767	2.033	2.257	2.475	2.673	2.765	2.499	1.841	0.886	-1.131	-3.173	-5.209	-7.175	-8.971	-10.542	-11.948	-13.292
5)	1.210	1.611	1.914	2.153	2.371	2.575	2.700	2.517	1.912	1.006	-1.019	-3.061	-5.098	-7.064	-8.860	-10.433	-11.848	-13.192
0)	1.006	1.443	1.788	2.053	2.274	2.477	2.624	2.517	1.975	1.121	-0.908	-2.949	-4.987	-6.953	-8.748	-10.324	-11.748	-13.092
5)	0.814	1.270	1.655	1.951	2.181	2.378	2.536	2.496	2.030	1.228	-0.796	-2.837	-4.875	-6.842	-8.636	-10.215	-11.648	-12.992
0)	0.638	1.099	1.515	1.843	2.088	2.280	2.439	2.453	2.072	1.325	-0.684	-2.725	-4.763	-6.730	-8.524	-10.105	-11.548	-12.892
5)	0.479	0.931	1.369	1.726	1.989	2.172	2.327	2.380	2.094	1.411	-0.572	-2.613	-4.651	-6.618	-8.412	-9.995	-11.439	-12.792
0)	0.342	0.769	1.216	1.596	1.878	2.066	2.210	2.286	2.093	1.485	-0.460	-2.500	-4.539	-6.506	-8.299	-9.883	-11.330	-12.692
5)	0.231	0.620	1.059	1.454	1.751	1.949	2.078	2.166	2.057	1.543	-0.349	-2.387	-4.427	-6.394	-8.185	-9.770	-11.221	-12.592
0)	0.143	0.490	0.905	1.303	1.610	1.819	1.941	2.036	1.994	1.584	-0.237	-2.274	-4.314	-6.281	-8.072	-9.657	-11.112	-12.492
5)	0.059	0.377	0.762	1.147	1.456	1.673	1.813	1.917	1.924	1.612	-0.125	-2.161	-4.201	-6.168	-7.958	-9.543	-11.003	-12.392
0)	-0.038	0.262	0.625	0.990	1.294	1.516	1.671	1.784	1.834	1.620	-0.022	-2.048	-4.088	-6.055	-7.844	-9.429	-10.893	-12.292
5)	-0.148	0.131	0.481	0.832	1.128	1.350	1.517	1.648	1.734	1.609	0.080	-1.935	-3.975	-5.941	-7.729	-9.315	-10.783	-12.189
0)	-0.249	-0.010	0.321	0.667	0.959	1.182	1.358	1.504	1.620	1.576	0.184	-1.822	-3.861	-5.827	-7.614	-9.201	-10.673	-12.080
5)	-0.327	-0.141	0.153	0.491	0.786	1.014	1.197	1.358	1.500	1.525	0.290	-1.709	-3.748	-5.713	-7.499	-9.085	-10.560	-11.971
0)	-0.380	-0.248	-0.007	0.308	0.609	0.845	1.034	1.205	1.371	1.453	0.397	-1.595	-3.634	-5.598	-7.383	-8.970	-10.446	-11.862
5)	-0.415	-0.328	-0.144	0.132	0.428	0.676	0.872	1.049	1.234	1.363	0.500	-1.482	-3.520	-5.484	-7.267	-8.854	-10.332	-11.753
0)	-0.438	-0.382	-0.251	-0.025	0.251	0.505	0.710	0.893	1.083	1.251	0.594	-1.370	-3.406	-5.369	-7.151	-8.738	-10.218	-11.644
5)	-0.453	-0.418	-0.329	-0.155	0.087	0.336	0.550	0.738	0.933	1.130	0.676	-1.267	-3.292	-5.254	-7.034	-8.622	-10.103	-11.534
0)	-0.441	-0.384	-0.257	-0.056	0.176	0.392	0.587	0.786	1.005	1.105	0.740	-1.164	-3.177	-5.138	-6.917	-8.505	-9.988	-11.424
5)	-0.468	-0.454	-0.419	-0.332	-0.175	0.030	0.241	0.441	0.645	0.867	0.780	-1.061	-3.063	-5.023	-6.800	-8.387	-9.872	-11.313
0)	-0.471	-0.462	-0.440	-0.383	-0.267	-0.095	0.102	0.303	0.511	0.733	0.794	-0.958	-2.949	-4.907	-6.683	-8.270	-9.756	-11.199
5)	-0.473	-0.467	-0.452	-0.415	-0.333	-0.197	-0.020	0.176	0.386	0.605	0.776	-0.855	-2.834	-4.792	-6.565	-8.152	-9.640	-11.085
0)	-0.475	-0.469	-0.458	-0.433	-0.376	-0.272	-0.120	0.063	0.269	0.486	0.734	-0.752	-2.720	-4.676	-6.448	-8.034	-9.523	-10.971
5)	-0.478	-0.472	-0.463	-0.444	-0.402	-0.323	-0.198	-0.034	0.161	0.373	0.677	-0.650	-2.611	-4.560	-6.330	-7.915	-9.406	-10.856
0)	-0.481	-0.476	-0.468	-0.452	-0.420	-0.357	-0.256	-0.113	0.065	0.265	0.610	-0.550	-2.507	-4.444	-6.212	-7.797	-9.289	-10.741
5)	-0.480	-0.474	-0.461	-0.434	-0.382	-0.299	-0.177	-0.019	0.163	0.360	0.531	-0.452	-2.404	-4.328	-6.094	-7.678	-9.171	-10.626
0)	-0.485	-0.484	-0.479	-0.469	-0.448	-0.405	-0.333	-0.229	-0.092	0.070	0.431	-0.351	-2.300	-4.212	-5.976	-7.559	-9.053	-10.510
5)	-0.487	-0.487	-0.484	-0.477	-0.462	-0.427	-0.365	-0.274	-0.155	-0.015	0.320	-0.253	-2.196	-4.096	-5.857	-7.439	-8.935	-10.394
0)	-0.490	-0.489	-0.488	-0.484	-0.474	-0.449	-0.397	-0.317	-0.212	-0.091	0.209	-0.161	-2.092	-3.980	-5.739	-7.320	-8.816	-10.277
5)	-0.492	-0.492	-0.491	-0.489	-0.483	-0.466	-0.428	-0.360	-0.267	-0.161	0.106	-0.090	-1.988	-3.863	-5.621	-7.200	-8.697	-10.160
0)	-0.494	-0.494	-0.494	-0.493	-0.489	-0.479	-0.453	-0.400	-0.321	-0.227	0.012	-0.039	-1.884	-3.758	-5.502	-7.081	-8.578	-10.043
5)	-0.497	-0.497	-0.497	-0.497	-0.494	-0.488	-0.473	-0.435	-0.370	-0.287	-0.074	-0.021	-1.782	-3.654	-5.384	-6.961	-8.459	-9.925
0)	-0.501	-0.500	-0.501	-0.500	-0.499	-0.495	-0.486	-0.461	-0.412	-0.342	-0.153	-0.028	-1.682	-3.549	-5.265	-6.841	-8.339	-9.807
5)	-0.504	-0.504	-0.504	-0.504	-0.503	-0.501	-0.496	-0.480	-0.445	-0.388	-0.225	-0.066	-1.582	-3.445	-5.147	-6.721	-7.219	-9.689
0)	-0.508	-0.509	-0.508	-0.508	-0.507	-0.506	-0.503	-0.494	-0.470	-0.426	-0.287	-0.120	-1.484	-3.340	-5.029	-6.601	-8.099	-9.570
5)	-0.512	-0.513	-0.513	-0.513	-0.512	-0.511	-0.509	-0.504	-0.488	-0.456	-0.341	-0.182	-1.386	-3.236	-4.910	-6.481	-7.979	-9.452

Table 2.1. Continued

$\log_{10} T(K)$	$\log_{10} \rho(\mathrm{g/cm^3})$																		
	-12	-11.5	-11	-10.5	-10	-9.5	-9	-8.5	-8	-7.5	-7	-6.5	-6	-5.5	-5	-4.5	-4	-3.5	-3
(7.70)	-0.517	-0.517	-0.517	-0.517	-0.517	-0.517	-0.517	-0.517	-0.517	-0.517	-0.517	-0.517	-0.517	-0.517	-0.517	-0.517	-0.517	-0.517	-0
(7.75)	-0.522	-0.522	-0.522	-0.522	-0.522	-0.522	-0.522	-0.522	-0.522	-0.522	-0.522	-0.522	-0.522	-0.522	-0.522	-0.522	-0.522	-0.522	-0
(7.80)	-0.528	-0.528	-0.528	-0.528	-0.528	-0.528	-0.528	-0.528	-0.528	-0.528	-0.528	-0.528	-0.528	-0.528	-0.528	-0.528	-0.528	-0.528	-C
(7.85)	-0.535	-0.535	-0.535	-0.535	-0.535	-0.535	-0.535	-0.535	-0.535	-0.535	-0.535	-0.535	-0.535	-0.535	-0.535	-0.535	-0.535	-0.535	-C
(7.90)	-0.542	-0.542	-0.542	-0.542	-0.542	-0.542	-0.542	-0.542	-0.542	-0.542	-0.542	-0.542	-0.542	-0.542	-0.542	-0.542	-0.542	-0.542	-0
(7.95)	-0.550	-0.550	-0.550	-0.550	-0.550	-0.550	-0.550	-0.550	-0.550	-0.550	-0.550	-0.550	-0.550	-0.550	-0.550	-0.550	-0.550	-0.550	-0
(8.00)	-0.559	-0.559	-0.559	-0.559	-0.559	-0.559	-0.559	-0.559	-0.559	-0.559	-0.559	-0.559	-0.559	-0.559	-0.559	-0.559	-0.559	-0.559	-C
(8.05)	-0.569	-0.569	-0.569	-0.569	-0.569	-0.569	-0.569	-0.569	-0.569	-0.569	-0.569	-0.569	-0.569	-0.569	-0.569	-0.569	-0.569	-0.569	-C
(8.10)	-0.580	-0.580	-0.580	-0.580	-0.580	-0.580	-0.580	-0.580	-0.580	-0.580	-0.580	-0.580	-0.580	-0.580	-0.580	-0.580	-0.580	-0.580	-0
(8.15)	-0.592	-0.592	-0.592	-0.592	-0.592	-0.592	-0.592	-0.592	-0.592	-0.592	-0.592	-0.592	-0.592	-0.592	-0.592	-0.592	-0.592	-0.592	-0
(8.20)	-0.605	-0.605	-0.605	-0.605	-0.605	-0.605	-0.605	-0.605	-0.605	-0.605	-0.605	-0.605	-0.605	-0.605	-0.605	-0.605	-0.605	-0.605	-0
(8.25)	-0.618	-0.618	-0.618	-0.618	-0.618	-0.618	-0.618	-0.618	-0.618	-0.618	-0.618	-0.618	-0.618	-0.618	-0.618	-0.618	-0.618	-0.618	-C
(8.30)	-0.632	-0.632	-0.632	-0.632	-0.632	-0.632	-0.632	-0.632	-0.632	-0.632	-0.632	-0.632	-0.632	-0.632	-0.632	-0.632	-0.632	-0.632	-0
(8.35)	-0.648	-0.648	-0.648	-0.648	-0.648	-0.648	-0.648	-0.648	-0.648	-0.648	-0.648	-0.648	-0.648	-0.648	-0.648	-0.648	-0.648	-0.648	-0
(8.40)	-0.665	-0.665	-0.665	-0.665	-0.665	-0.665	-0.665	-0.665	-0.665	-0.665	-0.665	-0.665	-0.665	-0.665	-0.665	-0.665	-0.665	-0.665	-0
(8.45)	-0.682	-0.682	-0.682	-0.682	-0.682	-0.682	-0.682	-0.682	-0.682	-0.682	-0.682	-0.682	-0.682	-0.682	-0.682	-0.682	-0.682	-0.682	-C
(8.50)	-0.700	-0.700	-0.700	-0.700	-0.700	-0.700	-0.700	-0.700	-0.700	-0.700	-0.700	-0.700	-0.700	-0.700	-0.700	-0.700	-0.700	-0.700	-0
(8.55)	-0.719	-0.719	-0.719	-0.719	-0.719	-0.719	-0.719	-0.719	-0.719	-0.719	-0.719	-0.719	-0.719	-0.719	-0.719	-0.719	-0.719	-0.719	-0
(8.60)	-0.739	-0.739	-0.739	-0.739	-0.739	-0.739	-0.739	-0.739	-0.739	-0.739	-0.739	-0.739	-0.739	-0.739	-0.739	-0.739	-0.739	-0.739	-0
(8.65)	-0.761	-0.761	-0.761	-0.761	-0.761	-0.761	-0.761	-0.761	-0.761	-0.761	-0.761	-0.761	-0.761	-0.761	-0.761	-0.761	-0.761	-0.761	-C
(8.70)	-0.784	-0.784	-0.784	-0.784	-0.784	-0.784	-0.784	-0.784	-0.784	-0.784	-0.784	-0.784	-0.784	-0.784	-0.784	-0.784	-0.784	-0.784	-0
(8.75)	-0.808	-0.808	-0.808	-0.808	-0.808	-0.808	-0.808	-0.808	-0.808	-0.808	-0.808	-0.808	-0.808	-0.808	-0.808	-0.808	-0.808	-0.808	-0
(8.80)	-0.833	-0.833	-0.833	-0.833	-0.833	-0.833	-0.833	-0.833	-0.833	-0.833	-0.833	-0.833	-0.833	-0.833	-0.833	-0.833	-0.833	-0.833	-0
(8.85)	-0.860	-0.860	-0.860	-0.860	-0.860	-0.860	-0.860	-0.860	-0.860	-0.860	-0.860	-0.860	-0.860	-0.860	-0.860	-0.860	-0.860	-0.860	-C
(8.90)	-0.887	-0.887	-0.887	-0.887	-0.887	-0.887	-0.887	-0.887	-0.887	-0.887	-0.887	-0.887	-0.887	-0.887	-0.887	-0.887	-0.887	-0.887	-C
(8.95)	-0.916	-0.916	-0.916	-0.916	-0.916	-0.916	-0.916	-0.916	-0.916	-0.916	-0.916	-0.916	-0.916	-0.916	-0.916	-0.916	-0.916	-0.916	-0
(9.00)	-0.944	-0.944	-0.944	-0.944	-0.944	-0.944	-0.944	-0.944	-0.944	-0.944	-0.944	-0.944	-0.944	-0.944	-0.944	-0.944	-0.944	-0.944	-0
(9.05)	-0.960	-0.960	-0.960	-0.960	-0.960	-0.960	-0.960	-0.960	-0.960	-0.960	-0.960	-0.960	-0.960	-0.960	-0.960	-0.960	-0.960	-0.960	-C
(9.10)	-0.969	-0.969	-0.969	-0.969	-0.969	-0.969	-0.969	-0.969	-0.969	-0.969	-0.969	-0.969	-0.969	-0.969	-0.969	-0.969	-0.969	-0.969	-0
(9.15)	-0.974	-0.974	-0.974	-0.974	-0.974	-0.974	-0.974	-0.974	-0.974	-0.974	-0.974	-0.974	-0.974	-0.974	-0.974	-0.974	-0.974	-0.974	-0
(9.20)	-0.978	-0.978	-0.978	-0.978	-0.978	-0.978	-0.978	-0.978	-0.978	-0.978	-0.978	-0.978	-0.978	-0.978	-0.978	-0.978	-0.978	-0.978	-0
(9.25)	-0.981	-0.981	-0.981	-0.981	-0.981	-0.981	-0.981	-0.981	-0.981	-0.981	-0.981	-0.981	-0.981	-0.981	-0.981	-0.981	-0.981	-0.981	-C
(9.30)	-0.982	-0.982	-0.982	-0.982	-0.982	-0.982	-0.982	-0.982	-0.982	-0.982	-0.982	-0.982	-0.982	-0.982	-0.982	-0.982	-0.982	-0.982	-C

$_{10}T(K)$	$\log_{10}\rho(g/cm^3)$																	
	-2.5	-2	-1.5	-1	-0.5	0	0.5	1	1.5	2	3	4	5	6	7	8	9	10
0)	-0.517	-0.518	-0.518	-0.518	-0.517	-0.517	-0.515	-0.512	-0.502	-0.480	-0.387	-0.246	-1.290	-3.131	-4.795	-6.362	-7.859	-9.332
5)	-0.522	-0.522	-0.524	-0.524	-0.523	-0.523	-0.522	-0.519	-0.513	-0.498	-0.426	-0.308	-1.197	-3.026	-4.690	-6.242	-7.739	-9.213
0)	-0.528	-0.528	-0.530	-0.530	-0.530	-0.529	-0.528	-0.527	-0.523	-0.513	-0.459	-0.365	-1.102	-2.925	-4.585	-6.122	-7.619	-9.093
5)	-0.535	-0.535	-0.537	-0.537	-0.537	-0.536	-0.536	-0.535	-0.533	-0.526	-0.486	-0.415	-1.009	-2.824	-4.480	-6.002	-7.498	-8.974
0)	-0.542	-0.542	-0.542	-0.544	-0.545	-0.545	-0.544	-0.543	-0.543	-0.538	-0.509	-0.460	-0.921	-2.724	-4.375	-5.883	-7.378	-8.853
5)	-0.550	-0.550	-0.550	-0.553	-0.553	-0.553	-0.553	-0.552	-0.552	-0.550	-0.528	-0.496	-0.848	-2.624	-4.270	-5.763	-7.257	-8.733
0)	-0.559	-0.559	-0.559	-0.563	-0.563	-0.563	-0.563	-0.563	-0.562	-0.560	-0.547	-0.528	-0.795	-2.523	-4.164	-5.643	-7.137	-8.613
5)	-0.569	-0.569	-0.569	-0.569	-0.569	-0.569	-0.569	-0.569	-0.569	-0.569	-0.569	-0.579	-0.824	-2.428	-4.059	-5.532	-7.017	-8.493
0)	-0.580	-0.580	-0.580	-0.580	-0.580	-0.580	-0.580	-0.580	-0.580	-0.580	-0.580	-0.588	-0.786	-2.329	-3.954	-5.427	-6.896	-8.372
5)	-0.592	-0.592	-0.592	-0.592	-0.592	-0.592	-0.592	-0.592	-0.592	-0.592	-0.592	-0.599	-0.760	-2.231	-3.854	-5.321	-6.776	-8.252
0)	-0.605	-0.605	-0.605	-0.605	-0.605	-0.605	-0.605	-0.605	-0.605	-0.605	-0.605	-0.611	-0.743	-2.133	-3.753	-5.216	-6.656	-8.131
5)	-0.618	-0.618	-0.618	-0.618	-0.618	-0.618	-0.618	-0.618	-0.618	-0.618	-0.618	-0.623	-0.732	-2.036	-3.653	-5.111	-6.536	-8.011
0)	-0.632	-0.632	-0.632	-0.632	-0.632	-0.632	-0.632	-0.632	-0.632	-0.632	-0.632	-0.636	-0.727	-1.940	-3.552	-5.005	-6.416	-7.890
5)	-0.648	-0.648	-0.648	-0.648	-0.648	-0.648	-0.648	-0.648	-0.648	-0.648	-0.648	-0.651	-0.727	-1.846	-3.452	-4.899	-6.296	-7.770
0)	-0.665	-0.665	-0.665	-0.665	-0.665	-0.665	-0.665	-0.665	-0.665	-0.665	-0.665	-0.668	-0.731	-1.748	-3.351	-4.793	-6.189	-7.649
5)	-0.682	-0.682	-0.682	-0.682	-0.682	-0.682	-0.682	-0.682	-0.682	-0.682	-0.682	-0.684	-0.737	-1.639	-3.250	-4.690	-6.084	-7.529
0)	-0.700	-0.700	-0.700	-0.700	-0.700	-0.700	-0.700	-0.700	-0.700	-0.700	-0.700	-0.702	-0.746	-1.525	-3.150	-4.589	-5.979	-7.409
5)	-0.719	-0.719	-0.719	-0.719	-0.719	-0.719	-0.719	-0.719	-0.719	-0.719	-0.719	-0.721	-0.756	-1.413	-3.050	-4.488	-5.873	-7.289
0)	-0.739	-0.739	-0.739	-0.739	-0.739	-0.739	-0.739	-0.739	-0.739	-0.739	-0.739	-0.740	-0.770	-1.314	-2.949	-4.388	-5.767	-7.168
5)	-0.761	-0.761	-0.761	-0.761	-0.761	-0.761	-0.761	-0.761	-0.761	-0.761	-0.761	-0.762	-0.786	-1.230	-2.849	-4.287	-5.662	-7.048
0)	-0.784	-0.784	-0.784	-0.784	-0.784	-0.784	-0.784	-0.784	-0.784	-0.784	-0.784	-0.785	-0.805	-1.165	-2.750	-4.186	-5.556	-6.931
5)	-0.808	-0.808	-0.808	-0.808	-0.808	-0.808	-0.808	-0.808	-0.808	-0.808	-0.808	-0.809	-0.826	-1.117	-2.650	-4.086	-5.449	-6.826
0)	-0.833	-0.833	-0.833	-0.833	-0.833	-0.833	-0.833	-0.833	-0.833	-0.833	-0.833	-0.834	-0.848	-1.086	-2.551	-3.985	-5.348	-6.721
5)	-0.860	-0.860	-0.860	-0.860	-0.860	-0.860	-0.860	-0.860	-0.860	-0.860	-0.860	-0.860	-0.872	-1.067	-2.453	-3.884	-5.247	-6.616
0)	-0.887	-0.887	-0.887	-0.887	-0.887	-0.887	-0.887	-0.887	-0.887	-0.887	-0.887	-0.887	-0.897	-1.052	-2.354	-3.783	-5.146	-6.510
5)	-0.916	-0.916	-0.916	-0.916	-0.916	-0.916	-0.916	-0.916	-0.916	-0.916	-0.916	-0.916	-0.924	-1.052	-2.243	-3.682	-5.046	-6.404
0)	-0.944	-0.944	-0.944	-0.944	-0.944	-0.944	-0.944	-0.944	-0.944	-0.944	-0.944	-0.944	-0.951	-1.056	-2.117	-3.582	-4.945	-6.298
5)	-0.960	-0.960	-0.960	-0.960-	0.960	-0.960	-0.960	-0.960	-0.960	-0.960	-0.960	-0.960	-0.966	-1.054	-1.986	-3.481	-4.844	-6.192
0)	-0.969	-0.969	-0.969	-0.969	-0.969	-0.969	-0.969	-0.969	-0.969	-0.969	-0.969	-0.969	-0.974	-1.050	-1.857	-3.380	-4.743	-6.087
5)	-0.974	-0.974	-0.974	-0.974	-0.974	-0.974	-0.974	-0.974	-0.974	-0.974	-0.974	-0.974	-0.979	-1.044	-1.738	-3.279	-4.642	-5.986
0)	-0.978	-0.978	-0.978	-0.978	-0.978	-0.978	-0.978	-0.978	-0.978	-0.978	-0.978	-0.978	-0.982	-1.039	-1.624	-3.179	-4.542	-5.886
5)	-0.981	-0.981	-0.981	-0.981	-0.981	-0.981	-0.981	-0.981	-0.981	-0.981	-0.981	-0.981	-0.985	-1.034	-1.535	-3.078	-4.441	-5.785
0)	-0.982	-0.982	-0.982	-0.982	-0.982	-0.982	-0.982	-0.982	-0.982	-0.982	-0.982	-0.982	-0.985	-1.029	-1.461	-2.978	-4.340	-5.684

2.4 Heat Conduction in Matter
at High Densities and Temperatures

Heat transfer in white dwarfs and neutron star shells is determined by degenerate electrons which may be relativistic; in the interiors of neutron stars the heat flux is dominated by that due to degenerate neutrons. The heat conduction coefficient in a gas may be evaluated from kinetic theory by solving the Boltzmann equation [222].

2.4.1 Solution of the Boltzmann Equation for Non-relativistic Gas

The solution of the Boltzmann equation for degenerate gas was first obtained in [611]. Let the material contain a mixture of light Fermi particles and heavy non-degenerate gas particles. The Boltzmann equation for a light gas in a comoving frame reads

$$
\frac{df}{dt} + v_i \frac{\partial f}{\partial r_i} + \left(F_i - \frac{dc_{0i}}{dt} \right) \frac{\partial f}{\partial v_i} - \frac{\partial f}{\partial v_i} v_k \frac{\partial c_{0i}}{\partial r_k} = J,
$$

$$
J = J_{\mathrm{nn}} + J_{\mathrm{n}N}
$$

$$
= B \int [f' f_1'(1-f)(1-f_1) - f f_1(1-f')(1-f_1')]
$$

$$
\times g_{\mathrm{nn}} W_{\mathrm{nn}}(\theta, g_{\mathrm{nn}}) d\Omega dc_{1i}
$$

$$
+ \int [f' f_N'(1-f) - f f_N(1-f')]
$$

$$
\times g_{\mathrm{n}N} W_{\mathrm{n}N}(\theta, g_{\mathrm{n}N}) d\Omega dc_{Ni}.
$$

(2.4.1)

Here $\frac{d}{dt} = \frac{\partial}{\partial t} + c_{0i} \frac{\partial}{\partial r_i}$; f, f_N are the distribution functions for light and heavy particles; the distribution functions after collisions are primed, $v_i = c_i - c_{0i}$, $v_{iN} = c_{Ni} - c_{0i}$ are the random velocities of light and heavy particles; c_i, c_{Ni} are the velocities of light and heavy particles with respect to a fixed frame, θ is the angle between the directions of relative velocities having the same absolute value g_{nn} or $g_{\mathrm{n}N}$ before ($\tilde{g}_i$) and after ($\tilde{g}_i'$) collision, $d\Omega = \sin\theta d\theta d\varphi$, $W(\theta, g)$ is the differential cross-section for the collision of particles with relative velocity g deflected by angle θ and moving after collision within the solid angle $d\Omega$; $c_{0i} = \frac{1}{\rho}(\rho_{\mathrm{n}} \langle c_i \rangle + \rho_N \langle c_{Ni} \rangle)$ is the mean mass velocity of the mixture; $\rho = \rho_{\mathrm{n}} + \rho_N$ is the matter density; and F_i is the acceleration due to external forces. The average values $\langle \varphi_i \rangle$, $\langle \varphi_{Ni} \rangle$ are defined by

$$
\langle \varphi_i \rangle = \frac{B}{n_{\mathrm{n}}} \int f \varphi_i dc_i, \quad B = \frac{2m^3}{h^3},
$$

$$
\langle \varphi_{Ni} \rangle = \frac{1}{n_N} \int f_N \varphi_{Ni} dc_{Ni}.
$$

(2.4.2)

The functions f and f_N are normalized so that

$$n_n = B \int f dc_i, \quad n_N = \int f_N dc_{Ni},$$

$$\rho = n_n m, \quad \rho_N = n_N m_N,$$

$$(2.4.3)$$

m and m_N are the light and heavy particle masses, respectively.

In the limit of small deviations from the equilibrium Fermi distribution we may seek a solution of the Boltzmann equation in the form

$$f = f_0[1 + \chi(1 - f_0)],$$

$$f_0 = \{1 + \exp[(mv^2 - 2\mu)/2kT]\}^{-1},$$

$$(2.4.4)$$

where μ is the Fermi gas chemical potential. The presence of anisotropy arising from temperature and pressure gradients in either component leads to a correction to the equilibrium distribution function which may be written in the form

$$\chi = -A_i \frac{\partial \ln T}{\partial r_i} - n_n D_i d_i \frac{G_{5/2}}{G_{3/2}},$$

$$d_i = \frac{\rho_N}{\rho} \frac{\partial \ln P_n}{\partial r_i} - \frac{\rho_n}{P_n} \frac{1}{\rho} \frac{\partial P_N}{\partial r_i}.$$

$$(2.4.5)$$

Here

$P_n = \dfrac{1}{3} n_n m \langle v^2 \rangle$ is the Fermi gas pressure,

P_N is the pressure due to other components of the mixture,

$$G_n \equiv G_n(x_0) = \frac{1}{\Gamma(n)} \int_0^\infty \frac{x^{n-1} dx}{1 + \exp(x - x_0)} = \frac{F_{n-1}(x_0)}{\Gamma(n)},$$

$$(2.4.6)$$

$$x = \frac{mv^2}{2kT}, \quad x_0 = \frac{\mu}{kT},$$

where $F_1(x_0)$ are determined above in (1.2.49).

Assume the following properties for the distribution function of heavy non-degenerate particles:

$$f_N = f_{N0}(1 + \chi_N),$$

$$\chi_N = -A_{Ni} \frac{\partial \ln T}{\partial r_i} - n_n D_{Ni} d_i \frac{G_{5/2}}{G_{3/2}},$$

$$\int f_{N0} dv_{Ni} = n_N, \quad \int v_{Ni} v_{Nk} f_{N0} dv_{Ni} = n_N \delta_{ik} \frac{kT}{m_N}.$$

$$(2.4.7)$$

The isotropic distribution function f_{N0} here can differ from Maxwellian. Substituting (2.4.4–6) into the Boltzmann equation (2.4.1) and using the transfer equations to eliminate the time derivatives gives equations [53] for A_i and $\mathcal{D}_i$ as follows

$$f_0(1 - f_0)\left(\frac{mv^2}{2kT} - \frac{5}{2}\frac{G_{5/2}}{G_{3/2}}\right)v_i = I_{nn}(A_i) + I_{nN}(A_i),$$

$$\frac{1}{n_n}f_0(1 - f_0)v_i = I_{nn}(\mathcal{D}_i) + I_{nN}(\mathcal{D}_i),$$

(2.4.8)

where $I_{nn}(R_i)$ and $I_{nN}(R_i)$ are the linearized collision integrals (2.4.1). Note that a classical treatment of the collision in (2.4.1) leads to the substitution (b - impact parameter)

$$gW(\theta, g)d\Omega \Rightarrow gb\,db\,d\varphi.$$

(2.4.9)

2.4.2 Neutron Gas Heat Conduction

Let the light particles be neutrons $m = m_n$, and the heavy particles be nuclei. As the mass fraction of neutrons can be important, their contribution to the density should be included. A neutron gas is close to perfect, and when immersed in the crystal lattice formed by nuclei may exist in the non-equilibrium layer of the neutron star shells (see Sects. 1.4.5 and 1.4.6). If the nuclei are much heavier than neutrons the exact form of the function f_N is inessential. The neutron gas heat conductivity has been evaluated in [607], and the heat conductivity of the neutron gas–nuclei mixture in [53].

Following [53] we search for a solution of (2.4.8) in the form of an expansion in polynomials $Q(x)$ orthogonal with respect to the weight function $f_0(1 - f_0)x^{3/2}$, analogous to the Sonin polynomials $S_{3/2}^{(n)}$ [222] for the non-degenerate case

$$A_i = [a_0Q_0(x) + a_1Q_1(x)]v_i,$$

$$\mathcal{D}_i = [d_0Q_0(x) + d_1Q_1(x)]v_i,$$

$$Q_0(x) = 1, \quad Q_1(x) = \frac{5}{2}\frac{G_{5/2}}{G_{3/2}} - x,$$

(2.4.10)

$$x = u^2, \quad u_i = \left(\frac{m_n}{2kT}\right)^{1/2}v_i.$$

In A_{Ni}, $\mathcal{D}_{Ni}$ from (2.4.7) we retain only the first terms of the expansion, determined by the condition that the corrections to the equilibrium functions should not contribute to the mean mass velocity

$$A_{Ni} = a_{0N}Q_0u_{Ni}, \quad n_na_0 + n_Na_{0N} = 0,$$

$$\mathcal{D}_{Ni} = d_{0N}Q_0u_{Ni}, \quad n_nd_0 + n_Nd_{0N} = 0,$$

(2.4.11)

$$u_{Ni} = (m_N/2kT)^{1/2}v_{Ni}.$$

Using the definition of the neutron heat flux together with that of the diffusion velocity $\langle v_i \rangle$, and recalling (2.4.4), (2.4.7), (2.4.10) gives [53]

$$q_i = \frac{1}{2} n_n m_n \langle v^2 v_i \rangle$$

$$= -\frac{5}{2} n_n m_n \left(\frac{kT}{m_n}\right)^2 \frac{G_{5/2}}{G_{3/2}}$$

$$\times \left\{ \left[a_0 - a_1 \left(\frac{7}{2}\frac{G_{7/2}}{G_{5/2}} - \frac{5}{2}\frac{G_{5/2}}{G_{3/2}}\right) \right] \frac{\partial \ln T}{\partial r_i} \right.$$

$$\left. + n_n \left[d_0 - d_1 \left(\frac{7}{2}\frac{G_{7/2}}{G_{5/2}} - \frac{5}{2}\frac{G_{5/2}}{G_{3/2}}\right) \right] \frac{G_{5/2}}{G_{3/2}} d_i \right\}, \qquad (2.4.12)$$

$$\langle v_i \rangle = -\frac{kT}{m_n} \left(a_0 \frac{\partial \ln T}{\partial r_i} + d_0 n_n \frac{G_{5/2}}{G_{3/2}} d_i \right), \qquad (2.4.13)$$

$$\langle v_{Ni} \rangle = -\frac{\rho_n}{\rho_N} \langle v_i \rangle.$$

Setting $\langle v_i \rangle = 0$ in (2.4.13), (2.4.12) gives an expression for the heat flux due to heat conduction

$$q_i = -\lambda_i \frac{\partial T}{\partial r_i} = -\frac{5}{2} n_n m_n \left(\frac{kT}{m_n}\right)^2 \frac{G_{5/2}}{G_{3/2}}$$

$$\times \left(\frac{7}{2}\frac{G_{7/2}}{G_{5/2}} - \frac{5}{2}\frac{G_{5/2}}{G_{3/2}}\right) \left(\frac{a_0 d_1}{d_0} - a_1\right) \frac{\partial \ln T}{\partial r_i}. \qquad (2.4.14)$$

Multiplying (2.4.8) by $BQ_0(x)$ and $BQ_1(x)$ and integrating over dv_i we obtain the set of equations for the coefficients a_k, d_k, namely

$$0 = b_{00}a_0 + b_{01}a_1,$$

$$-\frac{15}{4} n_n \left(\frac{7}{2}\frac{G_{7/2}}{G_{3/2}} - \frac{5}{2}\frac{G_{5/2}^2}{G_{3/2}^2}\right) = b_{10}a_0 + (a_{11} + b_{11})a_1, \qquad (2.4.15)$$

$$\frac{3}{2} = b_{00}d_0 + b_{01}d_1,$$
$$\qquad (2.4.16)$$
$$0 = b_{10}d_0 + (a_{11} + b_{11})d_1,$$

where

$$a_{11} = B^2 \int f_0 f_{01}(1 - f_0')(1 - f_{01}')Q_1(x)u_i[Q_1(x)u_i + Q_1(x_1)u_{1i}$$

$$- Q_1(x')u_i' - Q_1(x_1')u_{1i}']g_{nn}W_{nn}(\theta g_{nn})d\Omega dv_i dv_{1i}, \qquad (2.4.17)$$

$$b_{j1} = B \int f_0 f_{N0}(1 - f_0')[Q_1(x)u_i - Q_1(x')u_i']$$

$$\times g_{nN} W_{nN}(\theta, g_{nN}) d\Omega dv_i dv_{Ni}, \quad j = 0, 1. \tag{2.4.18}$$

Using (2.4.11) gives

$$b_{j0} = B \int f_0 f_{N0}(1 - f_0')[Q_j(x)u_i \left[u_i - u_i' - \frac{n_n}{n_N}\left(\frac{m_n}{m_N}\right)^{1/2} \right.$$

$$\left. \times (u_{Ni} - u_{Ni}') \right] g_{nN} W_{nN}(\theta, g_{nN}) d\Omega dv_i dv_{Ni}. \tag{2.4.19}$$

Conservation of momentum after collision yields

$$u_i - u_i' = -\left(\frac{m_N}{m_n}\right)^{1/2}(u_{Ni} - u_{Ni}'). \tag{2.4.20}$$

Assuming $m_N \gg m_n$ we have

$$|u_i| = |u_i'|, \quad u_i(u_i - u_i') = u^2(1 - \cos\theta),$$

$$g_{nN} = |v_i| = v. \tag{2.4.21}$$

Applying (2.4.10) and (2.4.21) allows us to obtain an explicit form for the coefficients b_{jk} from (2.4.18) and (2.4.19) [53]. As for a_{11} from (2.4.17), these coefficients can only be obtained explicitly in the limiting cases of weak and strong degeneracy. There is a simple interpolation for the intermediate range of degeneracy. Substituting the solutions (2.4.15) and (2.4.16) into (2.4.14) we find the heat conduction coefficient for the neutron gas in the form

$$\lambda_n = \frac{75}{64\pi^2} \frac{k}{\sqrt{2}} \left(\frac{kT}{m_n}\right)^{1/2} \frac{n_n^2}{n_N} \left(\frac{2\pi^2\hbar^2}{kTm_n}\right)^{3/2} \frac{G_{5/2}^2}{G_{3/2}^2}$$

$$\times \left(\frac{7}{2}\frac{G_{7/2}}{G_{5/2}} - \frac{5}{2}\frac{G_{5/2}}{G_{3/2}}\right)^2 \left[\frac{25}{4}\frac{G_{5/2}^2}{G_{3/2}^2}\tilde{\Omega}_{nN}(1)\right.$$

$$\left. - 5\frac{G_{5/2}}{G_{3/2}}\tilde{\Omega}_{nN}(2) + \tilde{\Omega}_{nN}(3) + \frac{32}{21}\pi^4\frac{n_n}{n_N}I\frac{\epsilon}{1+\epsilon}\right]^{-1}, \tag{2.4.22}$$

where

$$\tilde{\Omega}_{nN}(r) = \int_0^\infty dx \int_0^\pi d\theta f_0(1 - f_0)x^{r+1}(1 - \cos\theta)W_{nN}(\theta, x)\sin\theta,$$

$$I = \int_0^1 \frac{y^3 W_{nn}(y) dy}{(1 - y^2)^{1/2}}, \quad y = \frac{g}{(2x_0)^{1/2}},$$

$$g = \frac{1}{2} \left(\frac{m_n}{kT} \right)^{1/2} g_{nn},$$

$$W_{nn}(\theta, g) = \sum_{l=0}^{\infty} W_{nn}^{(l)}(g) P_l(\cos \theta),$$

$$W_{nn}(g) = \sum_{l=0}^{\infty} W_{nn}^{(l)}(g) [P_l(0)]^2, \qquad (2.4.23)$$

$$\epsilon = \frac{21\sqrt{2}}{8\pi^{13/2}} \left(\frac{2\pi^2 \hbar^2}{kT m_n} \right)^{3/2} n_n \frac{\sqrt{\pi} \Omega_{nn}^{(2)}(2)}{8I} \begin{cases} \gg 1 & \text{for } e^{x_0} \gg 1, \\ \ll 1 & \text{for } e^{x_0} \ll 1, \end{cases}$$

$$\Omega_{nn}^{(l)}(r) = \sqrt{\pi} \int_0^{\infty} e^{-g^2} g^{2r+3} dg \int_0^{\pi} (1 - \cos^l \theta) W_{nn}(\theta, g) \sin \theta d\theta.$$

Note that $\Omega_{nn}^{(l)}(r)$ must be multiplied by $2 \left(\frac{kT}{m_n} \right)^{1/2}$ to obtain $\Omega_{12}^{(l)}(r)$ from [222], where dimensional relative speed is used in the definition. In the case of strong neutron degeneracy

$$\lambda_n^{\text{deg}} = \frac{1}{12} \left(\frac{\pi}{3} \right)^{1/3} \frac{k^2 T}{\hbar} \frac{n_n^{2/3}}{n_N} \frac{I}{\bar{W}_{nN}(x_0)}$$

$$\times \left[1 + \frac{64}{21\pi} \left(\frac{\pi}{3} \right)^{1/3} \frac{m_n^2}{\hbar^4} \frac{(kT)^2}{n_n^{1/3} n_N} \frac{I}{\bar{W}_{nN}(x_0)} \right]^{-1}, \qquad (2.4.24)$$

where

$$\bar{W}_{nN}(x_0) = W_{nN}^{(0)}(x_0) - \frac{1}{3} W_{nN}^{(1)}(x_0),$$

$$W_{nN}(\theta, x) = \sum_{l=0}^{\infty} W_{nN}^{(l)}(x) P_l(\cos \theta),$$

$$n_n = \frac{8\pi}{3} \left(\frac{kT m_n}{2\pi^2 \hbar^2} \right)^{3/2} x_0^{3/2}. \qquad (2.4.25)$$

For a pure neutron gas we have

$$\lambda_{nn} = \begin{cases} \dfrac{75 k}{64} \left(\dfrac{kT}{m_n} \right)^{1/2} \dfrac{1}{\Omega_{nn}^{(2)}(2)} & \text{without degeneracy [222],} \\[2ex] \dfrac{7\pi \hbar^3 n_n}{256 T m_n^2 I} & \text{at strong degeneracy [607].} \end{cases} \qquad (2.4.26)$$

The scattering cross-section can be evaluated from experimental data on pn- and pp- scattering. The cross-section for elastic isotropic scattering at neutron energies ≤ 10 MeV can be written in the form [23]

$$W_{nn} = \frac{\hbar^2}{m_n} \frac{1}{E + E_{nn}}, \tag{2.4.27}$$

where E is the neutron energy on the centre-of-mass frame of reference. The value of E_{nn} is very small: $E_{nn} \leq 60$ KeV (neutron singlet state energy). Note also that at high energies $150 \leq E \leq 400$ MeV the cross-section W_{nn} is constant and equals $\approx 3.4 \; 10^{-27}$ cm^2 [23]. W_{nn} is approximated by

$$W_{nn} = \left((3.4 \times 10^{-27}) + \frac{4.15 \times 10^{-25}}{E + E_{nn}} \right) \text{ cm}^2, \tag{2.4.28}$$

where E and E_{nn} are measured in MeV

$$E = \frac{m_n g_{nn}^2}{4 \times 1.602 \times 10^{-6}} = \frac{2kT x_0 y^2}{1.602 \times 10^{-6}} \text{ MeV}. \tag{2.4.29}$$

Substituting (2.4.28) into (2.4.23) and (2.4.26) we find, upon integrating, that in the case of strong degeneracy

$$I = \left((2.3 \times 10^{-27}) + \frac{1.5 \times 10^{-25}}{\rho^{2/3}} \right) \text{ cm}^2,$$

$$\lambda_{nn} = \frac{9.2 \times 10^{18}}{1 + \frac{66}{\rho_{12}^{2/3}}} \frac{\rho_{12}}{T_9}, \quad \rho_{12} = \frac{n_n m_n}{10^{12}}, \quad T_9 = \frac{T}{10^9}. \tag{2.4.30}$$

Assume that

$$W_{nN} \approx A^{2/3} W_{nn} \text{ with } W_{nn} \text{ from (2.4.27) at}$$

$$E = \frac{mv^2}{2 \times 1.602 \times 10^{-6}} = \frac{kTx}{1.602 \times 10^{-6}} \text{ MeV}. \tag{2.4.31}$$

We then have from (2.4.24) that the neutron–nuclei mixture heat conduction coefficient is

$$\lambda_n = \frac{9.2 \times 10^{18} \rho_{12}/T_9}{1 + \frac{66}{\rho_{12}^{2/3}} + 2.5 \frac{\rho_{12}^{1/3} \rho_{12N}}{A_{100}^{1/3} T_9^2} \left(\frac{3}{2} + \frac{132}{\rho_{12}^{2/3}} \right)}, \tag{2.4.32}$$

$$\rho_{12N} = A m_n n_N/10^{12}, \quad A_{100} = A/100.$$

Relation (2.4.32) holds for strong degeneracy

$$x_0 \gg 1, \quad 5.8 \frac{\rho_{12}^{2/3}}{T_0} \gg 1.$$

Extrapolation of (2.4.30) to the neutron liquid with $\rho \geq 1.5 \times 10^{14}$ g cm^{-3} is in quantitative agreement with [357], where the heat conduction has been investigated using the nuclear matter theory.

2.4.3 Non-relativistic Electron Heat Conduction

Consider the electron gas kinetics including electron–ion collisions to the Lorentz approximation [222] and neglecting electron–electron collisions. Using (2.4.9), (2.4.20) and (2.4.21) the right-hand sides of (2.4.8) take[4] the form

$$I_{eN}(R_i) = B \int f_0 f_{N0}(1 - f_0')(R_i - R_i') vb\, db\, d\varphi\, dv_{Ni}. \qquad (2.4.33)$$

Setting $R_i = R(v)v_i$, multiplying (2.4.8) by v_i and using (2.4.3) and (2.4.6) gives

$$A(v) = \frac{x - \frac{5G_{5/2}}{2G_{3/2}}}{2\pi n_N \int_0^\infty (1 - \cos\theta) vb\, db}, \qquad (2.4.34)$$

$$\mathcal{D}(v) = \frac{1}{2\pi n_e n_N \int_0^\infty (1 - \cos\theta) vb\, db}. \qquad (2.4.35)$$

In [222] it is found for Coulomb interactions between electrons and ions with charge Z

$$\cos\theta = \frac{t_0^2 - 1}{t_0^2 + 1}, \qquad t_0 = \frac{bv^2 m_e}{Ze^2},$$

$$\Phi_{12} = \int_0^{b_{max}} (1 - \cos\theta) vb\, db = \frac{2Z^2 e^4}{m_e^2 v^3} \Lambda_v, \qquad (2.4.36)$$

$$\Lambda_v = \ln\left(\frac{b_{max} v^2 m_e}{Ze^2}\right).$$

Using (2.4.2), (2.4.4), (2.4.12) and (2.4.34–36) yields

$$q_i = -\frac{640k}{\Lambda} \frac{m_e (kT)^4)}{n_N Z^2 e^4 h^3} \left(G_5 - \frac{1}{2}\frac{G_{5/2}}{G_{3/2}} G_4\right) \frac{\partial T}{\partial r_i}$$
$$- \frac{128}{\Lambda} \frac{m_e (kT)^5)}{n_N Z^2 e^4 h^3} \frac{G_{5/2}}{G_{3/2}} G_4 d_i, \qquad (2.4.37)$$

$$\langle v_i \rangle = -\frac{128k}{\Lambda} \frac{m_e (kT)^3}{n_e n_N Z^2 e^4 h^3} \left(G_4 - \frac{5}{8}\frac{G_{5/2}}{G_{3/2}} G_3\right) \frac{\partial T}{\partial r_i}$$
$$- \frac{32}{\Lambda} \frac{m_e (kT)^4}{n_e n_N Z^2 e^4 h^3} \frac{G_{5/2}}{G_{3/2}} G_3 d_i. \qquad (2.4.38)$$

Setting $\langle v_i \rangle = 0$ and expressing d_i in terms of $\frac{\partial T}{\partial r_i}$, we have from (2.4.37) and (2.4.38)

---

[4] The classical cross-section for Coulomb collisions equals the quantum cross-section [144]; the electron quantities in these formulae should be substituted for the respective neutron quantities: P_e for P_n and so on.

$$q_i = -\frac{128k}{\Lambda}\frac{m_e(kT)^4)}{n_N Z^2 e^4 h^3}\left(5G_5 - 4\frac{G_4^2}{G_3}\right)\frac{\partial T}{\partial r_i} = \lambda_e \frac{\partial T}{\partial r_i}. \tag{2.4.39}$$

Recalling (2.3.6), (2.4.25) and the expansion of the functions G_n with the aid of (1.2.24)

$$G_n = \frac{1}{\Gamma(n)}\left[\frac{x_0^n}{n} + \frac{\pi^2}{6}(n-1)x_0^{n-2} + \dots\right], \tag{2.4.40}$$

we may write the heat conduction coefficient in the form

$$\lambda_e = -\frac{128k}{\Lambda}\frac{m_e(kT)^4)}{n_N Z^2 e^4 h^3}\left(5G_5 - 4\frac{G_4^2}{G_3}\right)$$

$$= \begin{cases} \frac{16\sqrt{2}}{\pi^{3/2}\Lambda}k\frac{n_e}{n_N}\left(\frac{kT}{e^2 Z}\right)^2\left(\frac{kT}{m_e}\right)^{1/2} & \text{(ND)} \\ \frac{1}{32\Lambda}\frac{k^2 T n_e^2 h^3}{m_e^2 n_N Z^2 e^4} & \text{(D)} \end{cases} \tag{2.4.41}$$

(the expressions for non-degenerate and degenerate electrons are denoted by ND and D, respectively). The effective frequency ν_{ei} of electron–ion collisions [90] used below is

$$\nu_{ei} = \frac{\int dp_i \nu(v)pv\left(-\frac{\partial f_0}{\partial \epsilon}\right)}{\int dp_i pv\left(-\frac{\partial f_0}{\partial \epsilon}\right)}$$

$$= \frac{4}{3}\sqrt{\frac{2\pi}{m_e}}\frac{Z^2 e^4 n_N \Lambda}{(kT)^{3/2}G_{3/2}}\frac{1}{1 + e^{-x_0}}$$

$$= \begin{cases} \frac{4}{3}\sqrt{\frac{2\pi}{m_e}}\frac{Z^2 e^4 n_N \Lambda}{(kT)^{3/2}} & \text{(ND)} \\ \frac{32\pi^2 Z^2 e^4 n_N \Lambda m_e}{3h^3 n_e} & \text{(D)}, \qquad \epsilon = \frac{pv}{2}, \end{cases} \tag{2.4.42}$$

where the electron momentum $p = m_e v$; $G_{3/2}$, x_0 are defined in (2.4.6), the distribution function f_0 is given by (2.2), and

$$\nu(v) = 2\pi n_N \Phi_{12} = \frac{4\pi Z^2 e^4 \Lambda_\nu n_N}{m_e^2 v^3}$$

is the frequency of collisions of electrons with velocity v and Φ_{12} given by (2.4.36). The Coulomb logarithm averaged over velocities is

$$\Lambda = \bar{\Lambda}_\nu = \ln\left(\frac{b_{\max}\bar{v}^2 m_e}{Ze^2}\right). \tag{2.4.43}$$

Using (2.4.2), (2.4.4), (2.4.6) and (2.4.25) gives

$$\bar{v^2} = \frac{3kT}{m_e} \frac{G_{5/2}}{G_{3/2}} = \begin{cases} \frac{3kT}{m_e} & \text{(ND)} \\ \frac{3h^2}{5m_e^2} \left(\frac{3n_e}{8\pi}\right)^{2/3} & \text{(D)}. \end{cases} \tag{2.4.44}$$

The parameter b_{max} is the charge screening radius and corresponds to the Debye radius in an ordinary plasma. The radius $r_{\mathcal{D}i}$ of non-degenerate ion screening is [176] [5]

$$r_{\mathcal{D}i}^2 = \frac{4\pi n_N e^2 Z^2}{kT}. \tag{2.4.45}$$

Evaluation of the electron screening reduces to the Thomas–Fermi equation [144] with finite temperature. Setting approximately [4, 90]

$$\begin{aligned} r_{\mathcal{D}e}^{-2} &= 4\pi n_e e^2 \frac{\int dp_i \left(-\frac{\partial f_0}{\partial \epsilon}\right)}{\int dp_i f_0} = \frac{4\pi n_e e^2}{m_e} \left(\frac{\bar{1}}{v^2}\right) \\ &= \frac{4\pi n_e e^2}{kT} \frac{G_{1/2}}{G_{3/2}} = \begin{cases} 4\pi n_e e^2/kT & \text{(ND)} \\ 16(3\pi^5)^{1/3} n_e^{1/3} e^2 m_e/h^2 & \text{(D)}. \end{cases} \end{aligned} \tag{2.4.46}$$

The general screening radius is

$$\frac{1}{b_{max}^2} = \frac{1}{r_{\mathcal{D}i}^2} + \frac{1}{r_{\mathcal{D}e}^2} = \frac{4\pi e^2}{kT} \left(n_N Z^2 + n_e \frac{G_{1/2}}{G_{3/2}}\right). \tag{2.4.47}$$

For strong degeneracy $x_0 \gg 1$, $r_{\mathcal{D}e} \gg r_{\mathcal{D}i}$ and only ion screening is essential, so $b_{max} \approx r_{\mathcal{D}i}$. At high temperatures and densities, including quantum effects changes the Λ value. Denoting

$$r_m = \frac{Ze^2}{m_e \bar{v^2}}, \tag{2.4.48}$$

we can rewrite (2.4.43) in the form

$$\Lambda = \ln \frac{b_{max}}{b_{min}}, \qquad b_{min} = r_m. \tag{2.4.49}$$

According to quantum laws the minimum impact parameter cannot be less than the de Broglie wavelength, or [246]

$$\begin{aligned} b_{min} > r_F &= \frac{h\sqrt{3/5}}{4\pi m_e \sqrt{\bar{v^2}}} = \frac{h}{4\pi\sqrt{5m_e kT}} \left(\frac{G_{3/2}}{G_{5/2}}\right)^{1/2} \\ &= \begin{cases} \frac{h}{4\pi\sqrt{5m_e kT}} & \text{(ND)} \\ \frac{h}{4\pi p_{Fe}} = \frac{n_e^{-1/3}}{2(3\pi)^{1/3}} & \text{(D)}. \end{cases} \end{aligned} \tag{2.4.50}$$

---

[5] Strictly speaking, the ion screening has a dynamic character here and the screening radius does not exactly equals $r_{\mathcal{D}i}$ in (2.4.45) (see [461]).

$b_{\min}$ is therefore generally determined by

$$
b_{\min} = \max\{r_m, r_F\} = \begin{cases} r_m & \text{for } \frac{v_{ch}}{c} < \frac{2\sqrt{5}e^2 Z}{3\hbar c} Z = \frac{Z}{92}, \\[2mm] r_F & \text{for } \frac{v_{ch}}{c} > \frac{Z}{92}, \end{cases} \qquad (2.4.51)
$$

where $v_{ch} = \sqrt{\overline{v^2}/3} = (kT/m_e)^{1/2}(G_{5/2}/G_{3/2})^{1/2}$. Thus, over a wide range of conditions where non-relativistic electrons exist it is necessary to take $b_{\min} = r_F$. Note that in the case of a nuclear mixture we have to transform (2.4.37–43) using the substitution

$$
n_N Z^2 \Rightarrow \langle n_N Z^2 \rangle = \sum_i n_{Ni} Z_i^2,
$$

$$
n_N Z^2 \Lambda \Rightarrow \langle n_N Z^2 \Lambda \rangle = \sum_i n_{Ni} Z_i^2 \Lambda_i, \qquad (2.4.52)
$$

where Λ_i is given by (2.4.43) with $Z = Z_i$. Electron–electron collisions can be included either by refusing the Lorentz approximation and calculating λ_e similarly to λ_n from Sect. 2.4.2, or by substituting phenomenologically $(\nu_{ei} + \nu_{ee})$ for ν_{ei} in (2.4.41). The latter is the more appropriate of these two alternative paths since in almost all cases of electron degeneracy it does not reduce the calculation accuracy for $e - i$ collisions dominating. Studies of the $e - e$ collision contribution [202, 461] show that it is negligible for $Z \gg 1$ and does not exceed 50% at $Z \sim 1$.

At high densities and strong electron degeneracy the Coulomb interaction between ions becomes important and influences $e - i$ collisions. We describe the electron heat conduction under these conditions in the following section.

2.4.4 The Degenerate Electron Heat Conduction Including Relativistic Effects and Effects of Non-perfect Ions

In most cases, relativistic electrons in stars are strongly degenerate. Under these conditions the heat conduction is calculated in [245, 246]. Writing λ_e in a form similar to (2.4.41) with the total electron mass m_* at the Fermi boundary (see Sect. 1.2) instead of m_e:

$$
m_* = (m_e^2 + p_{Fe}^2/c^2)^{1/2}, \qquad \text{with } p_{Fe} \text{ from (1.2.20)} \qquad (2.4.53)
$$

gives

$$
\lambda_e = \frac{\pi^2 k^2 T n_e}{3 m_* \nu_e}
$$

$$
= \frac{4.11 \times 10^{15}(\rho_6/\mu_Z)T_6}{[1 + (\rho_6/\mu_Z)^{2/3}]^{1/2}} \left(\frac{10^{16} s^{-1}}{\nu_e} \right) \text{ erg cm}^{-1}\text{ s}^{-1}\text{ K}^{-1}, \qquad (2.4.54)
$$

$$
\rho_6 = \rho/10^6 \text{ g cm}^{-3}, \qquad T_6 = T/10^6 \text{ K}.
$$

The quantity ν_e is the total frequency of electron collisions in the medium. In [246] ν_e is evaluated for a degenerate electron gas with

$$T < T_F = \frac{m_* - m_e}{k}c^2 = 5.93 \times 10^9 \frac{(\rho_6/\mu_Z)^{2/3}}{1 + [1 + (\rho_6/\mu_Z)^{2/3}]^{1/2}}, \qquad (2.4.55)$$

$$20Z^2 \ll \rho < 4 \times 10^{11} \text{ g cm}^{-3}.$$

The lower limit of ρ in (2.4.55) is the perfect electron gas condition ([145], see also (1.4.23)). At $T_m < T < T_F$, where

$T_m = Z^2e^2/kГ_m l_{WS}$ is the crystallization temperature,

$Г = Z^2e^2/kTl_{WS}$ is the gas parameter, $Г_m \approx 150$ (2.4.56)

$l_{WS} = (3/4\pi n_N)^{1/3}$ is the Wigner-Seitz cell radius (see (1.4.25)),

electron scattering by ions is the most important[6]. Calculating ν_{ei} similarly to Sect. 2.4.3 gives, analogously to (2.4.42)

$$\nu_{ei} = \frac{32\pi^2 m_* n_N Z^2 e^4}{3h^3 n_e}\Lambda_{ei}$$

$$= 1.77 \times 10^{16}\frac{n_N Z^2}{n_e}\left[1 + \left(\frac{\rho_6}{\mu_Z}\right)^{2/3}\right]^{1/2}\Lambda_{ei}. \qquad (2.4.57)$$

To determine the Coulomb logarithm we must keep in mind that relativistic corrections make the terms $\sim v_{Fe}^2/c^2$ important, and ion correlations lead to the maximum impact parameter being interpolated by $b_{max} = r_{\mathcal{D}i}^2 + l_{WS}^2/6$, where l_{WS} is given in (2.4.56) and $r_{\mathcal{D}i}$ is determined by (2.4.45)[7]. The value $b_{min} = r_F = h/4\pi p_{Fe}$ is similar to (2.4.50). An interpolation formula for Λ_{ei} valid at all temperatures $T_m < T < T_{Fe}$ is obtained in [246]:

$$\Lambda_{ei} = \ln\left[\left(\frac{2\pi Z}{3}\right)^{1/3}\left(\frac{3}{2} + \frac{3}{Г}\right)^{1/2}\right] - \frac{v_{Fe}^2}{2c^2} =$$

$$= \ln\left[(4Z)^{1/3}\left(1 + \frac{2}{Г}\right)^{1/2}\right] - \frac{(\rho_6/\mu_Z)^{2/3}}{2\left[1 + (\rho_6/\mu_Z)^{2/3}\right]}, \qquad (2.4.58)$$

$$v_{Fe}/c = (p_{Fe}/m_e c)\left(1 + \frac{p_{Fe}^2}{m_e^2 c^2}\right)^{-1/2} \quad \text{(see (1.2.3), (2.4.53)).}$$

At temperatures below the crystallization temperature $T < T_m$, instead of $e-i$ scattering electron scattering occurs as a result of interaction with thermal lattice oscillations called phonons. The frequency $\nu_{e,ph}$ of collisions between

---

[6] See footnote at p. 54

[7] At low densities, $\rho \sim 20Z^2$, it is important to take into account electron screening; this decreases b_{max} and causes ν_{ei} to decrease by about 40% [432].

electrons and phonons in classical $(T > \theta = 0.45\hbar\omega_{pi}/k)$[8] and quantum $(T < \theta)$ crystals is approximated to an accuracy of better than 10% by the expression [245, 246]

$$
\nu_{e,ph} = \frac{e^2}{\hbar v_{Fe}} \omega_{pi} x \left[\left(2 - \frac{v_{Fe}^2}{c^2} \right) \varphi^{(0)}(x) \right.
$$
$$
\left. + \frac{1}{\pi^2} \left(3\Lambda_{ph} - 1 + \frac{v_{Fe}^2}{2c^2} \right) \varphi^{(2)}(x) \right],
\tag{2.4.59}
$$

where the interpolation formulae

$$
x^2 \varphi^{(0)}(x) = 13[1 + (\theta/3.46T)^2]^{-1/2},
$$
$$
(x/\pi)^2 \varphi^{(2)}(x) = 13(\theta/5.1T)^2[1 + (\theta/4.17T)^2]^{-3/2},
\tag{2.4.60}
$$

are valid and
$$
\omega_{pi}^2 = \frac{4\pi n_N Z^2 e^2}{m_N}, \quad \text{where } \omega_{pi} \text{ is the ion plasma frequency and}
$$
$$
x = \frac{\hbar\omega_{pi}}{kT} = \frac{\theta}{0.45T},
\tag{2.4.61}
$$

$\Lambda_{ph} = \Lambda_{ei}$ from (2.4.58) with $\Gamma \gg 1$.

Formula (2.4.59) has been calculated with greater accuracy by a Monte-Carlo method in [546]. If there are impurity ions in the crystal, that is ions of another species introduced at lattice nodes, scattering by these ions can become important at low temperatures. The frequency of collisions $\nu_{e,imp}$ between electrons and impurity ions is given by [245]

$$
\nu_{e,imp} = \frac{4m_* e^2 \overline{(\Delta Z)^2} x_{imp} \Lambda_{imp}}{3\pi\hbar^3 Z},
\tag{2.4.62}
$$

where

$\overline{(\Delta Z)^2}$ is the mean-square deviation of charge

        due to impurities, and

x_{imp} is the mass concentration of impurities,   (2.4.63)

$$
\Lambda_{imp} = \ln\left(\frac{2p_{Fe}}{\hbar q_{TF}} \right) - \frac{1}{2} - \frac{v_{Fe}^2}{2c^2}.
$$

Here, p_{Fe} is defined by (1.2.20), $q_{TF} = \frac{\sqrt{3}\omega_{pe}}{v_{Fe}}$, $\omega_{pe}^2 = \frac{4\pi n_e e^2}{m_*}$, v_{Fe} is given in (2.4.58)[9]

---

[8] The quantity θ is defined by (1.4.38), where it is expressed in terms of $\omega_i = \omega_{pi}/\sqrt{3}$.

[9] The quantity q_{TF} has the meaning of reciprocal screening radius: $q_{TF} = r_{De}^{-1}$ (for the non-relativistic case see (2.4.46)).

According to calculations, $e - e$ collisions may be more important in the case of relativistic electrons than in a non-relativistic gas. The $e - e$ collision frequency is given by [245]

$$\nu_{ee} = \frac{3}{2\pi^3} \left(\frac{e^2}{\hbar c}\right)^2 \frac{(kT)^2}{\hbar m_* c^2} \left(\frac{2k_{Fe}}{q_{TF}}\right)^3 J(y)$$

$$= 5.08 \times 10^{11} T_6^2 (v_{Fe}/c)^{3/2} [1 + (\rho_6/\mu_Z)^{2/3}]^{-1/2} J(y) \ \text{s}^{-1},$$

(2.4.64)

where

$$k_{Fe} = \frac{p_{Fe}}{\hbar}, \qquad y = \frac{\sqrt{3} T_p}{T},$$

$$T_p = \frac{\hbar\omega_{pe}}{k} = 3.33 \times 10^8 \left(\frac{\rho_6}{\mu_Z}\right)^{1/3} \left(\frac{v_{Fe}}{c}\right)^{1/2},$$

$$J(y \geq 20) \approx 51.0 - \frac{995}{y} + \frac{1.51 \times 10^4}{y^2},$$

$$J(y \leq 1) \approx \frac{y^3}{3} \Lambda_{ee} \qquad \Lambda_{ee} = \ln \frac{2}{y}.$$

(2.4.65)

Intermediate values can be found from the following table:

y	0.5	1.5	3	7	10
J(y)	0.056	0.53	2.6	8.8	13.5

Thus the degenerate electron heat conductivity is determined by (2.4.54), where

$$\nu_e = \begin{cases} \nu_{ei} + \nu_{ee} & \text{for } T \geq T_m \\ \nu_{e,ph} + \nu_{e,imp} + \nu_{ee} & \text{for } T \leq T_m. \end{cases}$$

(2.4.66)

Using substitutions (2.4.52) we can obtain the corresponding formulae for ion mixtures. Other calculations of heat and electrical conductivities in the crystalline phase at $T < T_m$ have been performed in [431a].

2.4.5 Absorption of Radiation at High Temperatures and High Magnetic Fields

Regions with strong magnetic fields, where the opacity resulting from synchrotron and cyclotron self-absorption is important, can exist during accretion onto neutron stars and black holes. The opacity under these conditions has been estimated in [294]. It follows from Kirchhoff's law (2.1.13) that the total emissivity of matter ϵ_{tot} (erg g^{-1}cm^{-1}) and the Rosseland mean opacity κ are linearly related:

$$\epsilon_{tot} = A\sigma T^4 \kappa.$$

According to [89] and (2.3.15) the respective ϵ_{tot} and κ for bremsstrahlung emission due to non-relativistic electrons are

$$\epsilon_{ff} = 6 \times 10^{20} \rho T^{1/2},$$
$$\kappa_{ff} = 6.4 \times 10^{22} \rho T^{-7/2}, \tag{2.4.67}$$

so that

$$A = \frac{6}{6.4 \times 10^2 \sigma} \approx 170. \tag{2.4.68}$$

The electron emissivity in a magnetic field with component $B_\perp$ normal to the electron motion is calculated for a Maxwellian distribution[10] in [59]:

$$\epsilon_B = \begin{cases} 2.3 \, T B_\perp^2 & \text{for } kT \ll m_e c^2, \\ 3.2 \times 10^{-10} T^2 B_\perp^2 & \text{for } kT \gg m_e c^2. \end{cases} \tag{2.4.69}$$

From (2.4.68) and (2.4.69) the respective opacities are

$$\kappa_B = \begin{cases} 2 \times 10^2 B^2/T^3 & \text{for } kT \ll m_e c^2, \\ 3 \times 10^{-8} B^2/T^2 & \text{for } kT \gg m_e c^2. \end{cases} \tag{2.4.70}$$

In [294] κ_{ff} is obtained analogously using ϵ_{ff} (see, for example, [59]) for the relativistic Maxwellian electrons:

$$\epsilon_{ff} = 2 \times 10^{16} \rho T \ln \frac{kT}{m_e c^2},$$
$$\kappa_{ff} = 2 \times 10^{18} (\rho/T^3) \ln \frac{kT}{m_e c^2}. \tag{2.4.71}$$

The estimates of the present section concern a pure hydrogen non-degenerate plasma. Including pair production causes ϵ and κ in (2.4.70–71) to increase by a factor $\left(1 + \frac{2n_+ m_u}{\rho}\right)$. For equilibrium pairs n_+ is obtained from (1.2.9) and (1.2.17). If we compare (2.4.71) with (2.3.29), (2.3.31), we see that $\kappa_{ff} \ll \kappa_{es}$ always for non-degenerate electrons.

Problem 1. Find the rate of energy exchange between electrons and non-degenerate, non-relativistic nuclei.

Solution [441a]. The collision integral for charged particles is simpler than in the Boltzmann equation (2.4.1), because collisions involving small-angle deflections are of major importance and the expansion $f(p_i') = f(p_i) + \frac{\partial f}{\partial p_i}(p_i' - p_i)$ is valid in momentum space. The collision integral then takes the Landau form which, for the case of the ion kinetic equation, and relativistic and degenerate electrons reads [148a]

---

[10] Here, the Maxwellian distribution is used for relativistic electrons instead of the Fermi function because the concentration of pairs is not assumed to be in equilibrium.

$$J_{Ne} = -\frac{\partial S_\alpha}{\partial p_\alpha}, \qquad p_\alpha = m_N v_\alpha, \qquad v_\alpha \equiv v_{N\alpha}, \tag{1}$$

where the ion flux in momentum space due to collisions with electrons is

$$S_\alpha = \frac{4\pi Z^2 e^4 \Lambda}{h^3 m_N^3} \int \left[f(p) \frac{\partial f_e}{\partial p_\beta'} - f_e(p') \frac{\partial f}{\partial p_\beta} (1 - f_e(p')) \right] B_{\alpha\beta} d^3 p', \tag{2}$$

$$\alpha, \beta = 1, 2, 3.$$

We adopt here the normalization conditions (2.4.3). For electron–ion collisions

$$B_{\alpha\beta} = \frac{\gamma\gamma' \left(1 - \frac{\bar{v}\bar{v}'}{c^2}\right)^2}{c \left[\gamma^2 \gamma'^2 \left(1 - \frac{\bar{v}\bar{v}'}{c^2}\right)^2 - 1\right]^{3/2}} \left\{ \left[\gamma^2 \gamma'^2 \left(1 - \frac{\bar{v}\bar{v}'}{c^2}\right)^2 - 1 \right] \delta_{\alpha\beta} \right.$$

$$\left. - \gamma^2 \frac{v_\alpha v_\beta}{c^2} - \frac{\gamma'^2}{c^2} v_\alpha' v_\beta' + \frac{\gamma^2 \gamma'^2}{c^2} \left(1 - \frac{\bar{v}\bar{v}'}{c^2}\right) (v_\alpha v_\beta' + v_\alpha' v_\beta) \right\}. \tag{3}$$

We use in (2) and (3) the notation

$$p_\alpha' = \frac{m_e v_\alpha'}{\left(1 - \frac{v'^2}{c^2}\right)^{1/2}}, \qquad \varepsilon' = \frac{m_e c^2}{\left(1 - \frac{v'^2}{c^2}\right)^{1/2}}, \qquad v' \equiv v_e,$$

$$v_\alpha' = \frac{c^2 p_\alpha'}{\varepsilon'}, \qquad \gamma' = \frac{\varepsilon'}{m_e c^2}, \qquad \gamma = 1, \qquad |v_\alpha| \ll c. \tag{4}$$

Substituting (4) into (3) gives

$$B_{\alpha\beta} = \frac{\gamma'}{c (\gamma'^2 - 1)^{3/2}} \left[\left(\gamma'^2 - 1\right) \delta_{\alpha\beta} - \frac{\gamma'^2}{c^2} v_\alpha' v_\beta' \right]$$

$$= \frac{1}{v'} \left(\delta_{\alpha\beta} - \frac{v_\alpha' v_\beta'}{v'^2} \right). \tag{5}$$

The rate of energy exchange between ions and electrons in the unit of mass is

$$\frac{dE_N}{dt} = -\frac{1}{\rho} \int \varepsilon_N \frac{\partial S_\alpha}{\partial p_\alpha} d^3 p = \frac{1}{\rho} \int S_\alpha v_\alpha \, d^3 p. \tag{6}$$

With the electron distribution function $f_e(p')$ from (1.2.2), temperature T', the Maxwellian ion distribution function

$$f(p) \equiv f_N = n_N \left(\frac{m_N}{2\pi kT} \right)^{3/2} \exp \left(-\frac{p^2}{2 m_N kT} \right), \tag{7}$$

using the relations

$$\frac{\partial f}{\partial p_\beta} = -\frac{v_\beta}{kT} f, \qquad \frac{\partial f_e}{\partial p'_\beta} = \frac{\partial f_e}{\partial \varepsilon'} \frac{\partial \varepsilon'}{\partial p'_\beta} = -\frac{v'_\beta}{kT'} f_e(1 - f_e), \tag{8}$$

and the general property [148a] of the tensor $B_{\alpha\beta}$ in (3)

$$v_\beta B_{\alpha\beta} = v'_\beta B_{\alpha\beta}, \tag{9}$$

we obtain from (2), (5), and (6)

$$\frac{dE_N}{dt} = \frac{4\pi Z^2 e^4 \Lambda}{h^3 m_N^3 \rho} \int f f_e (1 - f_e) \left(\frac{1}{kT} - \frac{1}{kT'} \right)$$
$$\times \frac{v_\alpha v_\beta}{v'} \left(\delta_{\alpha\beta} - \frac{v'_\alpha v'_\beta}{v'^2} \right) d^3 p \, d^3 p'. \tag{10}$$

Integrating over $d^3 p$ in (10) with $f(p)$ from (7) gives [222]

$$\frac{1}{m_N^3} \int f v_\alpha v_\beta \, d^3 p = \frac{n_N kT}{m_N} \delta_{\alpha\beta}. \tag{11}$$

We then have from (10)

$$\frac{dE_N}{dt} = \frac{8\pi Z^2 e^4 \Lambda}{h^3 \rho} \frac{n_N}{m_N} \frac{T' - T}{T'} \int f_e(1 - f_e) \frac{d^3 p'}{v'}. \tag{12}$$

It follows from (4) that in the isotropic case

$$d\varepsilon' = v' dp', \qquad \frac{d^3 p'}{v'} = 4\pi p'^2 \frac{dp'}{v'} = 4\pi \frac{p'^2}{v'^2} d\varepsilon' = \frac{4\pi}{c^4} \varepsilon'^2 d\varepsilon'. \tag{13}$$

Integrating (12) with use of (1.2.2), (2.54), (2.4.3), and (13) gives

$$\frac{dE_N}{dt} = \frac{32\pi^2 Z^2 e^4 \Lambda k m_e^2}{h^3 \rho m_N} n_N(T' - T) Q(\beta' - \alpha'), \tag{14}$$

where

$$Q(x) = \frac{1}{1 + e^{-x}} + \frac{2}{\alpha'} \ln(1 + e^x) + \frac{2}{\alpha'^2} F_1(x), \tag{15}$$

α' and β' are given in (1.2.8) with T' instead of T, the Fermi function $F_1(x)$ is defined in (1.2.49). In the limit of strong degeneracy $\beta'^2 - \alpha'^2 \gg 1$ we have, using (1.2.24) and (1.2.29)[11]

$$Q(\beta' - \alpha') = (1 + y^2) \left(1 - \frac{1}{3} \frac{\pi^2}{y^2 \alpha'^2} \right),$$

$$y = \left(\frac{\rho}{9.74 \times 10^5 \text{ g cm}^{-3} \mu_Z} \right)^{1/3}. \tag{16}$$

---

[11]In [441a] this limit is given erroneously.

In the non-relativistic limit $\alpha' \sim \beta' \gg 1$

$$Q(\beta' - \alpha') = \frac{1}{1 + \exp(\alpha' - \beta')} \tag{17}$$

and in the non-degenerate, non-relativistic limit, when $\alpha' - \beta' \gg 1$, we obtain from (17), (1.2.48), and (1.2.50)

$$Q(\beta' - \alpha') = \frac{1}{4\sqrt{2}\pi^{3/2}} \frac{\rho}{\mu_Z m_n} \frac{h^3}{(m_e k T_e)^{3/2}}, \qquad T' \equiv T_e. \tag{18}$$

Here, μ_Z is defined in (1.2.17), and for a mixture of nuclei we have approximately

$$\left\langle \frac{Z^2 n_N}{\rho m_N} \right\rangle \approx \left\langle \frac{Z^2}{A^2 m_u^2} \right\rangle \approx \frac{1}{\mu_Z^2 m_u^2}. \tag{19}$$

We may obtain from (14) and (18) the Landau formula [148a] for a one-component plasma

$$\frac{3}{2} \frac{k}{m_N} \frac{dT}{dt} = \frac{4\sqrt{2\pi} Z^2 e^4 \Lambda k \sqrt{m_e} n_e}{m_N^2 (kT')^{3/2}} (T' - T) = \frac{3}{2} \frac{k}{m_N} \frac{T' - T}{\tau_{ie}},$$

$$n_e = Z n_N, \qquad \rho = m_N n_N,$$

$$m_N = A m_u, \qquad E_N = \frac{3}{2} \frac{kT}{m_N}, \tag{20}$$

$$\tau_{ie} = \frac{3(kT_e')^{3/2} m_N}{8\sqrt{2\pi m_e} Z^2 e^4 \Lambda n_e} = \frac{n_N}{n_e} \tau_{ei}.$$

Note that the treatment of energy exchange between two components in thermodynamic equilibrium is valid by virtue of the inequality between relaxation times [148a]

$$\tau_{ee} = 2\frac{m_e}{m_N}\tau_{ie} \ll \tau_{ii} = 2\sqrt{\frac{m_e}{m_N}}\tau_{ei} \ll \tau_{ei}. \tag{21}$$

Problem 2. Find the rate of energy exchange between two non-relativistic, non-degenerate Maxwellian gases.

Solution [23a]. When f and f_i are Maxwellian with concentrations, particle masses, and temperatures (n_1, m_1, T_1) and (n_2, m_2, T_2) respectively, the integration over $d^3 v_1$ with the weight function $\varepsilon_1 = \frac{1}{2} m_1 v_1^2$ in (2.4.1) gives

$$\frac{3}{2} k n_1 \frac{dT_1}{dt} = \int \int \varepsilon (f_1' f_2' - f_1 f_2) g b \, db \, d\varphi \, d^3 v_1 \, d^3 v_2. \tag{1}$$

Here, $f_N \equiv f_2$, and the normalization condition for f_1 is analogous to f_N in (2.4.3). The integral in (1) may be evaluated exactly for arbitrary ratios

m_1/m_2, T_1/T_2, n_1/n_2 by use of the generating function for the Sonin polynomials $S_{1/2}^{(p)}(x)$, analogously to the collision integrals in [222], where the
polynomials $S_{3/2}^{(p)}(x)$ and $S_{5/2}^{(p)}(x)$ have been used.

The result may be written as [23a]

$$\frac{3}{2}kn_1\frac{dT_1}{dt} = 16n_1n_2\frac{m_1m_2}{(m_1+m_2)^2}k(T_2 - T_1)\Omega_{12}^{(1)}, \tag{1}$$

where

$$\Omega_{12}^{(1)}(1) = \sqrt{\pi}\int_0^\infty e^{-Z^2}Z^4\int_0^\infty(1-\cos\theta)gb\,db\,dZ,$$

$$Z = \left[\frac{m_1m_2}{2k(m_1T_2 + m_2T_1)}\right]^{1/2}g, \tag{2}$$

and $\Omega_{12}^{(1)}$ coincides in definition with the values in [222]. The values of $\Omega_{12}^{(l)}(r)$
have been evaluated exactly in [222] for power-law interactions between par-
ticles and for elastic collisions. The Landau formula ((20), problem 1) follows
from (1), (2) in the limit of $m_1T_2/m_2T_1 \ll 1$, $m_1 \equiv m_e$, $m_2 \equiv m_N$.

2.5 Radiative Transfer in Moving Media

If there are large-scale motions with a velocity field v_i in the medium, then
the form of transfer equation (2.1.9) will not formally change. The interaction
between quanta and matter, however, depends on the matter velocity affect-
ing the quantum frequency owing to the Doppler effect. In this case some
significant complications affect the right-hand side of (2.1.9). These suggest
that instead one should rewrite the transfer equation in a co-moving system
at rest with respect to the matter. The right-hand side of (2.1.9) then does
not change, but the left-hand convective terms require transformation.

2.5.1 The Transfer Equation in the Co-moving System

In early papers [113, 162, 320] the transfer equation in the co-moving system
was obtained by a complicated "geometric" method. In [36, 159] the equation
was derived using a simpler method of scattering invariants (see below). The
quantities related to the co-moving system will be denoted by subscript "0" .
We assume $v \ll c$ and retain only first-order corrections $\sim v/c$ [320, 36, 159]:

$$\mu_0 = \mu - (1-\mu^2)\frac{v}{c}, \qquad \mu = \mu_0 + (1-\mu_0^2)\frac{v}{c},$$

$$\nu_0 = \nu(1 - \mu\frac{v}{c}), \qquad \nu = \nu_0(1 + \mu_0\frac{v}{c}), \tag{2.5.1}$$

$$\mu_0 = \cos\theta_0, \qquad \mu = \cos\theta.$$

The invariance of the distribution function $\tilde{n}_\nu$ of photons in phase space with respect to the Lorentz transformation (the photon momentum $p_\nu = h\nu/c$),

$$\tilde{n}_\nu = \frac{c^3 f_\nu}{h^3 \nu^2} = \frac{c^2 I_\nu}{h^4 \nu^3}, \tag{2.5.2}$$

yields the invariant

$$\frac{I_\nu}{\nu^3} = \frac{I_{\nu_0}}{\nu_0^3}. \tag{2.5.3}$$

Here, $dn_\nu = \tilde{n}_\nu p_\nu^2 dp_\nu \, dV \, d\Omega$, f_ν and I_ν are determined by (2.1.1) and (2.1.4). The invariant form of the left-hand side of (2.1.9) obtained from (2.5.3) gives the following relations (upon writing $\left(\frac{dt_0}{\nu_0} = \frac{dt}{\nu} \right)$)

$$\frac{j_{\nu_0} \rho_0}{\nu_0^2} = \frac{j_\nu \rho_0}{\nu^2} \tag{2.5.4}$$

$$(\alpha_\nu + \sigma_\nu)\rho\nu = (\alpha_{\nu_0} + \sigma_{\nu_0})\rho_0 \nu_0.$$

For the plane atmosphere, substituting the variables

$$(t, r, \mu, \nu, I_\nu) \Rightarrow (t_0, r_0, \mu_0, \nu_0, I_{\nu_0}),$$

in (2.1.9) and (2.1.11), and using (2.5.1), (2.5.3) and (2.5.4) we obtain, to first order in v/c,

$$\frac{1}{c}\frac{\partial I_{\nu_0}}{\partial t_0} + \mu_0 \frac{\partial I_{\nu_0}}{\partial r_0} + \frac{3\mu_0^2}{c}\frac{\partial v}{\partial r_0} I_{\nu_0} - \frac{\mu_0^2}{c}\frac{\partial v}{\partial r_0} \nu_0 \frac{\partial I_{\nu_0}}{\partial \nu_0}$$
$$- \frac{\mu_0(1 - \mu_0^2)}{c}\frac{\partial v}{\partial r_0}\frac{\partial I_{\nu_0}}{\partial \mu_0} = j_{\nu_0}\rho_0 - \alpha_{\nu_0}\rho_0 I_{\nu_0}. \tag{2.5.5}$$

The scattering omitted here can be added to the right-hand side similarly to (2.1.9); also used are Lagrangian coordinates

$$t_0 = t, \qquad m = \int_{r_*}^{r_0} \rho_0(x)\, dx, \qquad \frac{dm}{dr_0} = \rho_0, \tag{2.5.6}$$

where r_0 is the radius of a given Lagrangian point. Transition to the Eulerian system $(\tilde{t}, \tilde{r}_0)$ in (2.5.5) requires the substitution

$$r_0 = \tilde{r}_0, \qquad \frac{\partial}{\partial r_0} = \frac{\partial}{\partial \tilde{r}_0} \qquad \frac{\partial}{\partial t} = \frac{\partial}{\partial \tilde{t}} + v\frac{\partial}{\partial \tilde{r}_0}. \tag{2.5.7}$$

The transfer equation is written in the Eulerian form in [159]. For the case of spherical symmetry, using (2.1.12) instead of (2.1.11) and Lagrangian coordinates

$$t_0 = t, \qquad m = 4\pi \int_{r_*}^{r_0} \rho_0(x)x^2 \, dx, \qquad \frac{dm}{dr_0} = 4\pi\rho_0 r_0^2. \tag{2.5.8}$$

we find similarly to (2.5.5) [320]

$$
\frac{1}{c}\frac{\partial I_{\nu_0}}{\partial t} + \mu_0 \frac{\partial I_{\nu_0}}{\partial r_0} + \frac{3v}{cr_0}\left[1 - \mu_0^2\left(1 - \frac{\partial \ln v}{\partial \ln r_0}\right)\right] I_{\nu_0}
$$
$$
+ \frac{1-\mu_0^2}{r_0}\left[1 + \frac{\mu_0 v}{c}\left(1 - \frac{\partial \ln v}{\partial \ln r_0}\right)\right]\frac{\partial I_{\nu_0}}{\partial \mu_0} \qquad (2.5.9)
$$
$$
- \frac{\nu_0 v}{cr_0}\left[1 - \mu_0^2\left(1 - \frac{\partial \ln v}{\partial \ln r_0}\right)\right]\frac{\partial I_{\nu_0}}{\partial \nu_0}
$$
$$
= j_{\nu_0}\rho_0 - \alpha_{\nu_0}\rho_0 I_{\nu_0}.
$$

Using (2.5.7) we may obtain the spherically symmetrical transfer equation in Eulerian coordinates from (2.5.9). The derivation of the transfer equation for moving media is given in more detail in [159].

2.5.2 The Moment Equations in the Eddington Approximation

We now derive moment equations similar to Sect. 2.2. Multiplying (2.5.5) by μ and μ^2 and integrating we can write for the plane atmosphere, upon dropping the subscript "0"

$$
\frac{\partial S}{\partial t} + \frac{\partial F}{\partial r} - \frac{1}{\rho}\frac{\partial \rho}{\partial t}(P + S) = -\rho c\left[\int_0^\infty \alpha_\nu S_\nu d\nu - \alpha_p B(T)\right], \qquad (2.5.10)
$$

$$
\frac{1}{c}\frac{\partial F}{\partial t} + c\frac{\partial P}{\partial r} - \frac{2}{c\rho}\frac{\partial \rho}{\partial t}F = -\kappa\rho F. \qquad (2.5.11)
$$

The notation of Sect. 2.2 is used here together with the continuity equation in Lagrangian coordinates

$$
\frac{\partial \rho}{\partial t} + \rho\frac{\partial v}{\partial r} = 0. \qquad (2.5.12)
$$

We have used (2.5.12) to substitute $\left(-\frac{1}{\rho}\frac{\partial p}{\partial t}\right)$ for $\partial v/\partial r$. Setting $P = S/3$ and using (2.5.11) and (2.5.12) we obtain the closed system of moment equations corresponding to the Eddington approximation [36, 37].

Proceeding analogously in the case of spherical symmetry and using (2.5.9) gives

$$
\frac{\partial S}{\partial t} + \frac{1}{r^2}\frac{\partial}{\partial r}(r^2 F) - \frac{1}{\rho}\frac{\partial \rho}{\partial t}(P + S) + \frac{v}{r}(S - 3P)
$$
$$
= -\rho c\left[\int_0^\infty \alpha_\nu S_\nu d\nu - \alpha_p B(T)\right], \qquad (2.5.13)
$$

$$
\frac{1}{c}\frac{\partial F}{\partial t} + c\frac{\partial P}{\partial r} - c\frac{S - 3P}{r} - \frac{2}{c\rho}\frac{\partial \rho}{\partial t}F - \frac{2}{r}\frac{v}{c}F = -\kappa\rho F. \qquad (2.5.14)
$$

Here we use the continuity equation in the Lagrangian system to eliminate $\partial v/\partial r$, namely

$$\frac{\partial \rho}{\partial t} + \frac{\rho}{r^2}\frac{\partial}{\partial r}(vr^2) = 0, \qquad \frac{\partial v}{\partial r} = -\frac{1}{\rho}\frac{\partial \rho}{\partial t} - 2\frac{v}{\rho}. \tag{2.5.15}$$

Finally, in order to close the system (2.5.13), (2.5.14) we make use of the relation $P = fS$ with the Eddington variable factor f from (2.2.28). In order to solve problems of stationary outflow from stars it is easiest to put the moment equations in Eulerian coordinates. This requires substituting

$$\frac{\partial}{\partial t} \Rightarrow v\frac{\partial}{\partial r} \tag{2.5.16}$$

in (2.5.10), (2.5.11) and (2.5.13), (2.5.14). It should be noted that under conditions of radiative equilibrium the right-hand sides of (2.5.10) and (2.5.13) reduce to zero. In outflowing shells with a velocity gradient the possible transformation of radiative energy into kinetic energy causes the flux $F \neq \text{const}$.[12]. If the flux properties are not influenced by radiation field, then (2.5.10–14) can be solved for a given $\rho(r,t)$ and $v(r,t)$. In order to solve a self-consistent problem the set of equations should be completed by the continuity, motion and energy equations for matter considered below.

---

[12] In the case of spherical symmetry $Fr^2 \neq \text{const}$.

3 Convection

The temperature and density distribution in stars satisfying the static equilibrium and heat conduction equations may be unstable with respect to the development of non-symmetrical disturbances on various scales, resulting in mass motion called convection. Consider convection in optically thick media. Convection in atmospheres of small optical depth has some specific features and must be considered separately.

3.1 Conditions for the Onset of Convection. Mixing-Length Theory

3.1.1 A Criterion for Convective Instability

Consider a stellar region in which the energy transport is caused by heat conduction. Imagine a small volume element within this region and displace it outwards or inwards towards the centre of the star by a distance dr. If the displacements are slow the pressure inside the element equals that of the ambient medium. After being released this element strives either to return to its initial state or to move further from it. In the second case the medium is convectively unstable.

The next step is to assume that no energy is exchanged between the element and the medium. If under these conditions the element is displaced the density inside it is governed by the adiabatic law. Let the initial state of the element correspond to the point T_0, r_0, P_0. If the element is displaced by dr its parameters (subscripted e) and the medium parameters (subscripted m) will change by

$$\Delta T_e = \left(\frac{\partial T}{\partial P} \right)_S \frac{dP}{dr} dr,$$

$$\Delta T_m = \left(\frac{\partial T}{\partial P} \right)_S \frac{dP}{dr} dr + \left(\frac{\partial T}{\partial S} \right)_P \frac{dS}{dr} dr,$$

$$\Delta \rho_e = \left(\frac{\partial \rho}{\partial P} \right)_S \frac{dP}{dr} dr,$$

$$\Delta\rho_m = \left(\frac{\partial\rho}{\partial P}\right)_S \frac{dP}{dr}dr + \left(\frac{\partial\rho}{\partial S}\right)_P \frac{dS}{dr}dr. \tag{3.1.1}$$

The medium is convectively unstable if for any outward displacement and expansion of the element its density decreases more rapidly than that of the medium, and for any inward displacement and contraction the element density increases more rapidly. Under these circumstances the buoyancy force moves the element further from its initial position. We see that the medium is convectively unstable if

$$\Delta\rho_e > \Delta\rho_m \text{ or } \left(\frac{\partial\rho}{\partial S}\right)_P \frac{dS}{dr}dr < 0 \quad \text{for contraction,}$$

$$\Delta\rho_e < \Delta\rho_m \text{ or } \left(\frac{\partial\rho}{\partial S}\right)_P \frac{dS}{dr}dr > 0 \quad \text{for expansion.} \tag{3.1.2}$$

Using $(\partial\rho/\partial S)_P = -\gamma_2\rho^2 T/P$, where $\gamma_2 = \left(\frac{\partial\ln T}{\partial\ln P}\right)_S$ is given by (1.1.12), with $dr < 0$ for contraction and $dr > 0$ for expansion, we conclude from both inequalities (3.1.2) that in a convectively unstable star

$$dS/dr < 0 \tag{3.1.3}$$

or that the specific entropy decreases from the centre outwards.

Inequality (3.1.3) is a necessary but not sufficient condition for the development of convective instability because the adiabatic law is inevitably broken as the element moves through the medium. The element's adiabatic expansion in a medium with $dS/dr < 0$ causes its temperature to exceed that of the medium, and heat conduction, which tends to reduce temperature gradients, weakens the convective instability. In addition, viscosity may reduce the speed of the element and also contributes to enhancement of stability. The general criterion for convective instability including transfer phenomena was obtained by Rayleigh (1916) and has the form [312]

$$R = \frac{g\alpha_T\beta_T h^4}{k_T\nu} > R_{cr} \approx 10^3. \tag{3.1.4}$$

Here, g is the gravitational acceleration, h is the thickness of the given layer, $\alpha_T = -\frac{1}{\rho}\left(\frac{\partial\rho}{\partial T}\right)_P$ is the thermal expansion coefficient, $\beta_T = \left(\frac{\partial T}{\partial r}\right)_S - (dT/dr) = \Delta(\nabla T)$ is the excess of the temperature gradient over the adiabatic one, k_T is the temperature diffusion coefficient, and ν is the kinematic viscosity. The last two coefficients can be written in terms of the heat conduction coefficient λ and viscosity coefficient η as follows:

$$\lambda = \rho c_p k_T, \qquad \eta = \rho\nu. \tag{3.1.5}$$

The quantity R_{cr} is a function of the system geometry and the boundary conditions. For a layer with boundaries that exert no horizontal stress (free)

$R_{cr} = 27\pi^4/4 = 657.5$; for one free and one against which the fluid does not slip (rigid) $R_{cr} = 1101$; for two rigid boundaries $R_{cr} = 1708$ [312]. Using thermodynamic relations [145] we can also express the quantity $\beta_T = \Delta(\nabla T)$ in terms of dS/dr:

$$\Delta(\nabla T) = \left(\frac{\partial T}{\partial r}\right)_S - \frac{dT}{dr} = -\frac{T}{c_p}\frac{dS}{dr},$$

$$\left(\frac{\partial T}{\partial r}\right)_S = \left(\frac{\partial T}{\partial P}\right)_S \frac{dP}{dr}. \tag{3.1.6}$$

We see from the Rayleigh criterion (3.1.4) that development of convection requires a finite value for the negative entropy gradient. Under realistic conditions in stars with very large linear scales h, the coefficient before dS/dr in (3.1.4), (3.1.6) is so large that inequality (3.1.3) may also be used as a sufficient condition for the onset of convection. We thus obtain a condition known as the Schwarzschild criterion.

3.1.2 Effects of Inhomogeneous Chemical Composition

Changes in molecular weight μ (see (1.1.6), (1.1.7)) have been introduced into condition (3.1.3) in [147, 554] for chemically inhomogeneous compositions. This modification takes the form of a correction to the buoyancy force, arising from the fact that μ is no longer constant. The change in ρ is

$$\Delta\rho_m = \left(\frac{\partial\rho}{\partial P}\right)_{S,\mu}\frac{dP}{dr}dr + \left(\frac{\partial\rho}{\partial S}\right)_{P,\mu}\frac{dS}{dr}dr + \left(\frac{\partial\rho}{\partial\mu}\right)_{S,P}\frac{d\mu}{dr}dr. \tag{3.1.7}$$

From (3.1.1) and (3.1.7) inequalities (3.1.2) reduce to

$$\frac{dS}{dr} - \left(\frac{\partial S}{\partial\mu}\right)_{\rho,P}\frac{d\mu}{dr} < 0. \tag{3.1.8}$$

It follows from $(\partial S/\partial\mu)_{\rho,P} > 0$ [145] that convective instability is quenched by the molecular weight decreasing in the outward direction, so that the onset of convection corresponds to a finite negative value of dS/dr. This kind of quenching occurs during hydrogen burning in the centres of massive stars.

It is sometimes more convenient to rewrite criterion (3.1.8) using the gradients ∇T and ∇P given immediately by the stellar structure equations (see Chap. 6). Since the relation

$$\frac{dP}{dr} = \left(\frac{\partial P}{\partial T}\right)_{\rho,\mu}\frac{dT}{dr} + \left(\frac{\partial P}{\partial\rho}\right)_{T,\mu}\frac{d\rho}{dr} + \left(\frac{\partial P}{\partial\mu}\right)_{\rho,T}\frac{d\mu}{dr},$$

is valid along the star, we have

$$\frac{d\rho}{dr} = \left(\frac{\partial\rho}{\partial P}\right)_{T,\mu}\left[\frac{dP}{dr} - \left(\frac{\partial P}{\partial T}\right)_{\rho,\mu}\frac{dT}{dr} - \left(\frac{\partial P}{\partial\mu}\right)_{\rho,T}\frac{d\mu}{dr}\right]. \tag{3.1.9}$$

The convective instability criterion (3.1.2) including $\Delta\rho_e$ from (3.1.1) is now written in the form[1]

$$\frac{(\partial\rho/\partial P)_{S,\mu}}{(\partial\rho/\partial P)_{T,\mu}}\frac{dP}{dr} < \frac{dP}{dr} - \left(\frac{\partial P}{\partial T}\right)_{\rho,\mu}\frac{dT}{dr} - \left(\frac{\partial P}{\partial\mu}\right)_{\rho,T}\frac{d\mu}{dr}. \qquad (3.1.10)$$

Using Jacobian transformations (1.1.10) and the equality

$$\left(\frac{\partial T}{\partial P}\right)_{\rho,\mu}\left(\frac{\partial P}{\partial T}\right)_{S,\mu} + \left(\frac{\partial\rho}{\partial T}\right)_{S,\mu}\left(\frac{\partial T}{\partial\rho}\right)_{P,\mu} = 1,$$

following from them we have, from (3.1.10),

$$\frac{d\ln T}{d\ln P} > \left(\frac{d\ln T}{d\ln P}\right)_{S,\mu} + \left(\frac{\partial\ln T}{\partial\ln\mu}\right)_{P,\rho}\frac{d\ln\mu}{d\ln P}. \qquad (3.1.11)$$

The convective instability criterion in the form (3.1.11) is obtained in [147, 554]. For the perfect gas–radiation mixture

$$\left(\frac{\partial\ln T}{\partial\ln\mu}\right)_{P,\rho} = \frac{\beta_g}{4 - 3\beta_g}, \qquad \left(\frac{\partial\ln T}{\partial\ln P}\right)_{S,\mu} = \gamma_2 \quad \text{from} \quad (1.1.12).$$

Criterion (3.1.8) or (3.1.11) is obtained without taking account of heat exchange between the convective element and the medium. The heat exchange in the uniform medium leads to stabilization of convection (3.1.4), but in chemically non-uniform medium that is stable according to (3.1.8), but with decreasing entropy $dS/dr < 0$ it may lead to destabilization (molecular viscosity and diffusion are much less important in stars). When $dS/dr < 0$, then the ascending element even in the stable situation, is still hotter than the medium. When it loses its heat the restoring buoyancy force caused by positive $d\mu/dr$ increases and the element is returning to the origin with a larger speed than it has at the start. In the same way the descending element is cooler in the stable region with $dS/dr < 0$ and $d\mu/dr > 0$ and by gaining heat also returns with increasing speed. If we integrate over the period of unperturbed motion then the element will increase its kinetic energy, resulting in overstability of the situation. So the conditions when the Ledoux criterion (3.1.8) is fulfilled and the Schwarzschild (3.1.3) is not (intermediate regime) was claimed to be overstable in linear approximation (see [345] and references therein).

---

[1] We treat the quantity μ in this section as independent of T and ρ. In equilibrium, when $\mu = \mu(\rho, T)$, as in the case of ionization, for example, variability of μ is taken into account by derivatives with respect to T and ρ whereas the term involving $d\mu/dr$ cancels out.

However, even small non-linear effects could stabilize the situation. After a periodic trip of the convective element the temperature gradient decreases in the medium, and distribution of chemical composition becomes smoother. That means that non-linear effects tend to reduce the overstability, increasing the growth time. If this time starts to become larger than the average lifetime of the element, the overstability will become unimportant. It seems therefore that in the intermediate regime the small scale non-linear oscillations will be generated, but real convective motion will not be installed. The problem, in general, is still far from its solution. The analysis of non-linear effects and their stabilizing influence in a chemically non-uniform medium has been carried out in [378b].

Laboratory experiments in non-uniform fluids in the intermediate regime have shown that non-linear stabilization of the overstability leads to stratification of the fluid into thin layers consisting of overturning cells. There is an effective mixing inside each layer, which has uniform composition. The layers are divided from each other by convectively stable boundary layers with steep concentration gradients. It was supposed in [581b] that the same picture is developed in the intermediate regime zones (IRZ) in stars.

3.1.3 Heat Transfer by Convection: The Mixing-Length Approximation

The convective motion close to the boundary of the stable region is topologically complicated but sufficiently regular. Moving away from this boundary inwards in a convectively unstable region the velocities of the convective motion are found to increase (see (3.1.14)). The character of the motions in a liquid or gas is determined by the dimensionless Reynolds number

$$\mathrm{Re} = \frac{vL_d}{\nu}, \tag{3.1.12}$$

where L_d is usually taken as the pressure scale height $H_p = \frac{P}{|\nabla P|}$ (see (3.1.22)) and v is the speed of the convective motion. The motion is regular and laminar at small values of Re, while at large values the motion is chaotic, confused and turbulent. The critical Reynolds number depends on the geometry and boundary conditions and varies in the range $\mathrm{Re}_{cr} = 400 \div 10^4$. The characteristic scales in stars are so large that not far from the boundary of a stable region the convection is already turbulent. In the solar atmosphere, for example, the region with optical depth $\tau \approx 1$ [312] has $L_d \approx 5 \times 10^7$ cm, $\nu \approx 2 \times 10^3$ cm^2, $v \approx 2 \times 10^4$ cm s^{-1}, $\mathrm{Re} \approx 5 \times 10^8$ Re_{cr}, The regular pattern of convective motions becomes increasingly complicated with increasing Re and chaotic at $\mathrm{Re} > \mathrm{Re}_{cr}$.

A quantitative description of turbulent convection and ordinary hydrodynamic turbulence encounters substantial difficulties because we know little about the nature of chaotic and stochastic phenomena. More insight into these problems is connected with investigation of the stochastic nature of dynamic systems with a relatively small number of degrees of freedom. These investigations have resulted in a qualitative revision of the approach to evolution of the pattern of turbulence [198] and in the discovery of remarkable quantitative laws underlying the evolution of stochastic motion [211]. We are nevertheless far from being able to apply these achievements to the description of turbulent convection in stars.

The most widely used method for studying internal stellar structure is based on a simple quantitative description of convection called the mixing-length model [306] (see also [581a]). It is assumed that convective elements move upwards and downwards, rising or falling a length l called the mean mixing-length. After having passed a length l the convective element is assumed to dissolve completely in the surrounding medium. After having risen a height dr it has a temperature excess $\Delta(\nabla T)dr$ over its surroundings, corresponding to an enthalpy excess per unit volume $\rho c_p \Delta(\nabla T)dr$ since the energy exchange occurs at a constant pressure. Multiplying this by the velocity v of the convective element we obtain the convective energy flux in the form

$$F_{\text{conv}} = \Delta(\nabla T)\, dr\, c_p \rho v \quad (\text{erg cm}^{-2}\text{ s}^{-1}) . \tag{3.1.13}$$

To find the quantity v we make use of the condition that the work done by the buoyancy force acting over a distance dr is converted into kinetic energy. The mean buoyancy force per unit volume is $\frac{1}{2}\Delta(\nabla\rho)dr\, g$ and multiplying by dr we obtain

$$\frac{1}{2}\rho v^2 = \frac{1}{2}\Delta(\nabla\rho)\, dr^2 g. \tag{3.1.14}$$

The quantity $\Delta(\nabla\rho)$ can be found by calculations similar to those leading to (3.1.6), (3.1.11). We have

$$\Delta(\nabla T) = \left(\frac{\partial T}{\partial r}\right)_{S,\mu} - \frac{dT}{dr} = \left(\frac{\partial T}{\partial P}\right)_{S,\mu}\frac{dP}{dr} - \frac{dT}{dr}, \tag{3.1.15}$$

$$\Delta(\nabla\rho) = \frac{d\rho}{dr} - \left(\frac{\partial\rho}{\partial r}\right)_{S,\mu}$$

$$= -\left(\frac{\partial\rho}{\partial T}\right)_{P,\mu}\left[\Delta(\nabla T) + \left(\frac{\partial T}{\partial\mu}\right)_{P,\rho}\frac{d\mu}{dr}\right]. \tag{3.1.16}$$

Assuming that the mean heat flux corresponds to displacement of the element by half the mixing-length

$$\overline{dr} = \frac{1}{2}l, \tag{3.1.17}$$

gives

$$F_{\text{conv}} = c_p \rho \left[- \left(\frac{\partial \rho}{\partial T} \right)_{P,\mu} \frac{g}{\rho} \right]^{1/2}$$

$$\times \frac{l^2}{4} \left[(\Delta(\nabla T))^3 + (\Delta(\nabla T))^2 \left(\frac{\partial T}{\partial \mu} \right)_{P,\rho} \frac{d\mu}{dr} \right]^{1/2}. \tag{3.1.18}$$

The presence of convection is usually assumed to remove gradients in chemical composition, so we may set $d\mu/dr = 0$ in (3.1.18). The quantity c_p for the fully ionized gas–radiation mixture is given in (1.1.20), and

$$\left(\frac{\partial \rho}{\partial T} \right)_{P,\mu} = -\frac{\rho}{T} \left(1 + 4\frac{P_r}{P_g} \right) = -\frac{\rho}{T} \left(\frac{4}{\beta_g} - 3 \right). \tag{3.1.19}$$

In a spherically symmetrical star $g = Gm/r^2$, where m is the mass within radius r. Substituting this into (3.1.18) yields

$$F_{\text{conv}} = c_p \rho \left[\frac{Gm}{Tr^2} \left(\frac{4}{\beta_g} - 3 \right) \right]^{1/2} \frac{l^2}{4} (\Delta(\nabla T))^{3/2}. \tag{3.1.20}$$

An expression analogous to (3.1.20) has been used in [277, 295]. For a perfect gas we obtain the relation ($\beta_g = 1$)

$$F_{\text{conv}} = c_p \rho \left(\frac{Gm}{Tr^2} \right)^{1/2} \frac{l^2}{4} (\Delta(\nabla T))^{3/2}, \tag{3.1.21}$$

given in [229].

The mixing-length theory includes a free parameter l which is usually taken to be of the order of the pressure scale height

$$l = \alpha_p H_p = \alpha_p \frac{P}{|dP/dr|}, \tag{3.1.22}$$

where $\alpha_p \sim 1$ is a dimensionless coefficient determined by the condition of greatest consistency with observation data. Usually $\alpha_p = 0.5 - 2$.

In stellar convective cores with high density the efficiency of heat transfer by convection is very high and the quantity $\Delta(\nabla T)$ is so small that we can consider the temperature distribution as essentially adiabatic and replace the heat conduction equation by the relation

$$\frac{dT}{dr} = \left(\frac{\partial T}{\partial P} \right)_{S,\mu} \frac{dP}{dr}. \tag{3.1.23}$$

The modification of the mixing-length theory for IRZ in the picture when chemically uniform layers with overturning cells are separated by the stable radiative boundary layers with steep gradients, was carried out in [581b]. In the stratified medium convection carries heat inside the layers and radiation does so between them. The formula for heat flux F_{iz} in the IRZ was obtained in [581b] for a mixture of gas and equilibrium radiation

$$F_{iz} = \frac{1}{2}\left(\frac{4}{\beta_g} - 3\right)^{1/4} g^{1/4} dH_p^{-5/4} [\Delta(\nabla T)]^{5/4} \left(\frac{4acT^5 c_p}{3\kappa}\right)^{1/2}, \quad (3.1.24)$$

where d is the thickness of the layer, presumably $d < H_p$.

The mixing in the IRZ is not so effective, as in the convective case and does not make its composition uniform. The diffusion flux for this case with molecular diffusion between the layers, calculated in [581b] has a form

$$F_{ciz} = \frac{1}{2} g^{1/4} dH_p^{-5/4} \left(\frac{4}{\beta_g} - 3\right)^{5/4} [\Delta(\nabla T)]^{5/4} k_s^{1/2} C \frac{d\ln C}{d\ln \mu}, \quad (3.1.25)$$

where C (cm^{-3}) is concentration of the given element. The diffusion coefficient in the binary plasma can be calculated in the same way as the heat conduction coefficient from the Boltzmann equation (see Sect. 2.4), and has a form [222]

$$k_s = \frac{3}{16n}\left(\frac{2kT}{\pi m}\right)^{1/2}\left(\frac{2kT}{Z_1 Z_2 e^2}\right)^2 \frac{1}{\ln \Lambda}, \quad (3.1.26)$$

$$m = m_1 m_2/(m_1 + m_2), \qquad n = n_1 + n_2.$$

Numerical simulations of IRZ, appearing in the massive stars as semiconvective zones (see Chap. 9, Vol. 2), based on a 2-D model have been performed in [482b]. The simulations have been carried out in the layer with (see (3.1.11))

$$\gamma_2 = 0.282, \qquad \gamma_L = \gamma_2 + \frac{\beta_g}{4 - 3\beta_g}\frac{d\ln \mu}{d\ln P} = 0.528, \quad (3.1.27)$$

imitating the conditions in the semiconvective zone existing in $30M_\odot$ stars near the main sequence. The results of simulations appear to be dependent on the initial value of $\nabla = \frac{d\ln T}{d\ln P}$. For $\nabla = 0.445$ (strongly driven simulations) the development of overstability is accompanied by the growth of long waves, finally leading to almost full mixing, as in the real convection, and establishing $\nabla \approx \gamma_2$. For $\nabla = 0.390$ (weakly driven simulations) the short waves grow most rapidly, preventing the growth of the long waves before any such drastic mixing occurs. The growth of longer waves is halted by occasional episodes of weak overturning. Here the mixing is small and the resulting gradient is established $\nabla \approx \gamma_L$, as in the Ledoux criterion.

The formulae (3.1.24) and (3.1.25) contain the thickness of the layer d as a free parameter, so they can describe both regimes, corresponding to larger and smaller d.

Problem. Derive the equality $\left(\frac{\partial T}{\partial P}\right)_{\rho,\mu}\left(\frac{\partial P}{\partial T}\right)_{S,\mu} + \left(\frac{\partial \rho}{\partial T}\right)_{S,\mu}\left(\frac{\partial T}{\partial \rho}\right)_{P,\mu} = 1.$

Solution. Using (1.1.10) we write the equality in the forms (at $\mu=$ const.)

$$\frac{\partial(T,\rho)}{\partial(P,\rho)}\frac{\partial(P,S)}{\partial(T,S)} + \frac{\partial(\rho,S)}{\partial(T,S)}\frac{\partial(T,P)}{\partial(\rho,P)}$$

$$= \frac{\partial(T,\rho)}{\partial(T,S)} \left[\frac{\partial(P,S)}{\partial(P,\rho)} + \frac{\partial(\rho,S)}{\partial(\rho,P)} \frac{\partial(T,P)}{\partial(T,\rho)} \right]$$

$$= \left(\frac{\partial\rho}{\partial S} \right)_T \left[\left(\frac{\partial S}{\partial\rho} \right)_P + \left(\frac{\partial S}{\partial P} \right)_\rho \left(\frac{\partial P}{\partial\rho} \right)_T \right] = 1. \tag{1}$$

Consider the dependence of S on parameters in the form $S(\rho, P(\rho,T))$. Then the full differential dS is

$$dS = \left(\frac{\partial S}{\partial\rho} \right)_P d\rho + \left(\frac{\partial S}{\partial P} \right)_\rho \left(\frac{\partial P}{\partial\rho} \right)_T d\rho + \left(\frac{\partial S}{\partial P} \right)_\rho \left(\frac{\partial P}{\partial T} \right)_\rho dT,$$

so that

$$\left(\frac{\partial S}{\partial\rho} \right)_T = \left(\frac{\partial S}{\partial\rho} \right)_P + \left(\frac{\partial S}{\partial P} \right)_\rho \left(\frac{\partial P}{\partial\rho} \right)_T.$$

The left part of relation (1) is reduced to

$$\left(\frac{\partial\rho}{\partial S} \right)_T \left(\frac{\partial S}{\partial\rho} \right)_T,$$

which is equal to 1 with account of (1.1.10).

3.2 Non-local and Non-stationary Description of Convection

The convective elements are produced in an unstable region, receive an acceleration imparted by the buoyancy force, and penetrate into the convectively stable region beyond (overshooting). The heat flux at a given point depends not only on local values of $\Delta(\nabla T)$ and other parameters, but also at least on parameters in the region $r \pm l$ so that non-local and overshooting effects arise even in the mixing-length approximation. These effects may become important if there are several convective regions separated by radiative layers. The overshooting into these layers with $\Delta(\nabla T) \leq 0$ may have noticeable evolutionary consequences, for example mixing and homogenizing the chemical composition. Investigating non-stationary processes in stars (oscillations, heat flashes, collapse) also requires the description of convection to be non-stationary, i.e. to include both non-local-in-time effects together with non-local-in-space effects.

3.2.1 Generalization of the Mixing-Length Model

In order to include non-local and overshooting effects a simple generalization of the mixing-length model has been made in [475, 513, 576]. An iterative procedure has been developed in [475], allowing one to include these effects in evolutionary computations.

In the presence of convection the total heat flux is

$$F = F_{\text{rad}} + F_{\text{NL}} = \frac{L}{4\pi r^2}.$$ (3.2.1)

F_{rad} can be found from (2.2.32)

$$F_{\text{rad}} = -\frac{4acT^3}{3\kappa\rho}\frac{dT}{dr}$$ (3.2.2)

and F_{NL} is the non-local convective energy flux. Define the ratio

$$f = \frac{F_{\text{rad}}}{F}$$ (3.2.3)

which enables us to write

$$\frac{L(r)}{4\pi r^2} = F = -\frac{4\pi acT^3}{3\kappa\rho}\frac{1}{f}\frac{dT}{dr}.$$ (3.2.4)

Using (3.2.4) in the set of equations (6.1.1–7) with a fixed function $f_0(r)$ we can construct a single stellar model. Using this model and applying a non-local treatment [475, 576] we find the convective energy flux.

The temperature excess $\Delta T(r)$ inside a convective element which started its motion from radius r_i is written in the form

$$\Delta T(r) = \int_{r_i}^{r} \Delta(\nabla T)\,dr.$$ (3.2.5)

Similarly, the density excess of the medium is

$$\Delta\rho(r) = \int_{r_i}^{r} \Delta(\nabla\rho)\,dr.$$ (3.2.6)

The equation for determining the velocity of the convective element may then be written as

$$\frac{v^2}{2} = (1 - \nu_T)\int_{r_i}^{r} g\frac{\Delta\rho}{\rho}\,dr,$$ (3.2.7)

instead of (3.1.14) above. Here, ν_T is the fraction of the work done by the buoyancy force that is transformed into heat by frictional processes. The quantities $\Delta(\nabla T)$ and $\Delta(\nabla\rho)$ are given by (3.1.15) and (3.1.16). According to the mixing hypothesis we assume that after having passed a length l the convective element dissolves completely in the medium. The integration limits in (3.2.5), (3.2.7) will then, on average, be determined by

$$0 \leq r - r_i \leq l/2$$ (3.2.8)

for rising elements, and by

$$-l/2 \leq r - r_i \leq 0 \text{ with } r \geq 0 \tag{3.2.9}$$

for falling elements. In the second case, $\Delta T < 0$ and $\Delta \rho < 0$, but in (3.1.7) in both cases $v^2 > 0$. Denoting the quantities related to rising and falling elements by "+" and "−" respectively, we write the non-local convective heat flux in the form

$$F_{\text{NL}} = \frac{1}{2} c_p \rho \left(\Delta T_+ v_+ + |\Delta T_- v_-| \right). \tag{3.2.10}$$

The rising and falling elements are assumed here to have the same square. In [475] $\nu_T = 0.5$.

Using (3.2.2) and (3.2.10) we find the new value of f corresponding to the given model:

$$f_1(r) = \frac{F_{\text{rad}}}{F_{\text{NL}} + F_{\text{rad}}}. \tag{3.2.11}$$

Iterations are continued until the difference $|f_n - f_{n-1}|$ becomes smaller than the given accuracy ϵ. The parameter l can be found as the value of l from (3.1.22) at the point r.

Equation (3.2.7) with $\overline{\Delta \rho} = \frac{1}{2}\Delta(\nabla \rho)\Delta r$, and $\nu_T = 0$, reduces to (3.2.14). Using (3.2.5) and the relations $\overline{\Delta T_+} = \overline{|\Delta T_-|} = \Delta(\nabla T)\Delta r$, we find that (3.2.10) at $\Delta r = |r - r_i| = l/2$ coincides with the local value of the flux F_{conv} from (3.1.13) .

The above description assumes the rising elements to cross the boundary $\Delta(\nabla T) = 0$ at $r = r_\delta$ with a finite velocity and to overshoot a distance $r_v < r_\delta + l/2$, where their velocity reduces to zero. As they travel in the region $r_\delta < r < r_v$ they gradually lose their temperature excess, and at $r = r_\epsilon$ such that $r_\delta < r_\epsilon < r_v$ their temperature and density become equal to the temperature and density of the medium. In the region $r_\epsilon < r < r_v$ the buoyancy force slows down the element and the convective flux becomes negative. Calculations in [475, 576] show that the zone of overshooting extends over 10–15% of l and that the overshoot tends to the finite non-zero value $r_v - r_\delta \approx 0.06\,l$ in the limit of a vanishing convective zone. In [475] stellar models are constructed using the iterative procedure described above.

Another treatment of the non-local effects of convection, presented by Ulrich in 1969, is described in [612]. This treatment averages over convective elements which come to the given point from all other levels of the unstable zone. With the local values of the convective flux from (3.1.18) we derive the true non-local value of $F_{\text{NL}*}$ in the form

$$F_{\text{NL}*}(z_0) = \int_{-\infty}^{\infty} F_{\text{conv}}(z)\psi_0(z_0, z)\,dz,$$

$$\int_{-\infty}^{\infty} \psi_0(z_0, z)\,dz, = 1 \qquad dz = \frac{dr}{H_p}. \tag{3.2.12}$$

A weight function $\psi_0(z_0, z)$ is introduced in [612] in the form

$$\psi_0 = \frac{a_k}{2\alpha'} \exp\left(-\frac{z_0 - z}{\alpha'}\right), \qquad z_0 > z,$$

$$\psi_0 = (1 - a_k)\delta(z_0 - z), \qquad z_0 = z,$$

$$\psi_0 = \frac{a_k}{2\alpha''} \exp\left(-\frac{z - z_0}{\alpha''}\right), \qquad z_0 < z,$$

$$\alpha' = 1.2\alpha_p, \qquad \alpha'' = 0.6\alpha_p, \qquad a_k < 0.9,$$

(3.2.13)

where H_p and α_p are determined by (3.1.22). Function (3.2.13) allows for both an asymmetry in the contribution to F_{NL*} from rising ($z_0 > z$) and falling elements and a possible finite weight $(1 - a_k)$ of the local flux at a given point. The iterative procedure described above may also be applied to non-local flux computations with the aid of (3.2.12). Star envelope models with $M = 1.25M_\odot$ and $3M_\odot$ have been calculated in [242] using F_{NL*} from (3.2.12), and the solar convective flux has been evaluated in [513] with the aid of both F_{NL} from (3.2.10) and F_{NL*} from (3.2.12), the results of these evaluations having significant differences.

Non-local convection theory based on using the Boltzmann equation applied for convective fluid elements was developed in [383*, 383**]. A similar approach was used earlier by Spiegel [581+]. Numerical studies of convective overshooting, based on the 2-D modeling of fully compressible convection, have been performed in [553***].

3.2.2 Non-stationary Convection

Equations describing the convective energy and momentum transport under non-stationary conditions have been derived in [364]. The derivation depends on averaging the hydrodynamic equations over fluctuating variables and applying the mixing-length condition. For spherical symmetry, the following convection characteristics are introduced:

$$F_{conv} = \overline{\rho E w_r} \quad (\text{erg cm}^{-2}\ \text{s}^{-1}) \quad \text{is the convective energy flux,}$$

$$E_{conv} = 3\overline{\rho w_r^2}/2\overline{\rho} \quad (\text{erg g}^{-1}) \qquad \begin{array}{l}\text{is the mean specific} \\ \text{convective energy.}\end{array}$$

(3.2.14)

The mean modulus of the radial convective velocity is [364]

$$\overline{w}_r = -F_{conv}\left/ \left(\frac{\partial \overline{E}}{\partial \overline{\rho}}\right)_{\overline{P}} \overline{\rho}^2\right.$$

(3.2.15)

We define also the usual mixing length l from (3.1.22). If the distance r to the centre of the star or the thickness h of the convective zone is less than the mixing length from (3.1.22), then l equals the minimum of these three quantities. In the case of isotropic convection the equations for F_{conv} and E_{conv} have the form [364]

$$\frac{1}{r^2}\frac{D}{Dt}(r^2 F_{\text{conv}}) = -\frac{2}{3}\overline{\rho}E_{\text{conv}}\left(\frac{\partial\overline{E}}{\partial r} + \overline{P}\frac{\partial(1/\overline{\rho})}{\partial r}\right)$$

$$+ F_{\text{conv}}\left[\left(\frac{\partial\epsilon_n}{\partial\overline{E}}\right)_{\overline{P}} - 2\frac{\partial v_r}{\partial r} - \frac{|w_r|}{l} - \frac{16ac\overline{T}^3}{3\overline{\kappa}\,\overline{\rho}^2}\frac{(\partial\overline{T}/\partial\overline{\rho})_{\overline{P}}}{(\partial\overline{E}/\partial\overline{\rho})_{\overline{P}}\,l^2}\right], \tag{3.2.16}$$

$$\frac{DE_{\text{conv}}}{Dt} = \left[\frac{F_{\text{conv}}}{\overline{\rho}^3}\bigg/\left(\frac{\partial\overline{E}}{\partial\overline{\rho}}\right)_{\overline{P}}\right]\frac{\partial\overline{P}}{\partial r} - \frac{|w_r|^3}{l}$$

$$-\frac{2}{3}\overline{\rho}E_{\text{conv}}\frac{D(1/\overline{\rho})}{Dt}. \tag{3.2.17}$$

Here, $D/Dt = \partial/\partial t + v_r\partial/\partial t$; $\overline{\rho},\overline{T},\overline{E},\overline{P},v_r$ are the mean values of quantities entering into the hydrodynamic equations from Sect. 10.2 (Vol. 2) where the overbar is usually dropped, ϵ_n (erg g^{-1} s^{-1}) is the bulk energy release (loss) rate (nuclear reactions, neutrino radiation), and $\overline{\kappa}$ is the opacity. The corrections to the hydrodynamic equations due to convection then reduce to the substitutions

$$P \Rightarrow \overline{P} + \frac{2}{3}\overline{\rho}E_{\text{conv}},$$

$$E \Rightarrow \overline{E} + E_{\text{conv}} \tag{3.2.18}$$

$$F_{\text{rad}} \Rightarrow F_{\text{rad}} + F_{\text{conv}}\left[1 - \frac{\overline{P}}{\overline{\rho}^2}\bigg/\left(\frac{\partial\overline{E}}{\partial\overline{\rho}}\right)_{\overline{P}}\right] \approx F_{\text{rad}} + \overline{(\rho E + P)w_r}.$$

Equations (3.2.16) and (3.2.17) involve the convective instability condition $dS/dr < 0$, so changes in the chemical composition are not included. When $dS/dr > 0$ an oscillating decay of perturbations occurs instead of their increase. This description of the convective flux is evidently non-local since the flux is determined by the solution of a differential equation. Equations (3.2.16) and (3.2.17) have been used in [364] for calculations of the collapse and explosion of massive stars following their loss of stability.

Another way of dealing with non-stationary and non-local convection by adding differential equations for F_{conv} and E_{conv} has been examined in [345].

Less universal but more detailed convective models, including non-stationarity, anisotropy, and low optical depth, are often used for constructing model atmospheres of pulsating stars (see e.g. [261, 378]). These non-local convection theories are almost never used in stellar evolution calculations because the serious mathematical complications do not yield a significant increase in accuracy as compared to the local theory [306]. In any case, there is still the uncertainty caused by the necessity to choose either a theory parameter l as in (3.2.8) or a function ψ_0 as in (3.2.2).

3.2.3 Penetration and Overshooting

Overshooting is not the only, and, perhaps, not a principal phenomena on the boundary between convectively stable and unstable regions. As was pointed out in [646a] on the basis of examples from geophysical fluids and laboratory experiments, more important could be "penetrative convection", which designates the spreading of the convective region, as flows reach their fully developed regime. The superadiabatic region that results from such penetration is often much broader than the initial, linearly unstable layer. In addition, the flows intrude into the stable adjacent domain, where they weaken the subadiabatic stratification, to a degree which depends on the efficiency of the convective heat transport.

When the efficiency is low, the motions carry little heat, but they can transport chemicals and momentum to appreciable distances into the stable (radiative) region. The term "overshooting" is related to such inefficient penetration that does not alter the stable temperature gradient.

On the other hand, when the convective heat transport is efficient enough, the penetrative flows establish a nearly adiabatic stratification, and they extend the convective zone into the subadiabatic domain, well beyond the limit predicted by the Schwarzschild criterion. It must be stressed that the mixing-length treatment cannot account for penetrative convection, and its predictions differ widely, depending on the assumptions made (see Sect. 3.1.1).

Numerical modeling of the convection always exhibits substantial penetration of the convective layer into its stable enviroment, irrespective of the approximation made, provided the stratification is almost adiabatic in the unstable zone; then the nearly adiabatic stratification extends well beyond the limits of the unstable region. The only obvious manifestation of the change of the entropy slope from the unstable to the stable one, is the reversal of the direction of the convective heat flux from upwards to downwards.

Consider a convective envelope where convection is efficient enough to establish an adiabatic stratification and to transport most of the thermal flux from the star. This requires that the turn-over time of the convective motion is shorter than the thermal diffusion time. In the subadiabatic region where an adiabatic gradient is established due to penetration (overshooting), the radiative flux exceeds the total one, so the convective heat flux is negative. Between the penetrative convection and radiative regions a thermal boundary layer is formed, where the gradient returns to its radiative subadiabatic value and the convective flux returns to zero.

Estimations of the penetration length L_p made in [646a] gave

$$L_p \approx H_p/\lambda_p, \tag{3.2.19}$$

where $\lambda_p = \left(\frac{\partial \log \lambda}{\partial \log P}\right)_S$, and $\lambda = \frac{4acT^3}{3\kappa\rho}$ is a coefficient of a radiative heat conductivity. Note that at the base of the of the solar convective zone $\lambda_p = 1.8$, so the penetration length is about one half of the pressure height H_p

(3.1.22). Similar estimations for the thickness of the thermal boundary layer L_t gave

$$L_t \approx \left(\frac{\lambda t_d}{\rho c_p}\right), \quad t_d = \left(\frac{H_p}{g}\right)^{1/2}. \tag{3.2.20}$$

Here, t_d is a dynamical time scale. Estimations for the solar convection zone gave $L_t \approx 1$ km, which is well below H_p. It was shown in [646a] that the conditions for subadiabatic convection are certainly fulfilled in the interior of the star, both at the bottom of the deep enough convective envelope and at the edge of a convective core, due to the high efficiency of the convective heat transport. The situation is very different, however, above a convective envelope, since much stronger thermal diffusion leads there to important departures from adiabaticity. The overshooting may still be very substantial there, but the above estimations can no longer be applied. The Roxburgh integral constraints [553*, 553**] are valid in conditions of effective convective heat transport and small entropy gradients, when their application for estimation of the length of the convective penetration and overshooting give results similar to [646a] for the convective core. For the case of the bottom of the convective envelope these estimations are rather crude, because the regions of integration include upper parts of the envelope, where deviations from the adiabaticity are substantial.

3.3 Numerical Simulation of Convection

These various modifications of the mixing-length approximation are inadequate for investigating such astrophysical problems as chemical composition anomalies obtained from stellar spectra, solar granulation, and the production of non-thermal flow by convection. The treatment of such problems requires numerical simulations including complicated two- and three-dimensional non-stationary computations [207]. Problems of this kind are examined in hydrodynamics for modeling liquid motions coupled with heat flows. Important applications are also related to the modeling of motions of liquids in the earth and planetary atmospheres.

Theories of stellar structure are only slightly related to these computations. They do assist the understanding of some observational properties of stars at different stages of evolution, but any numerical simulation is usually applied to resolving a single astrophysical problem. Since universality is absent and great computational difficulties arise from applying the numerical simulation to convection, it is not presently in use and in all probability will not soon be applied to stellar evolution calculations. We now represent a qualitative description of some approximations used in the numerical simulation of convection.

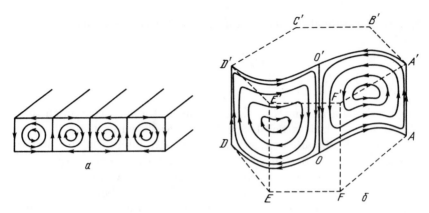

Fig. 3.1. Convective cellules: (a) two-dimensional "cylinder" convection, (b) three-dimensional cellular "torus" convection.

The simulation is based on the set of the hydrodynamical Navier–Stokes equations [150] including heat conduction and viscosity effects. In the case of laminar convection the transport properties are determined by the gas or radiation. Numerical investigations of laminar convection using two-dimensional models of the horizontal layer with solid boundaries ("cylinder convection") have revealed a cellular structure [83]. It should be noted that the picture of three-dimensional convection is more complicated [312] and is topologically equivalent to a line of flux winding around a torus called "torus convection" (see Fig. 3.1).

As pointed out in Sect. 3.1.1 convection within stars is usually turbulent. In numerical models of turbulent convection the small-scale convective motions are expressed in terms of the average quantities using turbulent viscosity and heat conductivity coefficients. Motions of maximum length-scale $\sim H_p$ and larger form regular structures similar to those observed in solar granulation [312], even under turbulent conditions. Investigations of turbulent convection by methods similar to [83] but with turbulent transport coefficients have been performed in [324, 581].

An important simplification of computations of turbulent convection by numerical simulation, called the "anelastic approximation", has been suggested by Ogura and Charney (1962), the equations for which were formulated for thermal convection by Gough [378a*]. In this approximation, P, ρ, T and other thermodynamic quantities are taken to be close to their equilibrium values and treated according to a linear approximation, whereas the velocity v and other dynamic quantities are considered exactly. This approximation allows the acoustic waves produced by convection, and which create computational difficulties but do not affect its principal characteristics, to be filtered out. Three-dimensional simulations of solar granulation have been computed in [338] by means of the anelastic approximation, two-dimensional simulations similar to [324] have also been computed in [405].

A simpler method but still involving long numerical computations relies on expanding all unknown variables as a given set of functions in the horizontal plane (and is called a local approximation). This approximation provides a set of ordinary equations for stationary convection. An anelastic modal description of stationary convection limiting the number of planform functions to one or two has been used in [480].

The simplifications provided by the anelastic modal approximation have allowed various ranges of parameters to be examined at the expense of considerably less computational work than direct two- and three-dimensional calculations [83, 324, 581, 338, 405]. The numerical simulation of convection up to now has made relatively little progress. We can hope, however, that further development of the theory will allow us to link it more closely to convective motions in stars.

4 Nuclear Reactions

Most of the energy radiated by stars is generated by transforming hydrogen into helium, and then helium into heavier elements. From the power engineering standpoint a star is equivalent to a giant nuclear reactor in which a controlled thermonuclear reaction takes place; the possibility of controlling this reaction requires thermal stability of the star (see Chap. 13, Vol. 2). The theory of nuclear reactions in stars was first investigated by M. Bethe, K. Weizsäcker, F. Hoyle, and E. Salpeter, while modern nuclear astrophysics is based largely on the studies of W. Fowler and his collaborators.

4.1 Nuclear Reaction Rates

Consider a reaction in which nuclei 1 and 2 combine to give nuclei 2 and 3

$$0 + 1 \rightarrow 2 + 3 + Q, \tag{4.1.1}$$

where

$$
\begin{aligned}
Q &= (M_0 + M_1 - M_2 - M_3)c^2 \\
&= \frac{931.478}{m_u}(M_0 + M_1 - M_2 - M_3) \ (\text{MeV}) \\
&= \frac{1.49232 \times 10^{-3}}{m_u}(M_0 + M_1 - M_2 - M_3) \ (\text{erg})
\end{aligned} \tag{4.1.2}
$$

is the thermal energy released per reaction. It increases by $2m_e c^2 = 1.0220\,\text{MeV} = 1.6374 \times 10^{-6}$ erg if the reaction yields a positron e^+ which annihilates with an electron of the medium. This is included in the expression for Q in reaction (4.2.1).

The main microscopic characteristic of (4.1.1) is its cross-section $\sigma(01)$. Let us introduce the following symbols[1]

$$\langle 01 \rangle \equiv \langle \sigma v \rangle_{01} \ (\text{cm}^3 \ \text{s}^{-1}) \tag{4.1.3}$$

---

[1] These are commonly used and are based for the most part on [360], as well as a content of this section.

where v is the relative velocity,

$$P_{01} = \frac{n_0 n_1}{1 + \delta_{01}} \langle 01 \rangle \ (\text{reactions cm}^{-3} \ \text{s}^{-1}), \tag{4.1.4}$$

where P_{01} is the reaction rate, δ_{01} is the Kronecker delta, and $n_i = \rho N_A \frac{x_i}{A_i} \ (\text{cm}^{-3}) \ (i = 0, 1, 2, 3)$, and where x_i are the fractional concentrations by weight of the nuclei i, $N_A = 6.02252 \times 10^{23} \ \text{g}^{-1}$ is the Avogadro number, $m_u = N_A^{-1} = 1.66043 \times 10^{-24}$ g is the atomic mass unit equal to $1/12$ of the ^{12}C (atom) mass, M_i are the atomic masses in reaction (13.1), and $A_i = M_i/m_u$;

$$[01] = \rho N_A \langle 01 \rangle \ (\text{s}^{-1}), \tag{4.1.5}$$

and

$$1/\tau_1(0) = \lambda_1(0) = -\frac{1}{n_0} \left(\frac{dn_0}{dt} \right)_1 = -\frac{1}{x_0} \left(\frac{dx_0}{dt} \right)_1$$
$$= n_1 \langle 01 \rangle = (x_1/A_1)[01] \ (\text{s}^{-1}), \tag{4.1.6}$$

where $\tau_1(0)$ is the mean lifetime of nucleus 0 before an interaction with nucleus 1, and $\lambda_1(0)$ is the probability per unit time for nucleus 0 to react with nucleus 1. The quantities $\tau_0(1)$ and $\lambda_0(1)$ are defined analogously. Equations (4.1.4) and (4.1.5) give

$$P_{01} = \rho N_A \frac{x_0 x_1}{A_0 A_1} \frac{[01]}{1 + \delta_{01}}. \tag{4.1.7}$$

The Kronecker $\delta_{01} = 1$ for the same particles ('0'='1') and $\delta_{01} = 0$ for different particles.

4.1.1 Cross-Sections and the Rates of Direct and Reverse Reactions

According to the reciprocity theorem (the principle of detailed balance), the cross-sections for the direct and reverse reactions in (4.1.1) are related by

$$\frac{\sigma(23)}{\sigma(01)} = \frac{1 + \delta_{23}}{1 + \delta_{01}} \frac{g_0 g_1}{g_2 g_3} \frac{A_0 A_1}{A_2 A_3} \frac{E_{01}}{E_{23}}, \tag{4.1.8}$$

where g_i are the statistical weights of different nuclei (see Sect. 1.3), and E_{01} and E_{23} are the energies of the nuclei in the centre-of-mass system before (E_{01}) and after (E_{23}) the reaction. After averaging over all velocities with a Maxwellian distribution we have

$$\frac{[23]}{[01]} = \frac{\langle 23 \rangle}{\langle 01 \rangle} = \frac{1 + \delta_{23}}{1 + \delta_{01}} \frac{g_0 g_1}{g_2 g_3} \left(\frac{A_0 A_1}{A_2 A_3} \right)^{3/2} \exp(-Q/kT). \tag{4.1.9}$$

If particle 3 is a photon, then $A_3 = 0$ and the time $\tau_\gamma(2)$ for photo-disruption of nucleus 2 may be written in terms of the quantity $\langle 01 \rangle$ for radiative capture:

$$\frac{1}{\tau_\gamma(2)} = \lambda_\gamma(2) = \frac{g_0 g_1}{(1 + \delta_{01}) g_2} \left(\frac{A_0 A_1}{A_2} \right)^{3/2} \left(\frac{m_u kT}{2\pi \hbar^2} \right)^{3/2}$$

$$\times \langle 01 \rangle \exp(-Q/kT)$$

$$= 0.98677 \times 10^{10} \frac{g_0 g_1}{(1 + \delta_{01}) g_2} \left(\frac{A_0 A_1}{A_2} \right)^{3/2}$$

$$\times \frac{T_9^{3/2}}{\rho} [01] \exp(-11.605 \, Q_6 / T_9) \, (\text{s}^{-1}), \qquad (4.1.10)$$

where Q_6 is the reaction energy expressed in MeV and $T_9 = T/10^9$ K. At high temperatures and densities, when reverse reactions are possible, the total reaction rate is

$$P_{01} - P_{23} = \frac{n_0 n_1}{1 + \delta_{01}} \langle 01 \rangle - \frac{n_2 n_3}{1 + \delta_{23}} \langle 23 \rangle. \qquad (4.1.11)$$

In equilibrium $P_{01} = P_{23}$, and using (13.9) gives

$$\frac{n_2 n_3}{n_0 n_1} = \frac{x_2 x_3}{x_0 x_1} \frac{A_0 A_1}{A_2 A_3} = \frac{g_2 g_3}{g_0 g_1} \left(\frac{A_2 A_3}{A_0 A_1} \right)^{3/2} \exp(Q/kT). \qquad (4.1.12)$$

Similarly, the total reaction rate for radiative capture is

$$P_{01} - P_{2\gamma} = \frac{n_0 n_1}{1 + \delta_{01}} \langle 01 \rangle - \frac{n}{\tau_\gamma(2)}, \qquad (4.1.13)$$

and equilibrium $(P_{01} = P_{2\gamma})$, using (13.10) yields

$$\frac{n_2}{n_0 n_1} = \frac{x_2 A_0 A_1}{\rho N_A x_0 x_1 A_2}$$

$$= \frac{g_2}{g_0 g_1} \left(\frac{A_2}{A_0 A_1} \right)^{3/2} \left(\frac{2\pi \hbar^2}{m_u kT} \right)^{3/2} \exp(Q/kT), \qquad (4.1.14)$$

or

$$\frac{x_2}{x_0 x_1} = 1.0134 \times 10^{-10} \rho T_9^{-3/2}$$

$$\times \left(\frac{g_2}{g_0 g_1} \right) \left(\frac{A_2}{A_0 A_1} \right)^{3/2} \exp \left(\frac{11.605 \, Q_6}{T_9} \right) \qquad (4.1.15)$$

(see also (1.3.3)). In atomic mass units $A_n = 1.008665$, $A_H = 1.007825$, $A_D = 2.014103$, $A_T = 3.016050$, $A_{^3He} = 3.016030$, $A_{^4He} = 4.002603$, while the other A_i are equal to the atomic weights to better than 3×10^{-3}. The rate ϵ (erg g^{-1} s^{-1}) of the energy release used in the theory of stellar structure is

$$\epsilon_{01} = \frac{P_{01}}{\rho} Q = 1.6021 \times 10^{-6} \frac{P_{01}}{\rho} Q_6$$

$$\times 9.6487 \times 10^{17} \frac{x_0 x_1}{A_0 A_1} \frac{[01]}{1 + \delta_{01}} Q_6 \quad (\text{erg g}^{-1}\,\text{s}^{-1}). \qquad (4.1.16)$$

For the reaction involving photodetachment $\epsilon_{2\gamma} < 0$ and

$$\epsilon_{2\gamma} = -9.6487 \times 10^{17} \frac{x_2}{A_2} \lambda_\gamma(2)\, Q_6 \quad (\text{erg g}^{-1}\,\text{s}^{-1}). \qquad (4.1.17)$$

Taking the reverse reaction into account we obtain the total energy release

$$\epsilon_{01} + \epsilon_{23} \quad \text{or} \quad \epsilon_{01} + \epsilon_{2\gamma}, \qquad (4.1.18)$$

where $\epsilon_{23} < 0$ is determined similarly to (4.1.16). The effect of neutrino escape is equivalent to a reduction in the thermal energy Q released in the reaction. For non-relativistic, non-degenerate nuclei the quantity $\langle \sigma v \rangle$ averaged over a Maxwellian distribution has the form

$$\langle \sigma v \rangle = \frac{(8/\pi)^{1/2}}{M^{1/2}(kT)^{3/2}} \int \sigma E \exp(-E/kT)\, dE$$

$$= 6.1968 \times 10^{-14} A^{-1/2} T_9^{-3/2}$$

$$\int \sigma_b E_6 \exp(-11.605\, E_6/T_9)\, dE \quad (\text{cm}^3\,\text{s}^{-1}), \qquad (4.1.19)$$

where E_6 is expressed in MeV and σ_b is expressed in barns (1 barn $= 10^{-24}$ cm^2),

$$M = \frac{M_0 M_1}{M_0 + M_1} \quad \text{is the reduced mass}$$

$$A = \frac{A_0 A_1}{A_0 + A_1} \quad \text{is the reduced atomic mass} \qquad (4.1.20)$$

$$E = Mv^2/2 \quad \text{is the energy in a centre-of-mass system.}$$

Using perturbation theory, the cross-section for reaction (4.1.1) can also be written in the form [212]

$$\sigma(01) = 2 \frac{g_2 g_3}{\pi \hbar^4} \frac{M' E_{23}}{v_{01} v_{23}} |\Omega H|^2, \qquad E_{23} = M' v_{23}^2/2. \qquad (4.1.21)$$

Here H is the matrix element for the transition and Ω is the volume to which the wave functions are normalized so that the product ΩH is independent of Ω, $v_{01} = v$ and v_{23} are the relative particle velocities before and after the reaction, and $M' = M_2 M_3/(M_2 + M_3)$ is the reduced mass of the particles after reaction. Evaluation of the matrix element, which is the same for direct and reverse reactions, encounters fundamental difficulties because of the absence of a rigorous, strong interaction theory. We have, approximately [212],

$$|\Omega H| = \left|\Omega \int dV \psi_{01}^* U \psi_{23}\right| \approx \overline{U} V_n \Omega \left|\overline{\psi_{01}\psi_{23}}\right| \quad (\text{erg cm}^3), \qquad (4.1.22)$$

where U (erg) is the energy, $\overline{U}$ is the mean interaction energy nearly equal to the depth of the potential well, V_n is the effective nuclear volume (volume of the strong interaction region), ψ_{01} and ψ_{23} ($\text{cm}^{-3/2}$) are the initial and final wave functions for the relative motions, and $\left|\overline{\psi_i\psi_f}\right|$ is the average of the product of wave functions over the nuclear volume. Theoretical estimates of the quantity $|\Omega H|$ are usually corrected by experimental data.

4.1.2 Non-resonant Reactions with Charged Particles

The reactions with charged particles are characterized by (quantum) tunnelling through a Coulomb barrier, since the kinetic energy of the reacting particles is usually much lower than the repulsive electrostatic energy $Z_0 Z_1 e^2 / r_{01}$, where $r_{01} = r_0 + r_1$, and r_0, r_1 are the nuclear radii. On isolating the slowly varying component of the energy, $S(E)$, the cross-section $\sigma(E)$ may be written in the form

$$\sigma = \frac{S(E)}{E} \exp\left(-\frac{2\pi Z_0 Z_1 e^2}{\hbar v}\right) \equiv \frac{S(E)}{E} \exp\left(-\sqrt{\frac{E_g}{E}}\right). \qquad (4.1.23)$$

The exponential in this expression is due to the product of the wave functions in (4.1.22) and determines the probability of tunnelling through the Coulomb barrier [212]. It is convenient to expand $S(E)$ as a Maclaurin's series:

$$S(E) = S(0) \left[1 + \frac{S'(0)}{S(0)} E + \frac{1}{2} \frac{S''(0)}{S(0)} E^2\right]. \qquad (4.1.24)$$

Substituting (4.1.3) and (4.1.24) into (4.1.19) and integrating by the saddle-point method [139] gives

$$\langle \sigma v \rangle = \frac{(8/\pi)^{1/2}}{M^{1/2}(kT)^{3/2}} \int S(E) \exp\left(-\frac{E_g^{1/2}}{E^{1/2}} - \frac{E}{kT}\right) dE$$

$$= \left(\frac{2}{M}\right)^{1/2} \frac{\Delta E_0}{(kT)^{3/2}} S_{ef} e^{-\tau}. \qquad (4.1.25)$$

From (4.1.3) and (4.1.5)

$$\langle 01 \rangle = \left[1.3006 \times 10^{-14} \left(\frac{Z_0 Z_1}{A}\right)^{1/3} S_{ef}\right] T_9^{-2/3} e^{-\tau} \quad (\text{cm}^3 \text{ s}^{-1}), \qquad (4.1.26)$$

$$[01] = \left[7.8327 \times 10^9 \left(\frac{Z_0 Z_1}{A}\right)^{1/3} S_{ef}\right] \rho T_9^{-2/3} e^{-\tau} \quad (\text{s}^{-1}), \qquad (4.1.27)$$

where

$$S_{ef} = S(0) \left[1 + \frac{5}{12\tau} + \frac{S'(0)}{S(0)} \left(E_0 + \frac{35}{36} kT \right) \right.$$

$$\left. + \frac{1}{2} \frac{S''(0)}{S(0)} \left(E_0^2 + \frac{89}{36} E_0 kT \right) \right] \quad (\text{MeV barn})$$

$$= S(0) \left[1 + 9.807 \times 10^{-2} W^{-1/3} T_9^{1/3} \right.$$

$$+ 0.1220 \frac{S'(0)}{S(0)} W^{1/3} T_9^{2/3} + 8.378 \times 10^{-2} \frac{S'(0)}{S(0)} T_9$$

$$+ 7.447 \times 10^{-3} \frac{S''(0)}{S(0)} W^{2/3} T_9^{4/3}$$

$$\left. + 1.300 \times 10^{-2} \frac{S''(0)}{S(0)} W^{1/3} T_9^{5/3} \right],$$

(4.1.28)

$$W = Z_0^2 Z_1^2 A, \tag{4.1.29}$$

$$E_0 = E_g^{1/3} \left(\frac{kT}{2} \right)^{2/3} = \left(\frac{\pi e^2 Z_0 Z_1 kT}{\hbar} \sqrt{\frac{M}{2}} \right)^{2/3}$$

$$= 0.12204 \, W^{1/3} T_9^{2/3} \quad (\text{MeV}),$$

(4.1.30)

$$\tau = 3 \frac{E_0}{kT} = 4.2487 \, W^{1/3} T_9^{-1/3}, \tag{4.1.31}$$

$$kT = 0.08617 \, T_9 = T_9/11.605 \quad (\text{MeV}), \tag{4.1.32}$$

$$\Delta E_0 = 4 \left(\frac{E_0 kT}{3} \right)^{1/2} = \frac{2^{11/6}}{3^{1/2}} \left(\frac{\pi e^2 Z_0 Z_1}{\hbar} \sqrt{M} \right)^{1/3} (kT)^{5/6}$$

$$= 0.23682 \, W^{1/6} T_9^{5/6} \quad (\text{MeV}). \tag{4.1.33}$$

The integral in (4.1.25) is evaluated by the saddle-point method, $E = E_0$ minimizes the absolute value of the exponent, and ΔE_0 is the effective width of the energy range contributing to this integral. The derivatives S' and S'' with respect to E in (4.1.28) are measured in barn and MeV^{-1} barn, respectively.

4.1.3 Resonant Reactions

The cross-section for reactions near a resonance is determined by the Breit–Wigner formula

$$\sigma = \frac{\pi\hbar^2}{2ME} \frac{\omega_r \Gamma_1 \Gamma_2}{(E - E_r)^2 + \Gamma^2/4}$$

$$= \frac{0.6566}{AE_{\text{MeV}}} \frac{\omega_r \Gamma_1 \Gamma_2}{(E - E_r)^2 + \Gamma^2/4} \quad \text{(barn)}, \tag{4.1.34}$$

where E_r is the resonant energy for the particles $0 + 1$ in a centre-of-mass system, Γ_1 is the partial width for decay of a resonant state with the formation of particles $0 + 1$, Γ_2 corresponds to particles $2 + 3$, $\Gamma = \Gamma_1 + \Gamma_2 + \ldots$ is the sum over all possible partial widths, $\omega_r = (1 + \delta_{01})g_r/g_0 g_1$, $g_r = 2J_r + 1$, and J_r is the resonant state spin. If the total width Γ of the resonance is much less than the effective dispersion in energies for the interacting particles

$$\begin{aligned} \Gamma &\ll \Delta E_0 \quad \text{from (4.1.33) for charged particles,} \\ \Gamma &\ll kT \quad \text{for reactions with neutrons,} \end{aligned} \tag{4.1.35}$$

then, substituting (4.1.34) into (4.1.19) gives, upon integrating,

$$\langle \sigma v \rangle = \left(\frac{2\pi\hbar^2}{MkT} \right)^{3/2} \frac{(\omega\gamma)_r}{\hbar} e^{-\frac{E_r}{kT}}. \tag{4.1.36}$$

From (4.1.3) and (4.1.5)

$$\begin{aligned} \langle 01 \rangle &= [2.557 \times 10^{-13} A^{-3/2} (\omega\gamma)_r] T_9^{-3/2} \\ &\quad \exp(-11.605 E_r/T_9) \quad (\text{cm}^3 \text{ s}^{-1}), \\ [01] &= [1.540 \times 10^{11} A^{-3/2} (\omega\gamma)_r] \rho T_9^{-3/2} \\ &\quad \exp(-11.605 E_r/T_9) \quad (\text{s}^{-1}). \end{aligned} \tag{4.1.37}$$

Here

$$(\omega\gamma)_r = \omega_r \gamma_r = (\omega\Gamma_1\Gamma_2/\Gamma)_r. \tag{4.1.38}$$

Using (4.1.34) and (4.1.10) gives the cross-section for resonant photo-disruption of a nucleus

$$\begin{aligned} \lambda_\gamma(2) &= \frac{g_r \gamma_r}{g_2 \hbar} \exp\left(-\frac{Q + E_r}{kT} \right) \\ &= \frac{(\omega\gamma)_r}{\omega_2 \hbar} \exp\left(-\frac{Q + E_r}{kT} \right) \quad (\text{s}^{-1}), \end{aligned} \tag{4.1.39}$$

where

$$\omega_2 = \frac{(1+\delta_{01})g_2}{g_0 g_1}. \tag{4.1.40}$$

It should be noted that the formulae for resonant reactions have the same form for both charged particles and neutrons. The quantity $(\omega\gamma)_r$ entering into the expressions for resonant reactions is to be found from experimental data. If the total cross-section $\int_r \sigma dE$ of the resonant reaction is known, then from (4.1.34) and (4.1.38) we obtain

$$(\omega\gamma)_r = \frac{ME_r}{\pi^2\hbar^2} \int_r \sigma dE. \tag{4.1.41}$$

If, on the contrary, it is easier to determine experimentally the quantity $\sigma_r = \sigma(E_r)$ and the total resonance width Γ_r, then $(\omega\gamma)_r$ should be obtained from the relation

$$(\omega\gamma)_r = \frac{\sigma_r M\Gamma_r E_r}{2\pi\hbar^2}. \tag{4.1.42}$$

Note that for reverse reactions to (4.1.1), the resonant energy E'_r is

$$E'_r = E_r + Q. \tag{4.1.43}$$

This quantity enters into (4.1.39).

4.1.4 Reactions at High Temperatures

At $T_9 \geq 1$ the interaction of light particles (protons, alpha particles and photons) with heavy nuclei proceeds through a large number of overlapping resonances. In this case the total value of $\langle\sigma v\rangle$ for these reactions is obtained by summing over a large number of formulae analogous to (4.1.36). The rates of the interactions between p, α, γ and nuclei with $19 \leq A \leq 40$ in the range of $1 \leq T_9 \leq 5$ have been calculated in [360, 361] and approximated by analytical formulae. The best agreement with experimental data is yielded by the formula

$$[01] = \rho N_A \langle 01 \rangle = CT^m \rho \exp(-E_{\text{th}}/kT), \tag{4.1.44}$$

where C, m, E_{th} are empirical parameters.

4.1.5 Non-resonant Reactions with Neutrons

The cross-section for inelastic neutron scattering at low energies is isotropic (S-wave) and $\sigma \sim 1/v$, so that $\sigma v = $ const. as $v \to 0$ [360]. It is convenient to express σv as a series in $v \sim \sqrt{E}$:

$$\sigma v = G(\sqrt{E}) = G(0)\left[1 + \frac{\dot{G}(0)}{G(0)}\sqrt{E} + \frac{1}{2}\frac{\ddot{G}(0)}{G(0)}E\right] \quad (\text{cm}^3 \text{ s}^{-1}). \tag{4.1.45}$$

Upon averaging, we have

$$\langle \sigma v \rangle = G(0) \left[1 + \frac{2}{\sqrt{\pi}} \frac{\dot{G}(0)}{G(0)} (kT)^{1/2} + \frac{3}{4} \frac{\ddot{G}(0)}{G(0)} kT \right]$$

$$= S(0) \left[1 + 0.3312 \frac{\dot{G}(0)}{G(0)} T_9^{1/2} + 0.06463 \frac{\ddot{G}(0)}{G(0)} T_9 \right], \tag{4.1.46}$$

where E is written in MeV, $\dot{G}(0)/G(0)$ in MeV$^{-1/2}$, $\ddot{G}(0)/G(0)$ in MeV^{-1} and the derivatives are calculated with respect to $E^{1/2}$. The quantity $S(0)$ is found from measurements of the cross-section σ_{th} for thermal neutron capture at $v_{\text{th}} = 2.2 \times 10^5$ (cm s^{-1}) and $E_{\text{lab}} = 2.53 \times 10^{-8}$ (MeV), so that

$$S(0) = (\sigma v)_{\text{th}} = 2.2 \times 10^{-19} \sigma_{\text{th}} \quad (\text{cm}^3 \text{ s}^{-1}), \tag{4.1.47}$$

where σ_{th} is written in barns. The coefficients $S(0)$, $S'(0)/S(0)$, $S''(0)/S(0)$ from (4.1.24) and (4.1.28), E_r and $(\omega\gamma)_r$ from (4.1.36–39), the parameters C and E_{th} with $m = 0$ from (4.1.44), $G(0)$, $\dot{G}(0)/G(0)$, $\ddot{G}(0)/G(0)$ from (4.1.45–46) and the values of Q and statistical weights of various reactions are listed in [360]. More accurate formulae for the same reactions and for many other reactions are given in [361, 389, 399, 638, 321]. Parameters in these formulae are chosen according to experimental data obtained specially for astrophysical purposes. The cross-sections for these reactions are listed in tables given in [650, 639].

4.2 Hydrogen, Deuterium and Helium Burning

The hydrogen burning phase is the longest one in the life of a star. This is caused by the large initial abundance of hydrogen in stars ($\sim 70\%$ by mass) and the high thermal yield for transformation of hydrogen into helium ($\sim 0.7\%$ mc^2); the released energy equals $\sim 70\%$ of the total energy released in converting hydrogen into the most stable element ^{56}Fe. The photon luminosity of stars on the main sequence, where hydrogen burns, is usually less than that of later evolutionary stages, and their neutrino luminosity is also essentially smaller because the central temperature is less than 4×10^7 K. Stars on the main sequence are therefore the most common, both in the Galaxy and throughout the Universe (see Chap. 9, Vol. 2).

The energy released per gram in helium burning is about an order of magnitude less than that in hydrogen burning, and the luminosity of stars in which helium burning is dominant is considerably higher. Hence the lifetime and number of these stars in the galaxy is much less than those of corresponding stars on the main sequence. The properties of such stars, which are giants and supergiants, are nevertheless well studied because of their high luminosity.

In addition to slow burning, thermal flashes may also occur in a few stars, in which fast, so-called explosive burning takes place at high temperatures $T_9 \sim 1$. These phenomena are associated with late evolutionary stages and in a matter accreting onto white dwarfs and neutron stars.

4.2.1 Proton–Proton Reaction and Deuterium Burning

Hydrogen burning resulting in helium formation requires the transformation of two protons into two neutrons via electron capture or positron emission together with simultaneous emission of neutrinos. These transformations are caused by reactions involving the weak interaction, which proceed much more slowly (see Chap. 5) than nuclear reactions resulting from the strong interaction. The fusion of two protons, leading to deuterium formation, is the first reaction of the proton–proton cycle being the main in stars with masses $M \leq M_\odot$ in which the central temperature is less than $\sim 1.5 \times 10^7$ K. Consider the several chains of the pp-cycle. The first reactions to occur [361, 359, 134, 320a] are

$$(1) \quad {}^1\text{H} + {}^1\text{H} \rightarrow {}^2\text{D} + e^+ + \nu, \quad Q_{\text{tot}} = 1.442, \quad Q_6 = 1.192, \qquad (4.2.1)$$

or

$$(1a) \quad {}^1\text{H} + e^- + {}^1\text{H} \rightarrow {}^2\text{D} + \nu, \quad Q_{\text{tot}} = 1.442, \quad Q_6 = 0.001. \qquad (4.2.1a)$$

Reaction (1) is much more probable than (1a) in the Sun and more massive stars. Deuterium burning corresponds to the reaction

$$(2) \quad {}^2\text{D} + {}^1\text{H} \rightarrow {}^3\text{He} + \gamma, \qquad Q_6 = 5.494. \qquad (4.2.2)$$

^{3}He takes part in two reactions:

$$(3) \quad {}^3\text{He} + {}^3\text{He} \rightarrow {}^4\text{He} + {}^1\text{H} + {}^1\text{H}, \qquad Q_6 = 12.860, \qquad (4.2.3)$$

or

$$(3a) \quad {}^3\text{He} + {}^4\text{He} \rightarrow {}^7\text{Be} + \gamma. \qquad Q_6 = 1.588. \qquad (4.2.3a)$$

^{7}Be is converted into ^{4}He in two different ways:

$$(3a_1) \quad {}^7\text{Be} + e^- \rightarrow {}^7\text{Li} + \nu \qquad Q_{\text{tot}} = 0.862, \quad Q_6 = 0.049,$$
$$\qquad \quad {}^7\text{Li} + {}^1\text{H} \rightarrow {}^4\text{He} + {}^4\text{He}, \qquad Q_6 = 17.346, \qquad (4.2.3a_1)$$

or

$$(3a_2) \quad {}^7\text{Be} + {}^1\text{H} \rightarrow {}^8\text{B} + \gamma, \qquad Q_6 = 0.137,$$
$$\qquad \quad \left. \begin{array}{l} {}^8\text{B} \rightarrow {}^8\text{Be}^* + e^+ + \nu \\[2mm] {}^8\text{Be}^* \rightarrow {}^4\text{He} + {}^4\text{He} \end{array} \right\} \quad Q_{\text{tot}} = 18.07, \quad Q_6 = 10.69. \qquad (4.2.3a_2)$$

Here Q_{tot} is the total energy release per reaction (in MeV), and Q_6 is the same minus the escaping neutrino energy. The total energy released in MeV per helium nucleus formed in the various chains is

$$Q_{\text{pp}} = \begin{cases} 26.23 & (1+2+3), \\ 23.85 & (1a+2+3), \\ 25.67 & (1+2+3a+3a_1), \\ 24.48 & (1a+2+3a+3a_1), \\ 19.1 & (1+2+3a+3a_2), \\ 17.91 & (1a+2+3a+3a_2). \end{cases} \tag{4.2.4}$$

The importance of each channel depends on the physical conditions in the star. The reaction $(1+2+3)$ is the principal reaction in the Sun and $Q_{\text{pp}} = 26.2$ (MeV). The rate of the pp-reaction is determined by the slowest reaction in the chain (4.2.1) and is proportional (see (4.1.4)) to the quantity [320a]

$$\begin{aligned} N_A\langle^1\text{H}p\rangle &= 4.01 \times 10^{-15} T_9^{-2/3} \exp(-3.380/T_9^{1/3}) \\ &\times (1 + 0.123\, T_9^{1/3} + 1.09\, T_9^{2/3} \\ &+ 0.938\, T_9)\ (\text{cm}^3\ \text{s}^{-1}\ \text{g}^{-1}), \quad T_9 \le 3. \end{aligned} \tag{4.2.5}$$

According to (4.1.16), the rate of energy release for $(1+2+3)$ is

$$\begin{aligned} \epsilon_{\text{pp}} &= 9.6487 \times 10^{17} \frac{x_{\text{H}}^2}{2} \rho N_A\langle^1\text{H}p\rangle \frac{Q_{\text{pp}}}{2} \\ &= 6.32 \times 10^{18} \frac{Q_{\text{pp}}}{26.2} x_{\text{H}}^2 \rho N_A\langle^1\text{H}p\rangle\ (\text{erg g}^{-1}\ \text{s}^{-1}). \end{aligned} \tag{4.2.6}$$

In young stars approaching the main sequence, where the temperature is not yet sufficient for hydrogen burning, deuterium or ^{3}He burning is possible. The reaction rate and energy release in (4.2.2) are [389]

$$\begin{aligned} N_A\langle^2\text{D}p\rangle &= 2.24 \times 10^3\, T_9^{-2/3} \exp(-3.72/T_9^{1/3}) \\ &\times (1 + 0.112\, T_9^{1/3} + 3.38\, T_9^{2/3} \\ &+ 2.65\, T_9)\ (\text{cm}^3\ \text{s}^{-1}\ \text{g}^{-1}), \end{aligned} \tag{4.2.7}$$

$$\begin{aligned} \epsilon_{2\text{D}p} &= 9.6487 \times 10^{17} \frac{x_{2\text{D}} x_{\text{H}}}{2} \rho N_A\langle^2\text{D}p\rangle Q_6 \\ &= 2.65 \times 10^{18} x_{2\text{D}} x_{\text{H}} \rho N_A\langle^2\text{D}p\rangle\ (\text{erg g}^{-1}\ \text{s}^{-1}). \end{aligned} \tag{4.2.8}$$

Similarly, for the reaction (4.2.3) we have [320a]

$$\begin{aligned} N_A\langle^3\text{He}^3\text{He}\rangle &= 6.04 \times 10^{10}\, T_9^{-2/3} \exp(-12.276/T_9^{1/3}) \\ &\times (1 + 0.034\, T_9^{1/3} - 0.522\, T_9^{2/3} - 0.124\, T_9 \\ &+ 0.353\, T_9^{4/3} + 0.213\, T_9^{5/3})\ (\text{cm}^3\ \text{s}^{-1}\ \text{g}^{-1}), \end{aligned} \tag{4.2.9}$$

$$\epsilon_{^3He^3He} = 9.6487 \times 10^{17} \frac{x_{^3He}^2}{18} \rho N_A \langle ^3He^3He \rangle Q_6$$

$$= 6.893 \times 10^{17} x_{^3He}^2 \rho N_A \langle ^3He^3He \rangle \quad (\text{erg g}^{-1}\,\text{s}^{-1}). \tag{4.2.10}$$

In order to include ^2D and ^3He burning in stellar evolution calculations of stars approaching the main sequence, all three reactions (4.2.5–10) must be considered and $Q_6 = 1.192$ must be substituted in (4.2.6) for $Q_{pp}/2$. When the concentrations of ^2D and ^3He reach small steady-state values near the main sequence, the energy release rate is determined by formulae (4.2.5–6).

4.2.2 Carbon and Other Hydrogen Burning Cycles

In the presence of ^{12}C and heavier elements, and at a sufficiently high temperature, hydrogen is converted into helium through various chains of reactions in which the heavy elements serve as catalysts. The lowest temperature cycle CN has the form [229, 361, 389, 259, 182, 320a]:

$$
\begin{aligned}
&(1) \ ^{12}C + {}^1H \rightarrow {}^{13}N + \gamma, &&Q_6 = 1.944 \ (\sim 10^6 \text{ years}), \\
&(2) \ ^{13}N \rightarrow {}^{13}C + e^+ + \nu, &&Q_{tot} = 2.22, \\
&&&Q_6 = 1.51 \ (\tau_{1/2} = 10 \text{ min}), \\
&(3) \ ^{13}C + {}^1H \rightarrow {}^{14}N + \gamma, &&Q_6 = 7.551 \ (\sim 3 \times 10^6 \text{ yr}), \\
&(4) \ ^{14}N + {}^1H \rightarrow {}^{15}O + \gamma, &&Q_6 = 7.297 \ (\sim 3 \times 10^8 \text{ yr}), \\
&(5) \ ^{15}O \rightarrow {}^{15}N + e^+ + \nu, &&Q_{tot} = 2.76, \\
&&&Q_6 = 1.76 \ (\tau_{1/2} = 12.4 \text{ s}), \\
&(6) \ ^{15}N + {}^1H \rightarrow {}^{12}C + {}^4He, &&Q_6 = 4.966 \ (\sim 10^5 \text{ yr}).
\end{aligned} \tag{4.2.11}
$$

The half-lives $\tau_{1/2}$ and reaction time scales for the Sun are given in parentheses [229]. According to [361], the energy released in the carbon cycle per helium nucleus is

$$Q_{CN} = 24.97 \ (\text{MeV}), \tag{4.2.12}$$

the neutrino energy not being included. The reaction rate in the Sun is determined by the slowest reaction (4) of the chain in (4.2.11); it is given by [320a]

$$
N_A \langle ^{14}Np \rangle = \left\{ 4.90 \times 10^7 \, T_9^{-2/3} \exp\left[-\frac{15.228}{T_9^{1/3}} - \left(\frac{T_9}{3.294} \right)^2 \right] \right.
$$

$$
\times (1 + 0.027 \, T_9^{1/3} - 0.778 \, T_9^{2/3} - 0.149 \, T_9 + 0.261 \, T_9^{4/3}
$$

$$
+ 0.127 \, T_9^{5/3}) + 2.37 \times 10^3 \, T_9^{-3/2} \exp(-3.011/T_9)
$$

$$
\left. + 2.19 \times 10^4 \exp(-12.530/T_9) \right\} \quad (\text{cm}^3 \, \text{s}^{-1} \, \text{g}^{-1}). \tag{4.2.13}
$$

The total energy release rate is

$$\epsilon_{CN} = 1.721 \times 10^{18} x_{14N} x_H \rho N_A \langle {}^{14}Np \rangle \quad (\text{erg g}^{-1}\,\text{s}^{-1}). \tag{4.2.14}$$

Here $Q_6 = Q_{CN}$ is taken from (4.2.12). The CN-cycle reactions dominate over those of the pp-cycle for stars roughly more massive than the Sun. With increasing temperature a higher Coulomb barrier can be overcome and the reactions of this cycle gain an advantage owing to the fact that the beta-decay reactions (2) and (5) from (4.2.11) are much faster than the pp-reaction (4.2.1). As the CN-cycle progresses, almost all the carbon is converted into nitrogen since reaction (4) from (4.2.11) is slow.

At temperatures $T \geq 2 \times 10^8$ K reaction (4) along with reactions (1), (3) and (6) from (4.2.11) become faster than the beta-decay (2), and this then starts to limit the overall reaction rate of the cycle. In this case, the capture of a proton onto a nucleus ^{13}N begins another cycle called the hot CNO-cycle:

$$^{12}C(p,\gamma)^{13}N(p,\gamma)^{14}O(e^+\nu)^{14}N(p,\gamma)^{15}O(e^+\nu)^{15}N(p,\alpha)^{12}C. \tag{4.2.15}$$

These reactions are written here in the more compact form used in nuclear physics. The ^{14}O half-life is 72 s in this cycle [232] and the slow reaction (2) from (4.2.11) ceases to limit the cycle as a whole. The rate of the $^{13}N(p,\gamma)^{14}O$ reaction is computed in [481] and corrected in [320a]:

$$
\begin{aligned}
N_A\langle {}^{13}Np \rangle = \Big\{ & 4.04 \times 10^7\, T_9^{-2/3} \exp[-15.202/T_9^{1/3} \\
& - (T_9/1.191)^2] \times (1 + 0.027\, T_9^{1/3} - 0.803\, T_9^{2/3} \\
& - 0.154\, T_9 + 5.00\, T_9^{4/3} + 2.44\, T_9^{5/3}) \\
& + (2.43 \times 10^5/T_9^{2/3}) \exp\left(-\frac{6.348}{T_9}\right) \Big\} \quad (\text{cm}^3\,\text{s}^{-1}\,\text{g}^{-1}).
\end{aligned} \tag{4.2.16}
$$

The energy $Q_6 = 4.628$ MeV is released in this reaction and the energy release per gram is

$$\epsilon_{13Np} = 3.434 \times 10^{17} x_{13N}\, x_H\, \rho N_A \langle {}^{13}Np \rangle \quad (\text{erg g}^{-1}\,\text{s}^{-1}). \tag{4.2.17}$$

Besides the reaction (6) from (4.2.11), the nucleus ^{15}N may also be involved in the $^{15}N(p,\gamma)^{16}O$ reaction. The production of ^{16}O launches a new chain. Together with (6) the reactions proceed as follows:

$$(6a) \quad ^{15}N(p,\gamma)^{16}O(p,\gamma)^{17}F(e^+\nu)^{17}O(p,\alpha)^{14}N \tag{4.2.18}$$

followed by reaction (4) from (4.2.11), or another cycle goes

$$^{15}N(p,\gamma)^{16}O(p,\gamma)^{17}F(e^+\nu)^{17}O(p,\gamma)^{18}F(e^+\nu)^{18}O(p,\alpha)^{15}N. \tag{4.2.19}$$

At high temperatures the reactions [462]

$$^{15}O(\alpha,\gamma)^{19}Ne\,(p,\gamma)^{20}Na\,(e^+\nu)^{20}Ne \qquad (4.2.20)$$

proceed together with (5) in (4.2.11). The production of ^{20}Ne acts as a catalyst for the NeNa-cycle [608]

$$^{20}Ne\,(p,\gamma)^{21}Na\,(e^+\nu)^{21}Ne\,(p,\gamma)^{22}Na\,(e^+\nu)$$

$$^{22}Ne\,(p,\gamma)^{23}Na\,(p,\alpha)^{20}Ne, \qquad (4.2.21)$$

while production of ^{24}Mg (together with(4.2.21)) through

$$^{23}Na\,(p,\gamma)^{24}Mg \qquad (4.2.22)$$

leads to yet another high temperature MgAl-cycle of hydrogen burning [361, 389]:

$$^{24}Mg\,(p,\gamma)^{25}Al\,(e^+\nu)^{25}Mg\,(p,\gamma)^{26}Al\,(p,\gamma)$$

$$^{27}Si\,(e^+\nu)^{27}Al\,(p,\alpha)^{24}Mg. \qquad (4.2.23)$$

At $T \geq 2 \times 10^8$ K, when the reaction rate for (4) in (4.2.11) no longer limits the rate of burning, in order to find the concentration of elements and the rate of energy release it is necessary to take into account the contribution of each of the reactions (4.2.11) separately, and also (4.2.15–23). The reaction rates of α- and p-capture from (4.2.20) are calculated in [462]. Formulae similar to (4.1.27) and (4.1.37) being too cumbersome, because of a large number of resonances, it is more convenient to reproduce Table 4.1 from [462] representing $N_A\langle\sigma v\rangle$ as a function of temperature. The heat release in these reactions is calculated from the data in [135]. We have for the reactions (p,γ) from (4.2.11) [320a]

$$^{12}C(p,\gamma)^{13}N, \qquad Q_6 = 1.944,$$

$$N_A\langle^{12}Cp\rangle = \left\{ 2.04 \times 10^7\,T_9^{-2/3}\exp\left[-\frac{13.690}{T_9^{1/3}} - \left(\frac{T_9}{1.500}\right)^2\right]\right.$$

$$\times \left(1 + 0.03\,T_9^{1/3} + 1.19\,T_9^{2/3} + 0.254\,T_9 + 2.06\,T_9^{4/3} + 1.12\,T_9^{5/3}\right)$$

$$+ 1.08 \times 10^5\,T_9^{-3/2}\exp\left(-\frac{4.925}{T_9}\right)$$

$$\left. + 2.15 \times 10^5\,T_9^{-3/2}\exp\left(-\frac{18.179}{T_9}\right)\right\} \quad (cm^3\,s^{-1}\,g^{-1}), \qquad (4.2.24)$$

$$\epsilon_{12Cp} = 1.563 \times 10^{17}\,x_{12C}\,x_H\,\rho N_A\langle^{12}Cp\rangle \quad (erg\,g^{-1}\,s^{-1});$$

Table 4.1. $N_A\langle\sigma v\rangle$, $\text{cm}^3\text{s}^{-1}\text{g}^{-1}$, and the energy release Q_6 per reaction in MeV for two capture reactions

T (10^9 K)	$N_A\langle^{15}\text{O}\alpha\rangle_\gamma$	$N_A\langle^{19}\text{Ne p}\rangle_\gamma$	T (10^9 K)	$N_A\langle^{15}\text{O}\alpha\rangle_\gamma$	$N_A\langle^{19}\text{Ne p}\rangle_\gamma$
0.1	5.02(-25)	1.01(-11)	1.5	3.57	31.82
0.16	–	3.47(-9)	2.0	17.6	78.7
0.2	8.81(-13)	3.40(-8)	3.0	99.2	167.0
0.4	7.42(-7)	1.64(-4)	4.0	267	–
0.6	1.80(-2)	4.82(-2)	5.0	573	–
0.8	1.01(-2)	0.814	10.0	2.36(+3)	–
1.0	0.129	4.16			
Q_6	3.53	2.19	Q_6	3.53	2.19

$^{13}\text{C}(\text{p},\gamma)^{14}\text{N}$, $Q_6 = 7.551$,

$$N_A\langle^{13}\text{Cp}\rangle = \left\{ 8.01 \times 10^7\, T_9^{-2/3} \exp\left[-\frac{13.717}{T_9^{1/3}} - \left(\frac{T_9}{2.0}\right)^2 \right] \right.$$

$$\times \left(1 + 0.03\, T_9^{1/3} + 0.958\, T_9^{2/3} + 0.204\, T_9 + 1.39\, T_9^{4/3} + 0.753\, T_9^{5/3} \right)$$

$$\left. + 1.21 \times 10^6\, T_9^{-6/5} \exp\left(-\frac{5.701}{T_9} \right) \right\} \quad (\text{cm}^3\ \text{s}^{-1}\ \text{g}^{-1}), \tag{4.2.25}$$

$$\epsilon_{13\text{Cp}} = 5.604 \times 10^{17} x_{13\text{C}}\, x_H\, \rho N_A\langle^{13}\text{Cp}\rangle \quad (\text{erg g}^{-1}\ \text{s}^{-1}).$$

The value of $N_A\langle 01\rangle$ for the reaction $^{14}\text{N}(\text{p},\gamma)^{15}\text{O}$ is given in (4.2.13) and replacing coefficient 1.721×10^{18} by 5.030×10^{17} in (4.2.14) yields $\epsilon_{14\text{Np}}$ instead of ϵ_{CN}. We have also

$^{15}\text{N}(\text{p},\alpha)^{12}\text{C}$, $Q_6 = 4.966$,

$$N_A\langle^{15}\text{Np}\rangle_\alpha = \left\{ 1.08 \times 10^{12}\, T_9^{-2/3} \exp\left[-\frac{15.251}{T_9^{1/3}} - \left(\frac{T_9}{0.522}\right)^2 \right] \right.$$

$$\times \left(1 + 0.027\, T_9^{1/3} + 2.62\, T_9^{2/3} + 0.501\, T_9 + 5.36\, T_9^{4/3} + 2.60\, T_9^{5/3} \right)$$

$$+ 1.19 \times 10^8\, T_9^{-3/2} \exp\left(-\frac{3.676}{T_9} \right)$$

$$\left. + 5.41 \times 10^8\, T_9^{-1/2} \exp\left(-\frac{8.926}{T_9} \right) \right\} \quad (\text{cm}^3\ \text{s}^{-1}\ \text{g}^{-1}), \tag{4.2.26}$$

$$\epsilon_{15\text{Np}_\alpha} = 3.194 \times 10^{17} x_{15\text{N}}\, x_H\, \rho N_A\langle^{15}\text{Np}\rangle_\alpha \quad (\text{erg g}^{-1}\ \text{s}^{-1}).$$

For the reactions (4.2.18), (4.2.19) we have [320a]

$$^{15}\text{N}(\text{p},\gamma)^{16}\text{O}, \qquad Q_6 = 12.128,$$

$$N_A\langle^{15}\text{Np}\rangle_\gamma = \left\{ 9.78 \times 10^8\, T_9^{-2/3} \exp\left[-\frac{15.25}{T_9^{1/3}} - \left(\frac{T_9}{0.45}\right)^2 \right] \right.$$

$$\times \left(1 + 0.027\, T_9^{1/3} + 0.219\, T_9^{2/3} + 0.042\, T_9 + 6.83\, T_9^{4/3} + 3.32\, T_9^{5/3} \right)$$

$$+ 1.11 \times 10^4\, T_9^{-3/2} \exp\left(-\frac{3.328}{T_9} \right)$$

$$+ 1.49 \times 10^4\, T_9^{-3/2} \exp\left(-\frac{4.665}{T_9} \right)$$

$$\left. + 3.80 \times 10^6\, T_9^{-3/2} \exp\left(-\frac{11.048}{T_9} \right) \right\} \quad (\text{cm}^3\ \text{s}^{-1}\ \text{g}^{-1}), \qquad (4.2.27)$$

$$\epsilon_{15\text{Np}_\gamma} = 7.801 \times 10^{17} x_{15\text{N}}\, x_H\, \rho N_A\langle^{15}\text{Np}\rangle_\gamma \quad (\text{erg}\ \text{g}^{-1}\ \text{s}^{-1});$$

$$^{16}\text{O}(\text{p},\gamma)^{17}\text{F}, \qquad Q_6 = 0.600,$$

$$N_A\langle^{16}\text{Op}\rangle = 1.5 \times 10^8\, T_9^{-2/3}$$

$$\times \left\{ 1 + 2.13\left[1 - \exp\left(-0.728\, T_9^{2/3} \right) \right] \right\}^{-1}$$

$$\times \exp\left(-\frac{16.692}{T_9^{1/3}} \right) \quad (\text{cm}^3\ \text{s}^{-1}\ \text{g}^{-1}), \qquad (4.2.28)$$

$$\epsilon_{16\text{Op}} = 3.618 \times 10^{16} x_{16\text{O}}\, x_H\, \rho N_A\langle^{16}\text{Op}\rangle \quad (\text{erg}\ \text{g}^{-1}\ \text{s}^{-1});$$

$$^{17}\text{O}(\text{p},\alpha)^{14}\text{N}, \qquad Q_6 = 1.191,$$

$$N_A\langle^{17}\text{Op}\rangle_\alpha = \left\{ 1.53 \times 10^7\, T_9^{-2/3} \exp\left[-\frac{16.712}{T_9^{1/3}} - \left(\frac{T_9}{0.565}\right)^2 \right] \right.$$

$$\times \left(1 + 0.025\, T_9^{1/3} + 5.39\, T_9^{2/3} + 0.940\, T_9 + 13.5\, T_9^{4/3} + 5.98\, T_9^{5/3} \right)$$

$$\left. + 2.92 \times 10^6\, T_9 \exp\left(-\frac{4.247}{T_9} \right) \right\} \quad (\text{cm}^3\ \text{s}^{-1}\ \text{g}^{-1}), \qquad (4.2.29)$$

$$\epsilon_{17\text{Op}_\alpha} = 6.760 \times 10^{16} x_{17\text{O}}\, x_H\, \rho N_A\langle^{17}\text{Op}\rangle_\alpha \quad (\text{erg}\ \text{g}^{-1}\ \text{s}^{-1});$$

$^{17}O(p,\gamma)^{18}F,\qquad Q_6 = 5.607,$

$$N_A\langle^{17}Op\rangle_\gamma = \left\{7.97 \times 10^7\, T_{9A}^{5/6}\, T_9^{-3/2}\, \exp\left(-\frac{16.712}{T_{9A}^{1/3}}\right)\right.$$

$$+\,1.51 \times 10^8\, T_9^{-2/3}\, \exp\left(-\frac{16.712}{T_9^{1/3}}\right)$$

$$\times\left(1 + 0.025\, T_9^{1/3} - 0.051\, T_9^{2/3} - 8.82 \times 10^{-3}\, T_9\right)$$

$$\left.+\,1.56 \times 10^5\, T_9^{-1}\, \exp\left(-\frac{6.272}{T_9}\right)\right\}\quad (\text{cm}^3\text{ s}^{-1}\text{ g}^{-1}),\qquad(4.2.30)$$

$T_{9A} = T_9(1 + 2.69\,T_9)^{-1},$

$\epsilon_{17\text{Op}_\gamma} = 3.183 \times 10^{17} x_{17\text{O}}\, x_H\, \rho N_A\langle^{17}Op\rangle_\gamma \quad (\text{erg g}^{-1}\text{ s}^{-1});$

$^{18}O(p,\alpha)^{15}N,\qquad Q_6 = 3.980,$

$$N_A\langle^{18}Op\rangle_\alpha = \left\{3.63 \times 10^{11}\, T_9^{-2/3}\, \exp\left[-\frac{16.729}{T_9^{1/3}} - \left(\frac{T_9}{1.361}\right)^2\right]\right.$$

$$\times\left(1 + 0.025\, T_9^{1/3} + 1.88\, T_9^{2/3} + 0.327\, T_9 + 4.66\, T_9^{4/3} + 2.06\, T_9^{5/3}\right)$$

$$+\,9.90 \times 10^{-14}\, T_9^{-3/2}\, \exp\left(-\frac{0.231}{T_9}\right)$$

$$+\,2.66 \times 10^4\, T_9^{-3/2}\, \exp\left(-\frac{1.670}{T_9}\right) + 2.41 \times 10^9\, T_9^{-3/2}\, \exp\left(-\frac{7.638}{T_9}\right)$$

$$\left.+\,1.46 \times 10^9\, T_9^{-1}\, \exp\left(-\frac{8.310}{T_9}\right)\right\}\quad (\text{cm}^3\text{ s}^{-1}\text{ g}^{-1}),\qquad(4.2.31)$$

$\epsilon_{18\text{O p}_\alpha} = 2.133 \times 10^{17} x_{18\text{O}}\, x_H\, \rho N_A\langle^{18}O\,p\rangle_\alpha \quad (\text{erg g}^{-1}\text{ s}^{-1}).$

The above formulae for (p,γ) and (p,α) reactions hold for $10^{-3} < T_9 < 10$, if no comments, and for densities at which electron screening is negligible (see Sect. 4.5). Indeterminate terms involving a factor (from 0 to 1) in [320a] are omitted everywhere in the reactions (4.2.29–31). We may write $N_A\langle01\rangle$ for the reactions (p,γ) and (p,α) from (4.2.21–23) and also for the reaction (α,γ) in the form

$$N_A\langle01\rangle = T_9^{-2/3}\, \exp\left[A - \left(\frac{TAU}{T_9^{1/3}}\right)\right.$$

$$\left.\times\left(1 + BT_9 + CT_9^2 + DT_9^3\right)\right]\quad (\text{cm}^3\text{ s}^{-1}\text{ g}^{-1}),\qquad(4.2.32)$$

Table 4.2. Parameters of $(p, \gamma), (p, \alpha)$ reactions for (4.2.32), Q_6 is the energy per reaction in MeV, T_{ran} is the application range

Reaction	τ	A	B	C	D	Q_6	T_{ran}
^{20}Ne(p,γ)^{21}Na	19.45	30.94	8.097(-2)	6.555(-1)	-4.272(-1)	2.431	$0 < T_9 < 1$
	18.57	28.01	2.428(-1)	-2.336(-2)	1.026(-3)	2.431	$1 < T_9 < 10$
^{21}Ne(p,γ)^{22}Na	19.46	31.61	2.391(-1)	-1.850(-2)	6.622(-4)	6.740	$0 < T_9 < 10$
^{22}Ne(p,γ)^{23}Na	19.48	31.41	1.802(-1)	-1.303(-2)	4.839(-4)	8.793	$0 < T_9 < 10$
^{23}Na(p,α)^{20}Ne	20.77	30.63	1.585(-2)	1.235(-3)	-1.294(-4)	2.376	$0 < T_9 < 10$
^{23}Na(p,γ)^{24}Mg	20.77	29.97	-6.586(-2)	2.804(-1)	-1.131(-1)	11.692	$0 < T_9 < 1$
	22.02	34.04	2.417(-1)	-2.162(-2)	9.298(-4)	11.692	$1 < T_9 < 10$
^{24}Mg(p,γ)^{25}Al	22.02	31.54	-3.333(-2)	6.356(-1)	-3.758(-1)	2.270	$0 < T_9 < 1$
	21.93	32.58	3.017(-1)	-3.410(-2)	1.671(-3)	2.270	$1 < T_9 < 10$
^{25}Mg(p,γ)^{26}Al	22.04	31.32	-8.352(-2)	3.614(-1)	-1.620(-1)	6.307	$0 < T_9 < 1$
	23.28	35.59	2.598(-1)	-2.450(-2)	1.097(-3)	6.307	$1 < T_9 < 10$
^{26}Al(p,γ)^{27}Si	23.26	31.81	-7.207(-2)	4.040(-1)	-1.990(-1)	7.465	$0 < T_9 < 1$
	24.17	35.29	2.528(-1)	-2.510(-2)	1.161(-3)	7.465	$1 < T_9 < 10$
^{27}Al(p,α)^{24}Mg	23.27	28.41	-2.716(-1)	3.533(-1)	-1.410(-1)	1.600	$0 < T_9 < 1$
	25.02	34.08	1.158(-1)	-1.657(-2)	8.537(-4)	1.600	$1 < T_9 < 10$

With this notation the corresponding numerical coefficients are shown in Table 4.2, taken from [638]. Further improvements for these reactions are presented in [320a]. Table 4.3 shows the parameters for the $(e^+\nu)$-decay: half-lives from [232], total energies of decay Q_{tot}, and mean energy release per reaction minus neutrino energy in MeV [135, 182].

While the time of ^{22}Na beta-decay from (4.2.21) is too large, another NeNa-cycle

$$^{20}\text{Ne}(p, \gamma)^{21}\text{Na}(e^+\nu)^{21}\text{Ne}(p, \gamma)^{22}\text{Na}(p, \gamma)$$

$$^{23}\text{Mg}(e^+\nu)^{23}\text{Na}(p, \alpha)^{20}\text{Ne} \tag{4.2.33}$$

may become more rapid at high temperatures. The ^{22}Na(p, γ)^{23}Mg reaction rate was calculated in [624]. Later derivation gave [320a]

$$N_A \langle^{22}\text{Na}\,p\rangle_\gamma = \left\{ 9.63 \times 10^{-5}\, T_9^{3/2} \exp\left(-\frac{0.517}{T_9}\right) \right.$$

$$\left. + 2.51 \times 10^4\, T_9 \exp\left(-\frac{2.013}{T_9}\right) \right\} \ (\text{cm}^3\ \text{s}^{-1}\ \text{g}^{-1}),$$

$$Q_6 = 7.578, \tag{4.2.34}$$

$$\epsilon_{^{22}\text{Na}\,p} = 3.324 \times 10^{17} x_{^{22}\text{Na}}\, x_H\, \rho N_A \langle^{22}\text{Na}\,p\rangle \ (\text{erg}\ \text{g}^{-1}\ \text{s}^{-1}).$$

The change in concentration of an element in the presence of beta-decay is

$$\frac{1}{x}\frac{dx}{dt} = -\frac{\ln 2}{\tau_{1/2}}. \tag{4.2.35}$$

Table 4.3. Parameters of beta–decay in hydrogen burning cycles

Reaction	$\tau_{1/2}$	Q_{tot}, MeV	Q_6, MeV
$^{13}N(e^+\nu)^{13}C$	10	2.22	1.51
$^{14}O(e^+\nu)^{14}N$	72	5.145	
$^{15}O(e^+\nu)^{15}N$	124	2.76	1.76
$^{17}F(e^+\nu)^{17}O$	70	2.76	1.82
$^{18}F(e^+\nu)^{18}O$	110	1.655	
$^{20}Na(e^+\nu)^{20}Ne$	0.38	13.89	
$^{21}Na(e^+\nu)^{20}Ne$	23	3.55	
$^{22}Na(e^+\nu)^{20}Ne$	2.6	2.84	
$^{25}Al(e^+\nu)^{20}Mg$	73	4.28	
$^{27}Si(e^+\nu)^{20}Al$	4.1	4.81	
$^{23}Mg(e^+\nu)^{20}Na$	12	4.06	

This equation together with (4.1.6) should be used for finding concentrations throughout the reactions. If an element appears and vanishes in more than one reaction, the right-hand sides of (4.1.6) and (4.2.35) should be replaced by the sums over corresponding reactions. The rate of energy release in β^+-decay of nuclei with atomic weight A_0 and concentration by mass x_0 is

$$\epsilon_{\beta^+} = N_A \frac{x_0}{A_0} Q \frac{\ln 2}{\tau_{1/2}} = 6.6880 \times 10^{17} \frac{x_0}{A_0} \frac{Q_6}{\tau_{1/2}} \quad \text{(erg g}^{-1}\text{ s}^{-1}\text{)}. \qquad (4.2.36)$$

The quantity $\ln 2/\tau_{1/2}$ in beta-decay is analogous to $(x_1/A_1)\rho N_A \langle 01 \rangle$ in the nuclear reaction $\langle 01 \rangle$, see (4.1.6) and (4.1.16).

Note that the presence of the isotopes ^{13}C, ^{17}O, ^{21}Ne, ^{25}Mg in the above reactions leads to neutron production in (α, n) reactions which is important for the production of heavy elements in s-processes (see Sect. 4.4). In addition to the above reactions many other (p, γ) and reverse reactions should be included in analyzing the case of explosive hydrogen burning. These reactions have been studied in [361, 389, 399, 638, 624, 320a].

4.2.3 Helium Burning

For stellar physics the most important of all helium burning reactions is the triple-alpha reaction leading to ^{12}C formation. This is caused by formation of an unstable nucleus of ^8Be owing to fusion of two alpha particles. The concentration of ^8Be nuclei in equilibrium is given by a relation similar to (1.3.3), namely

$$n_{^8Be} = n_{He}^2 \frac{A_{Be}^{3/2}}{A_{He}^3} \frac{h^3}{(2\pi m_u kT)^{3/2}} e^{-\frac{Q}{kT}},$$

$$Q = 92 \quad \text{(keV)}. \qquad (4.2.37)$$

This reaction leading to ^{8}Be formation is resonant with $E_r = Q$. The fusion of ^{8}Be nuclei and alpha particles also takes the form of a resonant reaction, with a reaction rate determined by (4.1.4) and (4.1.36). The fusion reaction of three particles 0, 1, 2 is characterized by the quantity $N_A^2 \langle 012 \rangle$ which is found to be [389]

$$N_A^2 \langle \alpha\alpha\alpha \rangle = 2.79 \times 10^{-8}\, T_9^{-3} \exp\left(-\frac{4.4027}{T_9}\right)$$

$$+ (0 \div 1) \times 1.35 \times 10^{-7}\, T_9^{-3/2} \exp\left(-\frac{24.811}{T_9}\right) \quad (\text{cm}^6\ \text{g}^{-2}\ \text{s}^{-1}), \quad (4.2.38)$$

$$Q_6 = 7.275.$$

The number of reactions proceeding in 1 cm^3 per second P_{012}, the mean lifetime of He nuclei $\tau_{3\alpha}(^4\text{He})$, and the energy released in the 3α-reaction $\epsilon_{\alpha\alpha\alpha}$, are given by [360]

$$P_{012} = \frac{n_0 n_1 n_2}{1 + \Delta_{012}} \langle 012 \rangle = \rho^3 N_A^3 \frac{x_0 x_1 x_2}{A_0 A_1 A_2} \frac{\langle 012 \rangle}{1 + \Delta_{012}}$$

$$= \frac{1}{384} \rho^3 N_A^3 x_{4\text{He}}^3 \langle \alpha\alpha\alpha \rangle \quad (\text{reactions s}^{-1}\ \text{cm}^{-3}), \quad (4.2.39)$$

$$\Delta_{012} = \delta_{01} + \delta_{12} + \delta_{02} + 2\delta_{012};$$

$$\frac{1}{\tau_{3\alpha}(^4\text{He})} = \frac{1}{n_{4\text{He}}} \frac{dn_{4\text{He}}}{dt} = \frac{3 P_{3\alpha}}{n_{4\text{He}}} = \frac{1}{2} n_{4\text{He}}^2 \langle \alpha\alpha\alpha \rangle,$$

$$\epsilon_{\alpha\alpha\alpha} = 9.6487 \times 10^{17} \frac{x_0 x_1 x_2}{A_0 A_1 A_2} \frac{\rho^2 N_A^2}{1 + \Delta_{012}} Q_6 \langle 012 \rangle$$

$$= \frac{9.6487 \times 10^{17}}{384} x_{4\text{He}}^3 \rho^2 N_A^2 Q_6 \langle \alpha\alpha\alpha \rangle$$

$$= 1.828 \times 10^{16} x_{4\text{He}}^3 \rho^2 N_A^2 \langle \alpha\alpha\alpha \rangle \quad (\text{erg g}^{-1}\ \text{s}^{-1}).$$

At $T \leq 2.8 \times 10^7$ K the non-resonant channel of the ^{8}Be$(\alpha, \gamma)^{12}$C reaction becomes more important than the resonant one. The non-resonant reaction is taken into account in [510]. This is equivalent, approximately with an accuracy $\sim 20\%$, to multiplying (4.2.38) and (4.2.39) by the function

$$f_{\text{He}}(T) = \left\{ 0.01 + 0.2 \frac{1 + 4 \exp\left[-(0.025/T_9)^{3.263}\right]}{1 + 4 \exp\left[-(T_9/0.025)^{9.227}\right]} \right\}^{-1}. \quad (4.2.40)$$

The rate of the ^{12}C$(\alpha, \gamma)^{16}$O reaction is comparable with the rate of the 3α reaction, so helium burning leads to formation of both ^{12}C and ^{16}O. Their relative production depends on the conditions under which the reaction is

Table 4.4. Parameters of (α, γ) reactions for (4.2.32), Q_6 is the energy per reaction in MeV, T_{ran} is the application range

Reaction	τ	A	B	C	D	Q_6	T_{ran}
$^{16}O(\alpha, \gamma)^{20}Ne$	39.76	41.17	-8.856(-3)	7.048(-2)	2.521(-2)	4.731	$0 < T_9 < 1$
	42.63	49.51	2.217(-1)	-1.750(-2)	6.622(-4)	4.731	$1 < T_9 < 10$
$^{20}Ne(\alpha, \gamma)^{24}Mg$	46.77	44.38	2.482(-2)	-2.855(-2)	3.987(-2)	9.315	$0 < T_9 < 1$
	48.75	49.62	1.087(-1)	-2.975(-3)	-2.077(-5)	9.315	$1 < T_9 < 10$
$^{24}Mg(\alpha, \gamma)^{28}Si$	53.32	47.06	1.344(-2)	1.954(-3)	7.228(-3)	9.986	$0 < T_9 < 1$
	54.77	50.89	6.885(-2)	3.133(-3)	-2.459(-4)	9.986	$1 < T_9 < 10$
$^{28}Si(\alpha, \gamma)^{32}S$	59.49	49.60	1.270(-2)	4.133(-3)	2.791(-3)	6.949	$0 < T_9 < 1$
	61.02	53.60	6.340(-2)	2.541(-3)	-2.900(-4)	6.969	$1 < T_9 < 10$
$^{32}S(\alpha, \gamma)^{36}Ar$	65.37	52.09	1.821(-2)	-5.033(-3)	5.584(-3)	6.642	$0 < T_9 < 1$
	66.69	55.41	4.913(-2)	4.637(-3)	-4.067(-4)	6.642	$1 < T_9 < 10$
$^{36}Ar(\alpha, \gamma)^{40}Ca$	71.01	54.48	2.676(-2)	-3.300(-2)	3.361(-2)	7.041	$0 < T_9 < 1$
	78.27	70.11	1.458(-1)	-1.069(-2)	3.790(-4)	7.041	$1 < T_9 < 10$
$^{40}Ca(\alpha, \gamma)^{44}Ti$	76.44	56.80	1.650(-2)	5.973(-3)	-3.889(-4)	5.128	$0 < T_9 < 10$
$^{44}Ti(\alpha, \gamma)^{48}Cr$	81.66	56.98	-8.364(-2)	2.085(-1)	-7.477(-2)	7.694	$0 < T_9 < 1$
	81.23	60.18	1.066(-1)	-1.102(-2)	5.324(-4)	7.694	$1 < T_9 < 10$
$^{48}Cr(\alpha, \gamma)^{52}Fe$	86.74	62.93	1.212(-1)	-1.340(-2)	-5.335(-4)	7.943	$0 < T_9 < 1$
	81.42	53.00	6.352(-2)	-5.671(-3)	2.848(-4)	7.943	$1 < T_9 < 10$
$^{52}Fe(\alpha, \gamma)^{56}Ni$	91.67	62.22	7.846(-2)	-7.430(-3)	3.723(-4)	8.001	$0 < T_9 < 10$
$^{56}Ni(\alpha, \gamma)^{60}Zn$	96.48	64.42	1.549(-2)	-4.664(-3)	4.888(-3)	2.704	$0 < T_9 < 1$
	104.92	79.65	8.188(-2)	-2.885(-3)	5.206(-5)	2.704	$1 < T_9 < 10$

proceeding, and varies with stellar mass. The rate of production of ^{16}O increases with increasing mass and may exceed the ^{12}C production (see Chap. 9, Vol. 2). We have for the $^{12}C(\alpha, \gamma)^{16}O$ reaction [361]

$$N_A\langle ^{12}C\,\alpha\rangle = 9.03 \times 10^7\, T_9^{-2}\frac{(1 + 0.621\, T_9^{2/3})^2}{(1 + 0.047\, T_9^{2/3})^2}$$

$$\times \exp\left[-\frac{32.120}{T_9^{1/3}} - \left(\frac{T_9}{5.863}\right)^2\right] + 2.74 \times 10^7\, T_9^{-2/3}$$

$$\times \exp\left(-\frac{32.120}{T_9^{1/3}}\right) + 1.25 \times 10^3\, T_9^{-3/2}\exp\left(-\frac{27.499}{T_9}\right)$$

$$+ 1.43 \times 10^{-2}\, T_9^5 \exp\left(-\frac{15.541}{T_9}\right) \quad (\text{cm}^3\ \text{s}^{-1}\ \text{g}^{-1}), \tag{4.2.41}$$

$$Q_6 = 7.162,$$

$$\epsilon_{12C\,\alpha} = 1.44 \times 10^{17} x_{12C}\, x_{4He}\, \rho N_A\langle ^{12}C\,\alpha\rangle \quad (\text{erg}\ \text{g}^{-1}\ \text{s}^{-1}).$$

The relations (4.2.38–41) are valid in the absence of electron screening at $10^{-3} < T_9 < 10$.

Other types of (α, γ) capture become important at evolutionary stages preceding a supernova explosion [363], and also in explosive helium burning. If we write $N_A\langle 01\rangle$ in the form (4.2.32), the coefficients for some reactions from [638] are given in Table 4.4. The parameters of various (α, γ), (α, n), (α, p) and reverse reactions that take part in explosive helium burning are given in [399, 638].

4.3 Reactions Involving Heavy Nuclei at High Temperatures

Reactions involving the direct fusion of heavy nuclei occur during late evolutionary stages of massive stars and in explosions leading to a supernova outburst. The energy release in ^{12}C, ^{16}O, ^{20}Ne, ^{24}Mg and ^{28}Si burning is comparable to that in the 3α reaction, but powerful neutrino emission arising from the high temperature causes the lifetime of the star on the ^{12}C and ^{16}O burning stage to be much less than on the ^{4}He burning stage. The likelihood of discovering such stars is small because of their short lifetime. Actually there is no known star in a calm state which can be reliably identified as radiating energy produced by the ^{12}C or other heavy element burning.

When Coulomb screening of the nuclei is negligible, the reaction rate calculations must include, according Sect. 4.1, resonant and non-resonant contributions. In [320a] relations are given for the following reactions:

$$^{12}C(^{12}C, \gamma)^{24}Mg, \qquad Q_6 = 13.933,$$

$$N_A\langle ^{12}C^{12}C\rangle = 4.27 \times 10^{26}\, T_{9A}^{5/6}\, T_9^{-3/2} \exp\left(-\frac{84.165}{T_{9A}^{1/3}} - 2.12 \times 10^{-3}T_9^3\right)$$

$$(\text{cm}^3\ \text{s}^{-1}\ \text{g}^{-1}), \tag{4.3.1}$$

$$T_{9A} = T_9(1 + 0.0396\, T_9)^{-1}.$$

The energy release rate is

$$\epsilon_{12C12C} = 4.67 \times 10^{16} x_{12C}^2\, \rho N_A\langle ^{12}C^{12}C\rangle \ \ (\text{erg}\ \text{g}^{-1}\ \text{s}^{-1}). \tag{4.3.2}$$

The reaction

$$^{12}C(^{16}O, \gamma)^{28}Si, \qquad Q_6 = 16.755,$$

$$N_A\langle ^{12}C^{16}O\rangle = 1.72 \times 10^{31}\, T_{9A}^{5/6}\, T_9^{-3/2} \exp\left(-\frac{106.594}{T_{9A}^{1/3}}\right)$$

$$\times \Big[\exp(-0.180\, T_{9A}^2)$$

$$+ 1.06 \times 10^{-3} \exp(2.562\, T_{9A}^{2/3})\Big]^{-1} \ \ (\text{cm}^3\ \text{s}^{-1}\ \text{g}^{-1}), \tag{4.3.3}$$

$$T_{9A} = T_9(1 + 0.055\, T_9)^{-1}, \quad T_9 \geq 0.5$$

releases the energy

$$\epsilon_{^{12}C^{16}O} = 8.42 \times 10^{16} x_{^{12}C}\, x_{^{16}O}\, \rho N_A \langle ^{12}C^{16}O \rangle \quad (\text{erg g}^{-1}\, \text{s}^{-1}). \tag{4.3.4}$$

The reaction

$$^{16}O(^{16}O, \gamma)^{32}S, \qquad Q_6 = 16.542,$$
$$N_A \langle ^{16}O^{16}O \rangle = 7.10 \times 10^{36}\, T_9^{-3/2}$$
$$\times \exp\left(-\frac{135.930}{T_9^{1/3}} - 0.629\, T_9^{2/3} - 0.445\, T_9^{4/3} + 0.0103\, T_9^2\right)$$
$$(\text{cm}^3\, \text{s}^{-1}\, \text{g}^{-1}) \tag{4.3.5}$$

has the energy release rate

$$\epsilon_{^{16}O^{16}O} = 3.12 \times 10^{16} x_{^{16}O}^2\, \rho N_A \langle ^{16}O^{16}O \rangle \quad (\text{erg g}^{-1}\, \text{s}^{-1}). \tag{4.3.6}$$

Reactions involving the direct fusion of ^{24}Mg and heavier nuclei require such a high temperature, $T \geq 3 \times 10^9$ K, for nuclei to pass through the Coulomb barrier that photodetachment of particles such as p, n and α from ^{12}C, ^{16}O, ^{20}Ne and their capture by other nuclei occur even before these reactions start. Thus the production of elements heavier than ^{32}S proceeds via (p, γ), (n, γ), (α, γ) captures. This fast ($\sim$ 3000 seconds) evolutionary stage is called the alpha-process or silicon burning stage [363, 302], because the most abundant source of ^{28}Si results from ^{12}C and ^{16}O burning in the stellar core. The rates of some (p, γ) and (α, γ) reactions are given in Sect. 4.2 and Tables 4.2 and 4.4. The rates of the photodetachment reactions reverse to fusion are determined by (4.1.10), the corresponding energy release given by (4.1.17). Characteristics of the reverse reactions involving four nuclei are given by (4.1.9). Taking into account [389] and (4.2.38) we find the ^{12}C photosplitting rate:

$$\frac{1}{\tau_\gamma(^{12}C)} = \lambda_\gamma(^{12}C) = 2.00 \times 10^{20}\, T_9^3\, \exp(-84.424/T_9) N_A^2 \langle \alpha\alpha\alpha \rangle. \tag{4.3.7}$$

If the interaction of two particles 0 and 1 results in three particles 2, 3 and 4 formation and a heat release Q, the relations between the direct and reverse reactions will then have the form

$$\frac{\langle 234 \rangle}{\langle 01 \rangle} = \frac{g_0 g_1}{g_2 g_3 g_4} \left(\frac{A_0 A_1}{A_2 A_3 A_4}\right)^{3/2} \left(\frac{2\pi \hbar^2}{m_u kT}\right)^{3/2}$$
$$\times \left(\frac{1 + \Delta_{234}}{1 + \delta_{01}}\right) \exp(-Q/kT), \tag{4.3.8}$$

and in equilibrium

$$\frac{n_2 n_3 n_4}{n_0 n_1} = \rho N_A \frac{x_2 x_3 x_4}{A_2 A_3 A_4} \frac{A_0 A_1}{x_0 x_1}$$

$$= \frac{g_2 g_3 g_4}{g_0 g_1} \left(\frac{A_2 A_3 A_4}{A_0 A_1}\right)^{3/2} \left(\frac{m_u kT}{2\pi \hbar^2}\right)^{3/2} \exp(Q/kT) \qquad (4.3.9)$$

or

$$\frac{x_2 x_3 x_4}{x_0 x_1} = 0.98677 \times 10^{10} \frac{T_9^{3/2}}{\rho} \left(\frac{g_2 g_3 g_4}{g_0 g_1}\right) \left(\frac{A_2 A_3 A_4}{A_0 A_1}\right)^{5/2}$$

$$\times \exp\left(\frac{11.605 \, Q_6}{T_9}\right). \qquad (4.3.10)$$

Considering these properties of the direct and reverse reactions one should bear in mind that both the resultant and initial nuclei may be excited, thereby enlarging the number of their possible states. At temperature T the population of the i-th excitation level is

$$P_k = \frac{g_k}{G} \exp(-E_k/kT),$$
$$G = \sum_k g_k \exp(-E_k/kT), \qquad (4.3.11)$$

where $g_k = 2J_k + 1$ is the statistical weight of the excited state with spin J_k and E_k is the excitation energy with respect to the ground state. $G(T)$ is the effective statistical weight of a nucleus including excitation states or a partition function. The importance of excitation states is determined by the normalized partition function

$$\mathcal{G} = G/g_0 = 1/P_0 \qquad (4.3.12)$$

reciprocal to the fractional content of nuclei in the ground state P_0. The excited states are taken into account by substituting partition function G_i for the respective statistical weights g_i. The difference between G_i and g_i or between $\mathcal{G}$ and unity become significant only at high temperatures ($T_9 \geq 2$ for ^{56}Fe [362]).

With increasing temperature, G from (4.3.11) formally diverges, but if we correctly calculate the continuous spectrum states making negative contributions, and take into account the limitations arising from nuclear disintegration, we shall find that, after passing over a maximum, G tends to zero as $T \to \infty$. The maximum value $G_{\max} \approx 33\,340$ is achieved for ^{56}Fe at $T_9 \approx 700$ [362]. Tables representing the partition functions for nuclei with $8 \leq Z \leq 86$ and $0 \leq T_9 \leq 10$ are given in [638].

4.4 Processes of Heavy Element Formation

By the time of the first star formation material in the expanding universe consisted mainly of hydrogen and helium, together with small amounts of deuterium, $x_{2D} \leq 10^{-4}$, helium-3, $x_{3He} \leq 3 \times 10^{-5}$, lithium, $x_{7Li} \approx 10^{-9}$ and negligible amounts of heavier elements [623, 567]. The low abundance of lithium and translithium elements is due to the absence of stable elements with $A = 5$ and 8. Elements with $A > 5$ and $A > 8$ can only form in reactions involving charged particles with $Z \geq 2$, and these have no time in the early universe to pass through the Coulomb barrier.

Carbon and transcarbon elements form in thermonuclear reactions in stellar interiors, in supernova explosions, and in (p, γ) and mostly (n, γ) captures [215]. Very heavy elements can also form in the ejection of material from neutron stars [61]. ^{6}Li, ^{9}Be, ^{10}B and ^{11}B are not produced in thermonuclear burning in stars. The observed mass fractions of these elements in stellar atmospheres are $\leq 10^{-8}$, a result explained by detachment reactions caused by interactions between fast particles in cosmic rays and heavy elements at the surface of stars and in supernova shells [170].

At the high temperatures achieved in supernova explosions material reaches an equilibrium with respect to nuclear reactions with kinetics determined by the beta-process (see Sect. 1.3), and this leads to the formation of iron peak elements [363, 509]. Neutron captures subsequently represent the principal mechanism for the formation of transiron elements [317].

4.4.1 Slow Captures (s-Process)

Neutron capture leads to the formation of nuclei that are unstable to e^--decay. If the neutron concentration is so small that the time between successive captures exceeds the time for beta-decay, then these captures are described as occurring in an s-process (slow process). During the s-process the growth of nuclei follows the valley of beta-stability (Fig. 4.1), and continues until formation of ^{209}Bi, which is the last stable element in the s-process chain. The neutron capture by ^{209}Bi results in the formation of an unstable nucleus ^{210}Bi which then decays into ^{206}Pb in the reactions ^{210}Bi$(\gamma, \alpha)^{206}$Te$(e^- \tilde{\nu})^{206}$Pb or ^{210}Bi$(e^- \tilde{\nu})^{210}$Po$(\alpha)^{206}$Pb.

The production in the s-process of a heavy element abundance close to that observed requires a neutron flux in the range $10^{15} - 10^{16}$ cm^{-2} s^{-1}, corresponding to their concentration $n_n \approx 5 \times 10^6 - 5 \times 10^7$ cm^{-3} at velocities $v_n \approx 2 \times 10^8$ cm s^{-1}, and a total time for the process $\geq 10^3$ yr [223]. $(\alpha, n) -$ captures onto nuclei with mass number $A = 4J + 1$ and charge $Z = 2j$ play a significant role in neutron production in stars [215]. Besides, the j strongly bound alpha particles in these nuclei, they also contain one loosely bound neutron. An interaction with a second alpha particle causes the neutron to detach easily and a nucleus consisting of $(j+1)$ strongly bound alpha particles

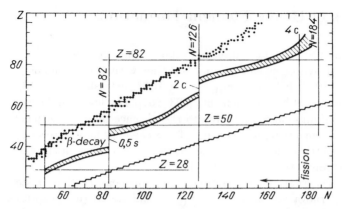

Fig. 4.1. Path of element formation on the (N, Z) diagram. The upper line shows s-process path along the beta-stability valley. The hatched band below corresponds to r-process. The times of achieving the corresponding concentration peaks are indicated. The values for r-process at $T_g = 1.0$, $\lg n_n = 24$ and initial iron nuclei $Z = 26$ are taken from calculations in [574]. The points mark stable nuclei which form in r-process. The low stepped line defines n-process. It corresponds to the limiting r-process at a high concentration of neutrons $\lg n_n > 30$, and its location does not change with increasing n_n. Beta-decay and fission accompanied by neutron emission cause nuclei formed in n-process to reach the stability valley.

is formed. As we have seen in Sect. 4.2.2, nuclei of ^{13}C, ^{17}O, ^{21}Ne and ^{25}Mg resulting from high-temperature hydrogen burning supplemented by reaction (4.2.20) act as targets for these reactions. The ^{13}C$(\alpha, n)^{16}$O reaction rate is given in [338b] by fitting of the experimental data

$$N_A\langle^{13}{\rm C}\,\alpha\rangle = 6.788 \times 10^{15}(1 + 0.485\,T_9^{1/3} - 7.948\,T_9^{2/3} + 10.725\,T_9)\,T_9^{-2}$$
$$\times \exp\left[-\frac{32.093}{T_9^{1/3}} - \left(\frac{T_9}{330.271}\right)^2\right] + T_9^{-3/2}[1016.988\,\exp(-6.259/T_9)$$
$$+ 3.474 \times 10^5 \exp(-8.430/T_9)], \tag{4.4.1}$$
$$Q_6 = 2.216, \quad 0.01 < T_9 < 1.0.$$

The rates of other neutron-producing reactions may be written in the form

$$N_A\langle 01\rangle = T_9^{-2/3}\exp\left[A - \frac{TAU}{T_9^{1/3}}(1 + BT_9 + CT_9^2 + DT_9^3)\right], \tag{4.4.2}$$

where A, B, C and TAU are given in Table 4.5 from [638], see also [320a].

The rate of the endothermic reaction ^{22}Ne $(\alpha, n)^{25}$Mg , which may be important for neutron production [332, 333] was investigated in [338b], see also [361]. The formula, based on fitting of experimental data has a form [338b]:

Table 4.5. Parameters of (α, n) reactions for (4.4.2), Q_6 is the energy release per reaction in MeV, T_{ran} is the application range

Reaction	τ	A	B	C	D	Q_6	T_{ran}
$^{17}O(\alpha,n)^{20}Ne$	39.92	41.42	1.673(-2)	2.304(-3)	-1.584(-4)	0.588	$0 < T_9 < 10$
$^{21}Ne(\alpha,n)^{20}Mg$	46.90	44.50	1.750(-2)	1.189(-3)	-7.053(-5)	2.554	$0 < T_9 < 10$
$^{23}Mg(\alpha,n)^{20}Si$	53.42	47.24	1.667(-2)	1.302(-3)	-7.781(-5)	2.654	$0 < T_9 < 10$

$$N_A\langle^{22}Ne\,\alpha\rangle = 1.266 \times 10^{21}\frac{t^{5/6}}{gT_9^{3/2}}$$

$$\times \exp\left[-\frac{50.899}{t^{1/3}} - \left(\frac{0.203}{t}\right)^{17.384}\right] + f_1 + f_2 + f_3 + f_4,$$

where

$$t = \frac{T_9}{1 - 0.027T_9}, \quad g = 1 + 129.22\exp\left(-\frac{5.056}{T_9}\right), \tag{4.4.3}$$

$$f_1 = \frac{7.425 \times 10^{-3}}{g}\exp\left(-\frac{3.989}{T_9^{4/3}}\right), \quad f_2 = -\frac{1.969 \times 10^{-14}}{g}\exp\left(-\frac{7.902}{T_9^{3/4}}\right),$$

$$f_3 = -\frac{1.094 \times 10^4}{g}\exp\left(-\frac{35.934}{T_9^{3/2}}\right), \quad f_4 = \frac{2.624}{g}\exp\left(-\frac{7.654}{T_9}\right),$$

$$Q_6 = -0.482, \quad 0.15 < T_9 < 1.6.$$

Reactions (4.4.1–3) occur at temperatures $T > 3 \times 10^8$ K during helium burning in the core of massive stars (see Chap. 9, Vol. 2). Beta-half-lives of nuclei produced by neutron capture on to stable nuclei vary from 14 s to 7×10^6 yr for ^{116}In and ^{107}Pd, respectively.

The cross-section for the capture of thermal neutrons involved in the beta-process scales as $\sigma_n \sim 1/v$ [212]. Table 4.6 gives $\sigma_{30} = \sigma_n$ (30 keV) from [506, 201] for nuclei involved in the s-process. These cross-sections have been verified from the best fit of the exact solution to the set of equations for the s-process to the observed element abundances in the solar system. The neutron exposition τ is serving as the argument in the s-process theory,

$$\tau = \int n_n v \, dt \quad (\text{cm}^{-2}) \tag{4.4.4}$$

and the unknown functions are given by

$$\psi_A(\tau) \equiv \sigma_A n_A(\tau)/n_1(0), \tag{4.4.5}$$

where $n_A(\tau)$ is the concentration of nuclei formed in the s-process which involve a single species of nucleus with initial concentration $n_1(0)$. For arbitrary but low neutron energies

$$\sigma(E) = \sigma_{30}\sqrt{\frac{30}{Q_{n,\text{KeV}}}}, \tag{4.4.6}$$

Table 4.6. Cross-section for capture of neutrons with energy $E = 30$ keV, σ_{30} (in mbarn=10^{-27} cm^2), by stable nuclei along the main s-process path and beta-half-lives $\tau_{1/2}$ for unstable nuclei formed in this process; $K = A - 55$

K	Z	Stable	σ_{30}	Unstable	$e^{\pm}$	$\tau_{1/2}$
1	26	^{56}Fe	13.5 ± 2			
2	26	^{57}Fe	30 ± 5			
3	26	^{58}Fe	15.9 ± 1.5	^{59}Fe	$-$	45 days
4	27	^{59}Co	72 ± 14	^{60}Co	$\gamma-$	90 days
5	28	^{60}Ni	31 ± 6			
6	28	^{61}Ni	135 ± 27			
7	28	^{62}Ni	26 ± 5	^{63}Ni	$-$	92 yr
8	29	^{63}Cu	49 ± 14	^{64}Cu	$-, +, K$	13 hours
9	28	^{64}Ni	23	^{65}Ni	$-$	2.6 hours
9	30	^{64}Zn	51.1	^{65}Zn	$+$	243 yr
10	29	^{65}Cu	42 ± 7	^{66}Cu	$-$	5.1 min
11	30	^{66}Zn	40			
12	30	^{67}Zn	160			
13	30	^{68}Zn	23 ± 3	^{69}Zn	$-$	57 min
14	31	^{69}Ga	130 ± 30	^{70}Ga	$-$	21 min
15	32	^{70}Ge	84	^{71}Ge	K	215 days
16	31	^{71}Ga	120 ± 30	^{72}Ga	$-$	14 hours
17	32	^{72}Ge	53.2			
18	32	^{73}Ge	330			
19	32	^{74}Ge	17 ± 5	^{75}Ge	$-$	82 min
20	33	^{75}As	490 ± 100	^{76}As	$-$	26 hours
21	34	^{76}Se	101			
22	34	^{77}Se	424			
23	34	^{78}Se	98 ± 14	^{79}Se	$\gamma-$	35 yr
24	35	^{79}Br	600 ± 150	^{80}Br	$+, -, K$	18 min
25	34	^{80}Se	20 ± 12	^{81}Se	$-$	19 min
25	36	^{80}Kr	163	^{81}Kr	K	$2 \cdot 10^5$ yr
26	35	^{81}Br	460 ± 80	^{82}Br	$-$	35 hours
27	36	^{82}Kr	127			
28	36	^{83}Kr	601			
29	36	^{84}Kr	25	^{85}Kr	$-$	11 yr
30	37	^{85}Rb	215 ± 20	^{86}Rb	$-$	19 days
31	38	^{86}Sr	74 ± 7			
32	38	^{87}Sr	109 ± 9			
33	38	^{88}Sr	6.9 ± 2.5	^{89}Sr	$-$	52 days
34	39	^{89}Y	21 ± 4	^{90}Y	$-$	64 hours
35	40	^{90}Zr	12 ± 2			
36	40	^{91}Zr	68 ± 8			
37	40	^{92}Zr	34 ± 6	^{93}Zr	$-$	$1.5 \cdot 10^6$ yr
38	40	^{93}Zr	81.3			
39	40	^{94}Zr	20 ± 2	^{95}Zr	$-$	65 days

Table 4.6. (continued)

K	Z	Stable	σ_{30}	Unstable	$e^{\pm}$	$\tau_{1/2}$
40	41			^{95}Nb	−	32 days
40	42	^{95}Mo	430 ± 50			
41	42	^{96}Mo	90 ± 10			
42	42	^{97}Mo	350 ± 50			
43	42	^{98}Mo	150 ± 40	^{99}Mo	−	67 hours
44	43			^{99}Tc	$\gamma-$	7 yr; $2.12 \cdot 10^5$ yr
44	44	^{99}Ru	640			
45	44	^{100}Ru	209 ± 6			
46	44	^{101}Ru	1011 ± 30			
47	44	^{102}Ru	189 ± 6	^{103}Ru	−	40 days
48	45	^{103}Rh	1072 ± 30	^{104}Rh	−	43 s
49	46	^{104}Pd	447 ± 22			
50	46	^{105}Pd	1189 ± 60			
51	46	^{106}Pd	382 ± 19	^{107}Pd	−	$7 \cdot 10^6$ yr
52	46	^{107}Pd	950			
53	46	(^{108}Pd)	345 ± 17	^{109}Pd	−	13 hours
54	47	^{109}Ag	620 ± 50	^{110}Ag	−	25 s
55	48	^{110}Cd	250 ± 30			
56	48	^{111}Cd	624			
57	48	^{112}Cd	233 ± 30			
58	48	^{113}Cd	569			
59	48	^{114}Cd	158 ± 25	^{115}Cd	−	54 hours
60	49	^{115}In	700 ± 45	^{116}In	−	14 s
61	50	^{116}Sn	100 ± 15			
62	50	^{117}Sn	420 ± 30			
63	50	^{118}Sn	63 ± 5			
64	50	^{119}Sn	348 ± 54			
65	50	^{120}Sn	50 ± 15	^{121}Sn	−	27 hours
66	51	^{121}Sb	740 ± 100	^{122}Sb	−	2.8 days
67	52	^{122}Te	270 ± 30			
68	52	^{123}Te	820 ± 30			
69	52	^{124}Te	150 ± 20			
70	52	^{125}Te	430 ± 20			
71	52	^{126}Te	82 ± 8	^{127}Te	−	9.4 hours
72	53	^{127}I	710 ± 35	^{128}I	−	25 min
73	54	^{128}Xe	232			
74	54	^{129}Xe	665			
75	54	^{130}Xe	143			
76	54	^{131}Xe	587			
77	54	^{132}Xe	90.9	^{133}Xe	−	5.3 days
78	55	^{133}Cs	700 ± 40	^{134}Cs	$\gamma-$	24 days
79	56	^{134}Ba	225 ± 35			
80	56	^{135}Ba	472			

Table 4.6. (continued)

K	Z	Stable	σ_{30}	Unstable	$e^{\pm}$	$\tau_{1/2}$
81	56	^{136}Ba	90 ± 20			
82	56	^{137}Ba	72.6			
83	56	^{138}Ba	4.22 ± 0.25	^{139}Ba	$-$	82 min
84	57	^{139}La	44 ± 4	^{140}La	$-$	40 hours
85	58	^{140}Ce	11.5 ± 0.6	^{141}Ce	$-$	33 days
86	59	^{141}Pr	111 ± 15	^{142}Pr	$-$	19 hours
87	60	^{142}Nd	67.7			
88	60	^{143}Nd	333			
89	60	^{144}Nd	67.4			
90	60	^{145}Nd	485			
91	60	^{146}Nd	105 ± 16	^{147}Nd	$-$	11 days
92	61			^{147}Pm	$-$	2.6 yr
92	62	^{147}Sm	1150 ± 90			
93	62	^{148}Sm	260 ± 50			
94	62	^{149}Sm	1620 ± 280			
95	62	^{150}Sm	370 ± 70	^{151}Sm	$-$	87 yr
96	63	^{151}Eu	3600 ± 500	^{152}Eu	$\gamma-$	9.3 hours
97	64	^{152}Gd	982	^{153}Gd	K	242 days
98	63	^{153}Eu	2700 ± 300	^{154}Eu	$\gamma-$	100 days
99	64	^{154}Gd	1164 ± 350			
100	64	^{155}Gd	2711 ± 813			
101	64	^{156}Gd	557 ± 166			
102	64	^{157}Gd	1464 ± 440			
103	64	^{158}Gd	540 ± 70	^{159}Gd	$-$	18 hours
104	65	^{159}Tb	2949 ± 340	^{160}Tb	$-$	72 days
105	66	^{160}Dy	1010			
106	66	^{161}Dy	2520 ± 270			
107	66	^{162}Dy	470 ± 50			
108	66	^{163}Dy	1600 ± 300			
109	66	^{164}Dy	180 ± 40	^{165}Dy	$-$	140 min
110	67	^{165}Ho	1170 ± 55	^{166}Ho	$-$	27 hours
111	68	^{166}Er	519 ± 156			
112	68	^{167}Er	1439 ± 432			
113	68	^{168}Er	243 ± 73	^{169}Er	$-$	9.4 days
114	69	^{169}Tm	2085 ± 290	^{170}Tm	$-$	130 days
115	70	^{170}Yb	790 ± 60			
116	70	^{171}Yb	1413 ± 424			
117	70	^{172}Yb	414 ± 124			
118	70	^{173}Yb	869 ± 261			
119	70	^{174}Yb	175 ± 52	^{175}Yb	$-$	101 hours
120	71	^{175}Lu	1460 ± 110	^{176}Lu	$\gamma-$	7 yr
121	72	^{176}Hf	640 ± 160			
122	72	^{177}Hf	1950			

Table 4.6. (continued)

K	Z	Stable	σ_{30}	Unstable	$e^{\pm}$	$\tau_{1/2}$
123	72	^{178}Hf	217 ± 27			
124	72	^{179}Hf	215 ± 25			
125	72	^{180}Hf	290 ± 80	^{181}Hf	$-$	43 days
126	73	^{181}Ta	865 ± 86	^{182}Ta	$-$	115 days
127	74	^{182}W	260 ± 30			
128	74	^{183}W	550 ± 50			
129	74	^{184}W	180 ± 20	^{185}W	$-$	75 days
130	75	^{185}Re	1530 ± 200	^{186}Re	$-$	90 hours
131	76	^{186}Os	467 ± 12			
132	76	^{187}Os	927 ± 19			
133	76	^{188}Os	413 ± 15			
134	76	^{189}Os	858			
135	76	^{190}Os	418 ± 63	^{191}Os	$-$	15 days
136	77	^{191}Ir	1900 ± 300	^{192}Ir	$-$	74 days
137	78	^{192}Pt	352	^{193}Pt	K	10^5 yr
138	78	^{193}Pt	1320			
139	78	^{194}Pt	386			
140	78	^{195}Pt	1040			
141	78	^{196}Pt	160 ± 40	^{197}Pt	$-$	18 hours
142	79	^{197}Au	610 ± 15	^{198}Au	$-$	2.7 days
143	80	^{198}Hg	411			
144	80	^{199}Hg	362			
145	80	^{200}Hg	69.5			
146	80	^{201}Hg	130			
147	80	^{202}Hg	50 ± 15	^{203}Hg	$-$	47 days
148	81	^{203}Tl	150 ± 30	^{204}Tl	$-$	3.8 yr
149	82	^{204}Pb	60 ± 16	^{205}Pb	K	$3 \cdot 10^7$ yr
150	82	^{205}Pb	57.7			
151	82	^{206}Pb	14 ± 1			
152	82	^{207}Pb	11.3 ± 0.7			
153	82	^{208}Pb	0.75 ± 0.09	^{209}Pb	$-$	3.3 hours
154	83	^{209}Bi	12.1 ± 4	^{210}Bi	$-, \alpha$	5 days

"$-$", "$+$" are the $e^{\pm}$ decays, K is the K-capture, γ indicates decays from an excited nuclear state, α is the α-decay. Parentheses in the third column denote nuclei with a long $\tau_{1/2}$.

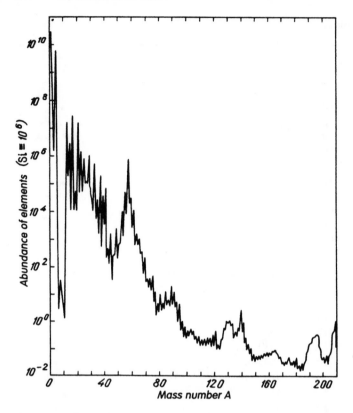

Fig. 4.2. Generalized curve of the elemental abundances in the solar system.

where $Q_{n,KeV}$ is the neutron energy in keV. More complicated formulae similar to (4.1.46) and including departures from the asymptotic law (4.4.6) are given in [399, 638] for the cross-sections of (n, γ) – reactions. Table 4.6 also gives half-lives $\tau_{1/2}$ from [232] for radioactive nuclei resulting from single-neutron capture onto a stable nucleus. At a high neutron exposition the intermediate nuclei formed in the s-process become close to a stationary state, when the number of nuclei forming with atomic weight A equals the number converted into an element with atomic weight $A + 1$:

$$\dot{n}_A = \langle \sigma_{A-1} v \rangle n_{A-1} - \langle \sigma_A v \rangle n_A. \tag{4.4.7}$$

At $\dot{n}_A = 0$, $\langle \sigma v \rangle \approx \sigma_T v_T$,

$$\frac{n_A}{n_{A-1}} = \frac{\sigma_{T,A-1}}{\sigma_{T,A}}. \tag{4.4.8}$$

We note that there must be an accumulation of nuclei that are weak neutron absorbers. Table 4.6 shows that nuclei ^{56}Fe, ^{58}Fe, ^{88}Sr, ^{90}Zr, ^{138}Ba, ^{140}Ge, ^{206}Pb and ^{207}Pb all have a small cross-section for capture and correspond to maxima on the observed curve of elemental abundances (Fig. 4.2). We can see that the relation (4.4.8) holds because the ratios $(n\sigma)$ for the stable isotopes of tin and samarium ^{148}Sm, ^{150}Sm, ^{116}Sn, and ^{118}Sn, ^{120}Sn are close to unity [223]:

$$\frac{n_{148}\sigma_{148}}{n_{150}\sigma_{150}} = 0.98 \pm 0.06,$$

$$\frac{n_{116}\sigma_{116}}{n_{118}\sigma_{118}} = 0.8 \pm 0.2, \qquad (4.4.9)$$

$$\frac{n_{120}\sigma_{120}}{n_{118}\sigma_{118}} = 0.9 \pm 0.3.$$

An important argument for s-process involvement in nucleosynthesis is provided by the discovery in stellar spectra of a line of technetium, which has no stable isotopes. ^{99}Tc has the longest half-life ($\tau_{1/2} = 2.12 \times 10^5$ yr) less than the lifetime of a star but still long enough for the isotope to remain in the s-process chain. Technetium forms in the ^{98}Mo $(n\gamma)^{99}$Mo $(e^- \tilde{\nu})^{99}$Tc reaction. An unstable but sufficiently long-lifetime nucleus can capture a neutron in the s-process before its beta-decay and thus lead to s-process branching. Neutron capture onto ^{99}Tc initiates one of the lateral chains.

4.4.2 Rapid Capture Processes
(r-Process), rbc- , n- and p- Processes

Elements with $A > 209$ and stable nuclei separated from the beta-stability valley by unstable isotopes will not be formed in the s-process (see Fig. 4.1). Instead, they are formed in the rapid r-process, characterized by such a high neutron density ($n_n \geq 10^{20}$ cm^{-3}) that a great many neutron captures may occur between two successive beta-decays [317]. In addition, a high temperature ($T \geq 10^9$ K) is needed in this case, so that not only n-captures but also neutron photo-detachments occur.

The path of element formation in the r-process goes through a region in the (N, Z) diagram containing neutron-rich, beta-unstable nuclei with properties that have not, in most cases, been studied experimentally. Theoretical estimates of the cross-sections for both n-captures $\sigma_{n\gamma}$ and the reverse photo-detachments $\sigma_{\gamma n}$ are therefore used in r-process calculations. The capture rate $\lambda_{n\gamma}$ is related to the photo-detachment rate $\lambda_{\gamma n}$ by (4.1.10). Theoretical techniques are also applied to calculations of beta-decay rates λ_β (see Chap. 5), fission–alpha-decay boundaries (essential for finding the maximum atomic weights achieved in r-processes), and binding energies and partition functions for neutron-rich nuclei (see Sects. 1.4.5–6). The r-process duration on the rapid dynamic stages of a supernova explosion is less than a few seconds or tens of seconds. When the neutron capture stage is terminated, the neutron-rich nuclei return to the stability valley via beta-decays. As a result, excited nuclei form. The delayed fission, alpha-decay and neutron evaporation associated with excited nuclei may significantly affect the final composition of material resulting from the r-process. Including delayed neutron evaporation smooths the large fluctuations in the concentration of even and odd nuclei arising in calculations of the r-process and which are not consistent with observed concentration ratios.

The kinetic equation for the time dependence of the concentration of nuclide $n_{A,Z}$ has the form

$$\frac{dn_{A,Z}}{dt} = - \lambda_{n\gamma}(A, Z)n_{A,Z} + \lambda_{\gamma n}(A + 1, Z)n_{A+1,Z}$$
$$+ \lambda_{n\gamma}(A - 1, Z)n_{A-1,Z} - \lambda_{\gamma n}(A, Z)n_{A,Z}$$
$$+ \lambda_{\beta\pm}(A, Z \pm 1)n_{A,Z-1} - \lambda_\beta(A, Z)n_{A,Z}. \qquad (4.4.10)$$

These equations should be supplemented by initial concentrations of seed nuclei and by the conditions of fission and alpha-decay of nuclei with large A so that fission products may be added to the amount of the corresponding nuclides. The set of equations (4.4.10) is often simplified by the assumption that equilibrium conditions hold with respect to the (n, γ), (γ, n) reactions. The isotopes of an element are in this case connected by the equilibrium equations (4.1.14), and each equation analogous to (4.4.10) refers to the sum over all isotopes of a given element with only beta-processes remaining on the right-hand side, r-processes under conditions of partial equilibrium are examined in [574]. Calculations including (n, γ) and (γ, n) reaction kinetics, time dependence of both $n_n(t)$ and $T(t)$, and delayed processes are performed in [151]. Overviews of earlier work are given in [609, 609a, 396].

The r-process calculations serve mostly to explain the observed abundances of r-process elements with three peaks at $A = 80, 130, 195$. Calculations performed up to now have not yet yielded a relevant height for all three peaks as occurring from the same physical phenomenon. This may be attributed to a non-satisfactory form of the functions $n_n(t)$, $T(t)$, or to large errors in the theoretically determined nuclear characteristics, but may also come from the fact that, in reality, different observed peaks form under different conditions.

Attributing the formation of very heavy elements near uranium to the r-process encounters serious difficulties. This would require a high concentration of neutrons, $n_n \approx 10^{24}$ cm^{-3}, which can hardly be achieved in supernova explosions. It has been suggested that these difficulties may be overcome by supplementing the r-process with one of proton photo-detachments (γ, p) occurring in equilibrium with respect to (n, γ) and (γ, n) reactions (rbc-process [179]). This strongly reduces the requirements on the neutron density since it is assumed to be governed by equilibrium processes.

In [60, 61, 287] the production of superheavy elements has been examined at a high neutron density $n_n \approx 10^{30}$ cm^{-3} and a moderate temperature $T \leq 10^8$ K, so that all nuclei are near the neutron drip line at $Q_n \approx 0$ (n-process), see Fig. 4.1. This situation may be realized in the non-equilibrium envelopes of neutron stars. Outbursts from these stars may lead to the appearance of superheavy elements in the interstellar medium, stars and planets [287, 61, 326] (see Chap. 13, Vol. 2)[2]. Only primary results are obtained in the study of n- and rbc-processes, and further investigations are necessary to determine the importance of these processes for the nucleosynthesis.

The processes of the neutron capture cannot explain the origin of about 30 nuclides on the proton-rich side of the valley of beta stability. These nuclides between ^{74}Se and ^{196}Hg are generally referred to as p-nuclei or bypassed nuclei. It was proposed in [317] that bypass nuclei were synthesized by (p, γ) or (γ, n) reactions on material which had already been synthesized by s- and r-processes. Operation of these reactions was called the p-process. A review of mechanisms of the origin of bypass nuclei is given in [459a].

4.5 Nuclear Reactions in Dense Matter

We assumed earlier that only two charged particles take part in a nuclear reaction (three in a 3α reaction) and ignored interactions with other nuclei and electrons. Coulomb interactions in a plasma reduce the height of the potential barrier, increasing in comparison with (4.1.23) the probability for particles to pass through it. The enhancement in the reaction rate caused by screening depends essentially on plasma parameters. At low densities and high temperatures, in the range of an almost perfect ion gas and for $\Gamma = Z^2 e^2 / akT \ll 1$ (see (2.4.56)), the mean energy of the Coulomb interactions between ions is much less than their kinetic energy, and weak screening occurs. With increasing density and decreasing temperature the mean Coulomb energy begins to exceed the kinetic energy, $\Gamma > 1$, but the energy of particles involved in the reaction is still greater than the mean Coulomb energy, so that we may treat them as free particles. A so-called strong screening takes place under these circumstances. Both types of screening have been studied in [557, 488], and also in a series of papers by Japanese authors (see reviews [431, 427]). On further reduction of temperature and increase of the density the mean Coulomb energy begins to exceed the energy of the particles involved in the reaction, and the role of the Coulomb interactions becomes dominant. Accordingly, the temperature dependence of the reaction rate vanishes, and the screening regime is called pycnonuclear (Π). It has been examined in [102, 557]. We shall characterize plasmas in which the reaction (4.1.1) with $Z_0 \geq Z_1$ occurs, by the dimensionless parameters

---

[2] The same term "n-process" is used in [298, 299] for another phenomenon: neutron capture, intermediate in rate between the s- and r-process, when $t_\beta \sim t_{n\gamma}$. This process is less effective than the r-process for heavy element production. In this book the term "n-process" has the same meaning as in [60, 61, 287].

$$\Gamma_Z = \frac{Z^2 e^2}{akT} = Z^{5/3} \frac{e^2}{kT} \left(\frac{4\pi}{3} n_e\right)^{1/3}$$

$$= Z^{5/3} \frac{5.7562 \times 10^8 \text{ K}}{T} \left(\frac{\rho}{\mu_Z \cdot 1.6203 \times 10^{10}}\right)^{1/3}, \tag{4.5.1}$$

$$a = \left(\frac{3Z}{4\pi n_e}\right)^{1/3},$$

(in (2.4.56) $a \equiv l_{WS}$)

$$\tau^3 = \frac{27\pi^2}{2} \frac{Am_u Z_0^2 Z_1^2 e^4}{kT\hbar^2} = \frac{A_0 A_1 Z_0^2 Z_1^2}{A_0 + A_1} \frac{7.6696 \times 10^{10} \text{ K}}{T}, \tag{4.5.2}$$

$$A = A_0 A_1/(A_0 + A_1),$$

$$\lambda = \frac{\hbar^2}{m_u e^2} \frac{1}{2A Z_0 Z_1} \left(\frac{n_e}{2\langle Z\rangle}\right)^{1/3}$$

$$= \frac{A_0 + A_1}{2A_0 A_1 Z_0 Z_1} \left(\frac{1}{\langle Z\rangle \mu_Z} \frac{\rho}{1.3574 \times 10^{11}}\right)^{1/3}. \tag{4.5.3}$$

The quantity μ_Z is determined by (1.2.17). Using μ_N from (1.2.16), the mean charge of plasma nuclei is

$$\langle Z\rangle = \frac{\mu_N}{\mu_Z}. \tag{4.5.4}$$

The parameter $\lambda^{3/2}$ is proportional to the ratio of the energy of the zero-point lattice fluctuations to the Coulomb energy of the interactions between nuclei at nuclear separations $\left(\sim \frac{\hbar^2}{m_u e^2}\right)$, whereas τ^3 is the ratio of the same Coulomb energy to the thermal energy. In the absence of screening and after averaging $\langle \sigma v\rangle$ over a Maxwellian distribution the quantity τ (coinciding with (4.1.31)) appears in the exponent of the relation (4.1.26) for the reaction rate. For $1 \leq \Gamma \leq \Gamma_m$ the matter is in a liquid state, while for $\Gamma \geq \Gamma_m \approx 150$ (see (1.4.37), (2.4.56) and the footnote to p. 54, Sect. 1.4) the ions form a crystal lattice in which they may oscillate with a frequency of the order of ω_{pi} from (2.4.61) combined with (2.4.52). The quantity

$$\xi = \frac{\hbar\omega_{pi}}{kT} \quad (= 0.075254\, \lambda^{3/2}\tau^3 \text{ at } Z_0 = Z_1 = Z) \tag{4.5.5}$$

characterizes the degree of excitation of the ionic lattice oscillations. For $\xi \gg 1$ the lattice is nearly in the ground state, while for $\xi \leq 1$ a large number of degrees of freedom are excited. The characteristic kinetic energy of reacting nuclei E_0 from (4.1.30) is within an order of magnitude

$$E_0 \sim kT_\tau \sim E_{\text{Coul}} \frac{\tau}{\Gamma}. \tag{4.5.6}$$

It follows from (4.5.1–5) that $\Gamma \sim \tau\xi^{2/3}$, so we have from (4.5.6)

$$E_0/E_{\text{Coul}} \sim \xi^{-2/3}. \tag{4.5.7}$$

Strong screening thus sets in at $\xi \ll 1$ from (4.5.5) and $\Gamma > 1$. The inequality $E_0 > E_{\text{Coul}}$ holds as a result of the large value of τ.

The Π-regime of screening occurs when $\Gamma \gg 1$ and $\xi \gg 1$, when penetration through the Coulomb barrier is due to zero-point fluctuations of the ionic lattice. We saw in Sect. 1.4.7 that zero-point fluctuations do not destroy crystalline structure under astrophysical conditions, which means that $\xi/\Gamma < 1$, therefore the strong inequality $\xi \gg 1$ holds only when $\Gamma > \Gamma_m$. According to [557], two Π- regimes may be distinguished:

pycnonuclear at $T > 0$ $(\Pi_{T>0})$

when $1 \ll \xi^{2/3} \ll \ln(1/\lambda)$

$$(4.5.8)$$

and

pycnonuclear at zero temperature $(\Pi_{T=0})$

when $\xi^{2/3} \gg \ln(1/\lambda) \gg 1$.

$$(4.5.9)$$

In the $\Pi_{T>0}$ regime, nuclei occupying high oscillation levels participate in the reaction, whereas at the $\Pi_{T=0}$ regime even reacting nuclei are in the ground oscillation state.

In all reactions occurring in stars during quasi-static evolutionary phases the Coulomb energy at nuclear separation is much more than the mean kinetic and zero-point energies. Therefore the inequalities

$$\tau \gg 1, \qquad \lambda \ll 1 \tag{4.5.10}$$

hold in all screening regimes, and the lifetime in nuclear burning in stars is very long compared to nuclear time scales. Areas of various screening regimes are shown in Fig. 4.3 on the $(\lg \rho, \lg T)$ diagram for homogeneous matter with $Z_0 = Z_1 = Z$. The horizontal line ab in Fig. 4.3 shows the limit of full ionization by pressure and is determined by (1.4.24) at $\theta_{12} = (3\pi^2)^{1/2} = 3.09367$. The vertical line bc corresponds to full temperature ionization and is

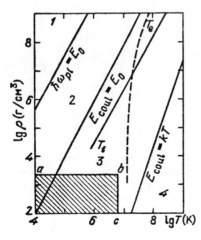

Fig. 4.3. Areas of various screening regimes (from [557]); at $T < T_e$ the electron screening is strong, at $T < T_t$ the electron screening calculations are not reliable (for ^{12}C, from [248]). 1 is the $\Pi_{T=0}$ area. 2 is the $\Pi_{T>0}$ area. 3 is the strong screening area. 4 is the weak screening area. Hatched is the non-complete ionization area.

determined at all densities by the equality $Ze^2/a_{Z_0} = 2kT$ where a_{Z_0} is the radius of the innermost electron orbit defined in Sect. 1.4. If the energy of electrostatic interaction between reacting particles $W(r)$ (r is the interparticle separation) differs from the Coulomb energy of pairwise interaction, then the probability for the particles to pass through the barrier should be expressed by the general formula

$$\sigma = \frac{S(E)}{E} \exp\left\{ -2\frac{\sqrt{2M}}{\hbar} \int_0^{r_{tp}} dr \left[\frac{Z_0 Z_1 e^2}{r} + H(r) - \frac{Mv^2}{2} \right]^{1/2} \right\} \quad (4.5.11)$$

rather than (4.1.23). Here M is the reduced mass given in (4.1.20), r_{tp} is the classical turning point where the integrand tends to zero, and

$$H(r) = W(r) - \frac{Z_0 Z_1 e^2}{r} \qquad \text{represent the part of the potential energy due to the screening.} \qquad (4.5.12)$$

At $H(r) = 0$ (4.5.11) yields (4.1.23)[3].

4.5.1 Weak Screening

When $\Gamma \ll 1$ the screening potential in a plasma should be calculated from the Debye–Hückel theory [176]. The nuclear and electron distributions in an electrostatic field with potential $\Phi(r)$ are described by the relations [145]

$$n_i(r) = n_i \exp\left[-Z_i e\, \Phi(r)/kT \right],$$

$$n_e(r) = \frac{1}{\pi^2} \left(\frac{kT}{c\hbar} \right)^3 \int_0^\infty \frac{x^2 dx}{1 + \exp\left[\sqrt{\alpha^2 + x^2} - (e\,\Phi(r)/kT) - \beta \right]}, \quad (4.5.13)$$

where n_i, n_e are the concentrations of ions and electrons, and $n_e(r) = n_e = \text{const}$ at $\Phi(r) = 0$. Assuming perturbations introduced by the potential $e\,\Phi(r)/kT \ll 1$ are small and using (1.2.10) and (1.2.70), we obtain, upon expanding,

$$n_i(r) = n_i \left[1 - Z_i e\, \Phi(r)/kT \right],$$

$$n_e(r) = n_e \left[1 + \frac{I_{5-} + I_{6-}}{I_{n-}} \frac{e\Phi(r)}{kT} \right]. \quad (4.5.14)$$

Poisson's equation for $\Phi(r)$ around a nucleus with charge Z_0 allowing for quasineutrality has the form

---

[3] The correct procedure consists of calculation of tunnelling in the field of random potential and subsequent averaging of the probabilities over the ensemble of surrounding particles, instead of calculating the tunnelling probability in the average potential of the screening field. The results of these "improved" calculations [515a] are given at the end of Sect. 4.5.3.

$$\Delta\Phi = -4\pi Z_0 e\, \delta^3(r) + \frac{4\pi e^2}{m_u kT}\rho\sum_i \frac{x_i}{A_i}\left(Z_i^2 + \frac{I_{5-}+I_{6-}}{I_{n-}}Z_i\right)\Phi \quad (4.5.15)$$

and may have the solution

$$\Phi = \frac{eZ_0}{r}\exp(-k_0 r) \approx \frac{eZ_0}{r} - eZ_0 k_0, \quad (4.5.16)$$

where

$$k_0^{-1} = \left[\frac{kTm_u}{4\pi e^2\rho}\frac{1}{\sum_i \frac{x_i}{A_i}\left(Z_i^2 + \frac{I_{5-}+I_{6-}}{I_{n-}}Z_i\right)}\right]^{1/2}$$

$$= 8.8924\times 10^{-12}\left(\frac{\mu_Z T}{\rho}\right)^{1/2}$$

$$\times\left(\mu_Z\sum_i \frac{x_i Z_i^2}{A_i} + \frac{I_{5-}+I_{6-}}{I_{n-}}\right)^{-1/2}. \quad (4.5.17)$$

The quantity k_0^{-1} corresponds to the Debye screening radius and coincides with the quantities introduced in (2.4.47) and (2.4.63) in the relevant limiting cases. For the case of weak screening, the addition to the potential energy is

$$H(r) = -e^2 Z_0 Z_1 k_0 = \text{const.} \quad (4.5.18)$$

Substituting (4.5.18) into (4.5.11) and integrating gives

$$\sigma = \frac{S(E)}{E}\exp\left\{-\frac{2Z_0 Z_1 e}{\hbar v\sqrt{1-\frac{2H}{Mv^2}}}\right\}. \quad (4.5.19)$$

Inserting (4.5.19) into (4.1.19), making use of the strong inequality $H/Mv^2 \ll 1$ and evaluating the integral by the saddle-point method shows that weak screening corresponds to multiplying $\langle\sigma v\rangle$ by a factor

$$e^{U_W} = e^{-\frac{H}{kT}} = e^{\frac{e^2 Z_0 Z_1 k_0}{kT}}. \quad (4.5.20)$$

Using (4.5.1) gives [557]

$$U_W = \sqrt{3}\frac{Z_1}{Z_0}\left(\frac{\mu_Z}{Z_0}\right)^{1/2}\eta\Gamma_{Z_0}^{3/2},$$

$$\eta^2 = \sum_i x_i\frac{Z_i^2}{A_i} + \frac{I_{5-}+I_{6-}}{I_{n-}}\frac{1}{\mu_Z}, \quad (4.5.21)$$

$$\frac{I_{5-}+I_{6-}}{I_{n-}} \to \begin{cases} 1 & \text{(ND)} \\ \frac{3\mu_{te}kT}{\mu_{te}^2-m_e^2 c^4} \xrightarrow{T\to 0} 0 & \text{(D)} \end{cases}$$

where the expansion (1.2.24) has been taken into account.

4.5.2 Strong Screening

In the strong screening regime at $\Gamma > 1$ and $\xi \ll 1$ from (4.5.7), the classical turning point $r_{tp} \ll a$. The screening calculations therefore imply that $H(r) \approx H(0)$ [557]. In the Wigner–Seitz approximation the quantity $H(0)$ is physically the difference between the total electrostatic energies of two cells after and before fusion. Using (1.4.26) and (4.5.1) gives

$$H(0) = 0.9kT(\Gamma_{Z_1} + \Gamma_{Z_2} - \Gamma_{Z_1+Z_2}). \tag{4.5.22}$$

Similarly to (4.5.20), we obtain that the effect of including strong screening in calculations corresponds to multiplying $\langle \sigma v \rangle$ by a factor

$$e^{U_{so}} = e^{-H/kT} = e^{0.9(\Gamma_{Z_1}+\Gamma_{Z_2}-\Gamma_{Z_1}-\Gamma_{Z_2})} = e^{\beta\tau}. \tag{4.5.23}$$

The quantity τ is determined by (4.1.31) and (4.5.2), while for β, using (4.5.1), we have [557] from (4.5.23)

$$\beta = \frac{1.8}{3(3\pi)^{1/3}} \frac{(Z_0 + Z_1)^{5/3} - Z_0^{5/3} - Z_1^{5/3}}{(AZ_0^2 Z_1^2)^{1/3}} \left(\frac{\rho}{\mu_Z}\right)^{1/3} \frac{1}{(kT)^{2/3}} \left(\frac{\hbar e}{m_u}\right)^{2/3}$$

$$= \frac{(Z_0 + Z_1)^{5/3} - Z_0^{5/3} - Z_1^{5/3}}{(AZ_0^2 Z_1^2)^{1/3}} \left(\frac{4.2579 \times 10^7 K}{T}\right)^{2/3} \left(\frac{\rho/\mu_Z}{1.6203 \times 10^{10}}\right)^{1/3}, \tag{4.5.24}$$

within an order of magnitude $\beta \sim \xi^{2/3} \ll 1$. Throughout the regions of strong and weak screening we may use, with uncertainty $\leq 30\%$, the interpolation formula

$$U_{SW} = \frac{U_{so}U_W}{\sqrt{U_{so}^2 + U_W^2}}. \tag{4.5.25}$$

The corrections to (4.5.23) in the region of $\beta \leq 1$ calculated in [557] are based on the relation for the potential energy $H(r)$ in the crystal lattice approximation, for an arbitrary interionic separation and nuclear motion directed towards the nearest neighbor:

$$H(r) = \frac{Z^2 e^2}{b}[-1.1547 - 1.1602(1 - \eta) + 1.0394(1 - \eta)^2$$

$$- 2.5690(1 - \eta)^3 + 1.6971(1 - \eta)^4]. \tag{4.5.26}$$

Here $b = \left(\frac{8\pi}{3}\right)^{1/3} a = 2.0310a = (N/2)^{-1/3}$ is the body-centred lattice constant (see (1.4.4), (4.5.1)), $\eta = r/d$ and $d = \frac{\sqrt{3}}{2}b = 0.866b$ is the nearest-neighbor separation in the same lattice. Integrating (4.5.11) and (13.19) numerically and using (4.5.26) yields [557] the strong screening factor $e^{U_{Sp}}$ with

$$U_{Sp} = \tau \left[\beta - \frac{(0.84Z_0 + 3.46Z_1)\beta^3 + 95.6Z_1\beta^9}{(Z_0 + Z_1) + (1.45Z_0 + 32.8Z_1)\beta^3} \right], \tag{4.5.27}$$

$$Z_1 \le Z_0, \qquad \beta \le 0.5.$$

We may also write out the relation for the potential energy from [431, 427] including terms up to fifth-order:

$$H(r) = -\frac{Z^2 e^2}{b} [1.1547 + 1.1547(1 - \eta) - 0.9935(1 - \eta)^2$$
$$+ 4.3385(1 - \eta)^3 - 5.3868(1 - \eta)^4 + 1.8728(1 - \eta)^5]. \tag{4.5.28}$$

In these expressions for $H(r)$ it is assumed that the distribution for surrounding charges adjusts instantaneously to the configuration of reacting nuclei. In reality, some retardation due to inertia is always present, and the strong screening factor is somewhat lower. It has been shown in [557] that the static approximation yields an estimate for $H(0)$ less than (4.5.22) by a factor of 1.337 (see (4.5.37)).

4.5.3 Strong Screening Calculations Based on Condensed Body Models

More accurate calculations of the strong screening factor compared to [557] have been made in [433, 434]. The dependence of the screening potential on r has been included in (4.5.11), and the fact also taken into account that tunelling between $r = r_{tp}$ and $r = 0$ is so fast that it conserves the charge distribution corresponding to $r = r_{tp}$[4]. An analysis of Coulomb liquids by Monte-Carlo methods has resulted [434] in the following relation for the correction to the potential energy $H(r)$ due to the interaction:

$$H(r) = kT \left(1.25\Gamma_{01} - 0.39 \frac{2r}{a_0 + a_1} \right), \tag{4.5.29}$$

$$0.5 \le 2r/(a_0 + a_1) \le 1.6,$$

where

$$\Gamma_{01} = \frac{2Z_0 Z_1 e^2}{(a_0 + a_1)kT}, \tag{4.5.30}$$

$$a_0 = \left[\frac{3Z_0}{4\pi(Z_0 n_0 + Z_1 n_1)} \right]^{1/3} = \left(\frac{3Z_0}{4\pi n_e} \right)^{1/3}, \tag{4.5.31}$$

$$a_1 = (3Z_1/4\pi n_e)^{1/3}.$$

---

[4] See footnote on p. 104.

For $Z_0 = Z_1 = Z$ the relation (4.5.30) reduces to (4.5.1). According to [433, 434], the dependence of the screening potential on r (in (4.5.11)) implies replacing $H(r)$ by a quantity $\psi(r_1, (r_{tp}+r)/2)$ equal, for $r < r_{tp}$, to the strong screening correction to the interaction energy created by the configuration of screening charges at $r = r_{tp}$. Statistical methods allow one to express $\psi(r, r')$ approximately in terms of $H(r)$ [433]:

$$\psi(r, r') - \psi(r', r') = r' \frac{dH(r')}{dr'} \frac{r}{r'} \left(1 - \frac{r}{r'}\right). \tag{4.5.32}$$

Substituting $\psi(r, r')\Big|_{r' = \frac{r+r_{tp}}{2}}$ from (4.5.32) for $H(r)$ in (4.5.11) and performing the integration (13.19) yields an approximate formula for U_S in the form

$$U_S = 1.25\, \Gamma_{01} - 0.095\tau \left(\frac{3\Gamma_{01}}{\tau}\right)^2. \tag{4.5.33}$$

The relations (4.5.22) and (4.5.33) imply different definitions of the parameter Γ (see (4.5.1) and (4.5.31)), which reduce to the same at $Z_1 = Z_2 = Z$: $\Gamma_{01} = \Gamma_{Z_1} = \Gamma_{Z_2} = \Gamma$, $\Gamma_{Z_1+Z_2} = 2^{5/3}\Gamma$. We then have from (4.5.23)

$$U_{S0} \approx 0.9\,(2^{5/3} - 2)\Gamma = 1.0573\,\Gamma, \tag{4.5.34}$$

which is less by $\sim 18\%$ than the first linear term in (4.5.33). The relation (4.5.33) applies only when the internuclear separation exceeds the turning point radius r_{tp} for nuclei which minimize the exponent in (4.1.25) and have energy $E = E_0 \approx \tau kT/3$. Taking $r_{tp} = \frac{Z^2 e^2}{E_0} = \frac{3Z^2 e^2}{\tau kT}$ gives $r_{tp}/a = 3\Gamma/\tau$ and consequently the condition for validity of (4.5.33), namely

$$3\Gamma_{01}/\tau \le 1. \tag{4.5.35}$$

Yakovlev and Schalybkov [248] give the following formula derived from Monte-Carlo calculations [578], holding for strong and moderate screening and valid at $\Gamma \ge 1$:

$$U_S = f(\Gamma_{Z_1+Z_2}) - f(\Gamma_{Z_1}) - f(\Gamma_{Z_2}),$$

$$f(\Gamma) = 0.897744\, \Gamma - 3.80172\, \Gamma^{1/4} + 0.75824\, \Gamma^{-1/4} \tag{4.5.36}$$
$$+ 0.81487 \ln \Gamma + 2.5820.$$

The function $f(\Gamma)$ for moderate screening $\Gamma \le 1$ is given in (4.5.48).

The screening potential based on further Monte-Carlo simulations was used in [515a] for calculation of strong screening making use of the correct averaging procedure, when averaging is performed on the reaction rates, calculated in the random field potential. The enhancement factor $E_{ij} = \exp(U_{ij})$ of the reaction between nuclei with charges Z_i and Z_j $(i, j = 1, 2; Z_2 > Z_1)$ is written in the form:

$$E_{ij} = E_{ij}^{IS} \exp\left[\frac{\Gamma_{ij} B_{ij}^2}{2h_{ij}} \Delta_{ij}\right.$$

$$\left. + (0.18528 - 0.03863 \ln \Gamma_{ij}) \Gamma_{ij} \left(\frac{3\Gamma_{ij}}{\tau_{ij}}\right)^3 \Delta_{ij}\right] \tag{4.5.36a}$$

where

$$E_{ij}^{IS} = \exp(U_{ij}^{IS})$$

$$U_{ij}^{IS} = \left[1.148 - 0.00944 \ln \Gamma_{ij} - 0.000168 (\ln \Gamma_{ij})^2\right] \Gamma_{ij}$$

$$- \frac{20h_{ij}}{32} \Gamma_{ij} \left(\frac{3\Gamma_{ij}}{\tau_{ij}}\right)^2 \tag{4.5.36b}$$

$$+ \left[-0.18528 + 0.03863 \ln \Gamma_{ij} + 0.01095 \left(\frac{3\Gamma_{ij}}{\tau_{ij}}\right)\right] \Gamma_{ij} \left(\frac{3\Gamma_{ij}}{\tau_{ij}}\right)^3,$$

$$B_{ij} = 0.456 - 0.0130 \ln \Gamma_{ij},$$

$$h_{ij} = \frac{(Z_i^{1/3} + Z_j^{1/3})^3}{16(Z_i + Z_j)},$$

$$\Delta_{ij} = 0.007x \left(\frac{Z_2}{Z_i} - 1\right) \left(\frac{Z_2}{Z_j} - 1\right).$$

Here, Γ_{ij} is identical to Γ_{01} from (4.5.30) and τ_{ij} is identical to τ from (4.5.2) if $(0 \equiv i, \ 1 \equiv j)$; $x = n_2/(n_1 + n_2)$ is the molar fraction of species "2".

4.5.4 Pycnonuclear Screening

At $\beta \gg 1$ $(3\Gamma/\tau \gg 1)$ the energy E_0 of nuclei involved in the reaction is much less than the mean Coulomb energy E_{coul}.

We cannot treat these nuclei as free, contrary to the case of the strong and weak screening regimes. The quantity $Mv^2/2$ in the relation for tunnelling (4.5.11) must be replaced by the energy of fluctuations in the lattice. It is not convenient to describe Π-screening by a factor representing an increase in the reaction rate because this gives relations for the reaction rates that are quite different in form from the formulae of Sect. 4.1. In addition to the potential (4.5.26), the potential

$$H_S(r) = \frac{Z^2 e^2}{b} \left[-1.1547 - 1.1602\,(1 - \eta) + 1.0394\,(1 - \eta)^2\right.$$

$$\left. - 0.4001\,(1 - \eta)^3 + 0.0692\,(1 - \eta)^4\right], \tag{4.5.37}$$

corresponding to the static approximation in which the surrounding nuclei and the centre of mass of reacting nuclei are taken as fixed during the reaction, has been used in [557] to examine Π-screening.

In the $\Pi_{T=0}$ regime the ions occupy the lowest fluctuation levels with energy $E_0 \sim \hbar\omega_{\text{pi}} \ll E_{\text{coul}}$. Therefore departures of the nuclei from their equilibrium locations are small. The $\Pi_{T=0}$ reaction rate is given in the WKB approximation by [557]

$$P_0 = \frac{2x_0 x_1}{1 + \delta_{01}} 10^{46} \lambda^{7/4} \frac{\rho}{\mu_N} (2A)^2 Z_0^2 Z_1^2 S(E)$$

$$\times \begin{pmatrix} 3.90 \\ 4.76 \end{pmatrix} \exp\left[-\frac{1}{\sqrt{\lambda}} \begin{pmatrix} 2.638 \\ 2.516 \end{pmatrix} \right] \quad \text{(reactions cm}^{-3}\text{s}^{-1}\text{)}. \qquad (4.5.38)$$

Here, the dimensions of s are MeV barn (see (4.1.23) and (4.1.24)). The quantities A, μ_N, and λ are given by (4.1.20), (1.2.16), and (4.5.3) respectively, the upper line referring to the static approximation with the potential function (4.5.37), and the lower line to the full relaxation approximation with the potential function (4.5.26). It should be noted that at $T \neq 0$ the probability for nuclei in the lattice to occupy excitation levels with energy $(E - E_0)$ is $\exp[-(E - E_0)/kT]$. The relation for P, the $\Pi_{T>0}$ reaction rate, is obtained in [557] on summing over the excited states,

$$\frac{P}{P_0} - 1 = \begin{pmatrix} 0.0430 \\ 0.0485 \end{pmatrix} \lambda^{-1/2} \left[1 + \begin{pmatrix} 1.2624 \\ 2.9314 \end{pmatrix} e^{-8.7833\,\beta^{3/2}} \right]^{-1/2}$$

$$\times \exp\left\{ -7.272\,\beta^{3/2} + \lambda^{-1/2} \begin{pmatrix} 1.2231 \\ 1.4331 \end{pmatrix} e^{-8.7833\,\beta^{3/2}} \right.$$

$$\times \left. \left[1 - \begin{pmatrix} 0.6310 \\ 1.4654 \end{pmatrix} e^{-8.7833\,\beta^{3/2}} \right] \right\}, \qquad (4.5.39)$$

where also the upper and lower lines refer to the static and full relaxation approximations, respectively.

Exact numerical solution to the Schrödinger equation for S-wave scattering between nearest-neighbor nuclei in the lattice potentials was formulated in [447+] through Monte-Carlo sampling. Thermal effects were evaluated by taking account of excited nuclear states. The results agree well with [557], except in the vicinity of the melting condition where the enhancement factor may be $\sim 10^{21}$ times larger than in [557].

4.5.5 Electrons in the Strong-Screening Regime

Electron screening is due to the fact that non-uniformity in the electron density distribution alters the interaction energy of reacting nuclei. Relation (4.5.21) allows for electron screening in the weak screening regime, while Sects. 4.5.2–4 deal only with ion screening. Electron screening in the strong-screening regime has been calculated in [177, 181, 428, 247]. The computations performed in [247, 248], for the nearly classical ion motion in a plasma, yield the following screening factor:

Table 4.7. Parameters from the relation (4.5.42) for electron screening of nuclear reactions [248]

Z	A_0	A_1	A_2	B_0	B_1	B_2	C_0	C_1	C_2
1	7.924(-3)	2.613(-2)	0.5280	0.2020	1.069(-1)	1.616	-2.935(-2)	-8.771(-2)	1.343
2	1.518(-2)	4.045(-2)	0.6262	0.2356	2.122(-2)	7.022	-5.274(-2)	-1.206(-2)	7.006
6	3.105(-2)	2.810(-2)	1.179	0.2602	1.160(-2)	10.96	-6.682(-2)	-7.733(-3)	9.321
8	3.528(-2)	2.563(-2)	1.330	0.2654	9.349(-3)	13.20	-7.014(-2)	-6.200(-3)	11.27
12	4.115(-2)	2.270(-2)	1.536	0.2715	6.283(-3)	19.22	-7.388(-2)	-3.926(-3)	17.65
18	4.683(-2)	1.994(-2)	1.753	0.2761	4.662(-3)	24.45	-7.668(-2)	-2.894(-3)	22.92

$$U_{Se} = f_e(\Gamma_{Z_1+Z_2}, Z_{12}, \beta) - f_e(\Gamma_{Z_1}, Z_1, \beta) - f_e(\Gamma_{Z_2}, Z_2, \beta),$$
$$f_e(\Gamma, Z, \beta) = \eta Z^{2/3}[A(Z,\beta)\Gamma + B(Z,\beta)\Gamma^{1/4} + C(Z,\beta)],$$

(4.5.40)

where

$$\eta = \kappa^2 l_e^2 = 3\left(\frac{12}{\pi}\right)^{1/3}\frac{e^2}{\hbar v_{Fe}} \approx \frac{0.03422}{\beta},$$
$$\beta = \frac{v_{Fe}}{c} = \frac{y}{\sqrt{1+y^2}}.$$

(4.5.41)

In (4.5.41) $y = \frac{p_{Fe}}{m_e c}$ from (1.2.21), $l_e = \left(\frac{3}{4\pi n_e}\right)^{1/3}$, $\kappa = \left(\frac{4\pi n_e e^2}{kT}\right)^{1/2}$ is the Debye radius of classical electrons from (2.4.46), and $Z_{12} = Z_1 + Z_2$. The functions $A(Z,\beta)$, $B(Z,\beta)$ and $C(Z,\beta)$ are approximated by

$$A(Z,\beta) = A_0(Z) + A_1(Z)\ln[1 + (1-\beta^2)A_2(Z)],$$
$$B(Z,\beta) = B_0(Z) + B_1(Z)\ln[1 + (1-\beta^2)B_2(Z)],$$
$$C(Z,\beta) = C_0(Z) + C_1(Z)\ln[1 + (1-\beta^2)C_2(Z)].$$

(4.5.42)

The coefficients $A_i(Z)$, $B_i(Z)$, $C_i(Z)$ are given in Table 4.7 from [248]. The electron screening factor should be added to the ion screening factor U_S in the exponent of the screening factor $\exp(U_S + U_{Se})$. The line T_e corresponding to $\exp(U_{Se}) = 2$ is plotted in Fig. 4.3 for carbon ^{12}C burning, so that electron screening is strong for $T < T_e$. The calculations performed in [247, 248] are no longer valid to the left of the line T_t.

4.5.6 Screening of Resonant Reactions

The major contribution to the mean value of $\langle \sigma v \rangle$ for resonant reactions comes from the energy region around the resonance energy $E = E_r$, instead of energies near $E = E_0$. Taking account of the shift in the energy levels due to the screening potential field, we have to substitute $E_r + H(0)$ for $Mv^2/2$

in (4.5.11). Evaluating the integral involving $H(r)$ in (4.5.11) and integrating (4.1.19) then gives the screening factor for the strong screening regime [557]

$$
\begin{aligned}
U_{Sr} = \tau\beta - \frac{\pi}{\sqrt{\lambda\epsilon_r}} &\left[\frac{1.30\, Z_0 + 1.14\, Z_1}{(Z_0 + Z_1)\epsilon_r^3} \right. \\
&\left. + \frac{28.2\, Z_0 - 34.4\, Z_1}{(Z_0 + Z_1)\epsilon_r^6} + \frac{150\, Z_1}{(Z_0 + Z_1)\epsilon_r^9} \right],
\end{aligned}
\tag{4.5.43}
$$

where

$$
\begin{aligned}
\epsilon_r = E_r \Big/ \left(\frac{Z_0 Z_1 e^2}{b} \right) \\
= \frac{1}{Z_0 Z_1} \left(\frac{Z_0 \mu_Z \times 1.3574 \times 10^{11}}{\rho} \right)^{1/3} \frac{E_r}{49.600\ \text{keV}},
\end{aligned}
\tag{4.5.44}
$$

$Z_0 \geq Z_1$.

The quantities λ, τ, β and b are determined in (4.5.2), (4.5.3), (4.5.24) and (1.4.4). If the radiative partial width for the outgoing channel is much less than the total width of the resonance, or the resonance energy is unusually large the strong screening factor reduces to

$$
U_{Sr} = \tau\beta = U_{S0}.
$$

The strong screening factor for the $3\alpha \rightarrow {}^{12}$ C reaction, which should be taken into account in (4.2.38), is given by

$$
\begin{aligned}
U_{S3\alpha} = 2.916\, \Gamma_2 - 3\ln(1 + 0.3\Gamma_2 + 0.266\Gamma_2^{3/2}) \\
+ \ln(1 + 1.87\Gamma_2 + 4.15\Gamma_2^{3/2}),
\end{aligned}
\tag{4.5.45}
$$

where $\Gamma_2(Z = 2)$ is determined by (4.5.1). The relations (4.5.43) and (4.5.45) hold for

$$
\epsilon_r > \frac{3.04\, Z_0 + 1.26\, Z_1}{Z_0 + Z_1}.
\tag{4.5.46}
$$

At the stage of ^{8}Be formation, this inequality reduces to $\rho < 5.5 \times 10^9$ g cm^{-3} for $E_r = 92$ keV from (4.2.37). For $U_{S3\alpha}$, we find in [248] the following relation calculated from the Monte-Carlo computations in [578] and valid for $\Gamma \gg 1$

$$
\begin{aligned}
U_{S3\alpha} = f(\Gamma_{123}) - 3f(\Gamma_2) = 2.90892\, \Gamma_2 + 5.3965\, \Gamma_2^{1/4} \\
- 1.7950\, \Gamma_2^{-1/4} - 1.62974\ \ln\Gamma_2 - 3.67196,
\end{aligned}
\tag{4.5.47}
$$

where $\Gamma_{123} = 3^{5/3}\Gamma_2$, and $f(\Gamma)$ is determined by (4.5.36). For moderate screening $\Gamma \leq 1$, $f(\Gamma)$ in (4.5.47) should be replaced by the function [248]

$$f(\Gamma) = \frac{\Gamma^{3/2}}{\sqrt{3}} - 0.14554\,\Gamma^{2.016}, \tag{4.5.48}$$

which yields the interpolation formula

$$U_{WS3\alpha} = 2.11\,\Gamma_2^{3/2} - 1.35\,\Gamma_2^{2.016}. \tag{4.5.49}$$

The function (4.5.48) is smoothly fitted at $\Gamma = 1$ to $f(\Gamma)$ from (4.5.36) and may be used in (4.5.36) for finding U_{WS} for moderate screening in the two-particle reaction.

The enhancement factor obtained in [515a] for the 3α reaction using the correct averaging procedure has a form

$$E_{3\alpha} = E_{\mathrm{HeHe}}E_{\mathrm{BeBe}}, \tag{4.5.50}$$

with E_{ij} from (4.5.36a) and

$$\Delta_{\mathrm{HeHe}} = 0.028x, \quad \Delta_{\mathrm{BeHe}} = 0.007x. \tag{4.5.51}$$

Problem. Find the $\Pi_{T=0}$ reaction rate for a spatial potential approximated by

$$W(r) = Z_0 Z_1 \left(\frac{e^2}{r} + \frac{e^2}{2r_0 - r} - \frac{2e^2}{r_0} \right). \tag{1}$$

Solution [102, 227]. The classical equilibrium state for the potential (1) occurs at $r = r_0$. For a small displacement ($|r - r_0| \ll r_0$)

the frequency $\omega = 2\sqrt{Z_0 Z_1 e^2}/\sqrt{Mr_0^3}$,

the zero-point energy $E_0 = \dfrac{\hbar\omega}{2} = \hbar\sqrt{\dfrac{Z_1 Z_2 e^2}{Mr_0^3}}$, \hfill (2)

the classical turning point $r_{tp} = r_0 - \left(\dfrac{\hbar^2 r_0^3}{4Z_0 Z_1 e^2 M} \right)^{1/4}$.

The exponential for the barrier penetration in (4.5.11) is given, for small ϵ, on writing $Mv^2/2 = E_0$, by

$$D = \exp\left(-\frac{4}{\epsilon} + \ln\sqrt{\frac{2}{\epsilon}} \right),$$

$$\epsilon = 2\left(1 - \frac{r_{tp}}{r_0} \right)^2 = (\hbar^2/Z_0 Z_1 e^2 Mr_0)^{1/2}. \tag{3}$$

Here, D is defined to be the ratio of transmitted to incident fluxes:

$$D = \frac{(v|\psi^2|)_{trans}}{(v|\psi^2|)_{inc}}. \tag{4}$$

Consider $|\psi^2|$ values averaged over angle variables. The probability of a reaction between two given nuclei is [227]

$$p = (v|\psi^2|)_{trans} \times R_n^2 P_n = (v|\psi^2|)_{inc} \times D R_n^2 P_n \quad (s^{-1}), \tag{5}$$

where P_n is the probability of a nuclear reaction between two particles separated by a nuclear radius R_n. The quantity $S(E) = E R_n^2 P_n$ defined similarly to (4.1.23) depends only on properties of the given nuclei. Using $S(E)$, the reaction probability reads

$$p = (v|\psi^2|)_{inc} \times D \frac{S(E)}{E} \quad (s^{-1}). \tag{6}$$

In the quasiclassical approximation, the wave function ψ in the spatial potential well (1) is [144]

$$\psi = \sqrt{\frac{2}{\pi}} \frac{1}{(r - r_0)\sqrt{r_0 - r_{tp}}}$$
$$\times \sqrt{\frac{v_0}{v}} \sin\left(\frac{1}{\hbar} \int_r^{r_0} \sqrt{2m(E_0 - W)}\, dx + \frac{\pi}{4}\right), \tag{7}$$

where $v_0 = \sqrt{2E_0/M}$ is the relative velocity of nuclei at the equilibrium point $r = r_0$. Replacing $\sin^2 F$ from (7) by its mean value $1/2$ we have for the flux of colliding nuclei at the turning point $r = r_{tp}$

$$(v|\psi^2|)_{inc} = \frac{1}{\pi} \frac{v_0}{(r_0 - r_{tp})^3}. \tag{8}$$

Recalling that the number of reactions per second in 1 cm^3 in a body-centred lattice [360, 557] is

$$P = \frac{8\rho N_A}{1 + \delta_{01}} \frac{x_0 x_1}{\mu_N} p \quad (cm^{-3}\, s^{-1}) \tag{9}$$

and using (3), (6) and (8) with $E = E_0$ gives

$$P = \frac{8\rho N_A}{1 + \delta_{01}} \frac{x_0 x_1}{\mu_N} \frac{4\sqrt{2}}{\pi} \left(\frac{Z_0^3 Z_1^3 e^6 M^3}{\hbar^{10} r_0^5}\right)^{1/4}$$
$$\times \exp\left(-\frac{4\sqrt{M r_0 Z_0 Z_1 e^2}}{\hbar}\right) S(E_0). \tag{10}$$

Introducing the parameter λ from (4.5.3) and making use of $r_0 = 2\langle a \rangle = \left(\frac{6\langle Z \rangle}{\pi n_e}\right)^{1/3}$ where $\langle a \rangle$ is the mean radius of the W–S cell, we have

$$\lambda = \frac{\hbar^2}{2m_u A Z_0 Z_1 e^2} \left(\frac{3}{\pi}\right)^{1/3} \frac{1}{r_0}, \tag{11}$$

$$P = \frac{2\rho x_0 x_1}{(1 + \delta_{01})\mu_N} \times 4.8 \times 10^{44} (2A)^2 Z_0^2 Z_1^2 \lambda^{5/4} S \exp\left(-\frac{2.85}{\sqrt{\lambda}}\right). \tag{12}$$

Similarly to (4.5.38), the dimensions of S are MeV $\times$ barn. The approximate formula (12) yields a lower value for P than the more exact formula (4.5.38).

5 β-Processes in Stars

5.1 Fundamentals of the Weak Interaction Theory

The reactions involving neutrinos are called weak interaction reactions (see (4.2.1), (4.2.11), (4.2.15), (4.2.18–21), and (4.2.23)). They owe their name to their small interaction cross-section that is substantially lower than the cross-section for electromagnetic and nuclear (strong) interactions. The small interaction cross-section results in a high penetrative capacity of neutrinos. They escape from the centre of the Sun and other stars and can easily make their way through the Earth along its diameter. Observations of solar neutrinos might provide a unique possibility to determine experimentally physical conditions in central regions of the Sun. The experiments headed by R. Davis continued for more than 20 years and based on the reaction

$$^{37}\text{Cl} + \nu_e \rightarrow {}^{37}\text{Ar} + e^- \tag{5.1.1}$$

give a result somewhat below theoretical expectations [260, 259b, 335b]:

$$\begin{aligned} E\nu_{\text{obs}} &= 2.28 \pm 0.69 \,\text{SNU} \\ E\nu_{\text{theor}} &= 7.4 \pm 2.8 \,\text{SNU}. \end{aligned} \tag{5.1.2}$$

The observational errors here correspond to the 3σ level, $1(\text{SNU}) = 1(\text{solar neutrino unit}) = 10^{-36}$ counts per second per ^{37}Cl atom. The disagreement between theory and experiment seems not statistically reliable enough to draw any radical conclusions, so uncertainty still persists in this problem. This disagreement could be attributed to the resonant neutrino oscillations in matter discussed in [160, 274], see also review [260a]. There are some hints in favour of the existence of $\nu_\mu \rightarrow \nu_e$ oscillations from anomalous cosmic ray data [567a] and from experiments with neutrino beams in Los Alamos [567b].

Another possibility for the solution of the solar neutrino problem suggested in [335b] is based on using of revised formulae for rates of nuclear reactions (4.2.3a) and (4.2.3a$_2$), which determine the production of the boron neutrino, mainly registered in the Davis experiment.

Low energy cross-sections for calculation of the reactions in the Sun were extrapolated from measurements at much higher energies using a parameterization (4.1.23) which in [335b] was claimed to be not good enough. Improvements in the Coulomb barrier penetration factor based on taking account of the dynamical effects in the screening process, and substruction of the higher partial waves contribution which give their input only at higher energies, considerably decrease the cross-sections compared to the values used in [260b]. The value of $S(0)$ (see (4.1.24)) is decreased from 0.56 keV·b in [260b] to 0.45 keV·b in [335b] for the reaction $^3\text{He}(\alpha, \gamma)^7\text{Be}$, and from 22.4 eV·b to 17.0 eV·b for the reaction $^7\text{Be}(p, \gamma)^8\text{B}$. This leads to a relative decrease of the energy production in the channel (4.2.3a) compared to (4.2.3) and to a decrease of the boron neutrino flux. Its value obtained in [335b] is equal to 2.77 SNU, which is in good agreement with the experimental results of Kamiokande (2.87 ±0.51 SNU) which measured only neutrino from ^8B. There is still a difference from the Davis experiment which measured neutrinos from ^8B (4.2.3a$_2$) and ^7Be (4.2.3a$_1$) with a small contribution (0.2 SNU) from pep. The average values over 23 years of the Davis experiment (5.1.2) are even less than from that from Kamiokande. After repairs in 1986 the Davis data become higher, resulting in 2.8 ± 0.9 SNU (3σ) since 1987. Its statistical significance is considered in [259b, 576a] as less reliable than the full average (5.1.2).

Solar neutrino measurements based on a $^{71}\text{Ga}(\nu, e^-)^{71}\text{Ge}$ experiment are provided now in the Russia–USA experiment SAGE at Baksan, and the European Gallium experiment GALLEX at Grand Sasso. The threshold in this experiment is low enough (0.233 MeV) permitting the main solar neutrino flux from the reaction (4.2.1) to be determined (see [259a]). The results do not have good enough statistics: 87 ±16 SNU (GALLEX, 1σ) and $85^{+22}_{-32} \pm$ 20 (SAGE, 1σ), and they are not in contradiction with different theoretical models within 3σ level [259b, 335b]. The conclusions of [335b] have been criticized (see [259b] and response in [335c]).

Using the revised formulae for screening and nuclear reaction rates of the p–p chain, as well as electrostatic corrections to the pressure in the equation of state with reasonable uncertainties in [342a], for construction of the solar model, called "nuclear minimal standard model", results in an important conclusion (similar to [335b]), that "even if such a model does not completely solve the solar neutrino enigma, it may considerably reduce the discrepancy between experiments and predictions and could be compatible with all the present experiments within 3σ."

The experimental results to the year 1999 are approximately 67 ± 11 SNU (SAGE), and 77.5 ± 11 SNU (GALLEX) [248*, 248**]. In combination, these results could still be within the "nuclear minimal standard model" (see also [366a]). Another interpretation, like neutrino oscillations in vacuum or in matter, leading to transitions between different types of neutrino, is also possible [248**], demanding non-zero rest masses of neutrinos. Additional restrictions to the model come from Super-Kamiokaude measurements [366b].

The particles involved in weak interactions and not involved in strong ones are called leptons. Neutral leptons (neutrinos) have their charged counterparts involved in electromagnetic interactions. All leptons have a spin 1/2 and obey Fermi statistics with a distribution function over phase space cells similar to (1.2.2). We know today three species of neutrino and three charged leptons [172, 544a, 346a]:

electron ν_e, e ($m_e = 0.5110$ MeV $= 9.11 \times 10^{-28}$ g,

$m_{\nu_e} < 5.1$ eV),

muon ν_μ, μ ($m_\mu = 105.7$ MeV $= 1.88 \times 10^{-25}$ g, $\qquad\qquad$ (5.1.3)

$m_{\nu_\mu} < 0.27$keV),

τ-neutrino ν_τ, τ-lepton ($m_\tau = 1.784$ MeV, $\quad m_{\nu_\tau} < 31$ MeV).

Here, we give experimental restrictions to masses[1]. All leptons in (5.1.3) have their antiparticles. Stars emit electron neutrinos in the hydrogen burning phase, while in late evolutionary stages ν_e energy losses may exceed photon losses [363] (see Chap. 9, Vol. 2). Emission of all species of neutrinos becomes very important in supernova explosions (see Chap. 10, Vol. 2).

The measurements of the shape of the Z^0 "resonance" curve have shown that there are only three types of the light neutrino, and three corresponding quarks (see below) doublets [553a].

The weak interaction (β-process) theory is developed in a fairly detailed form, almost as the electromagnetic field theory is, but it differs from the latter in that it is a purely quantum theory with no classical limit. Fundamentals of the β-interaction theory are considered necessary for calculations of neutrino processes in stars.

5.1.1 Nuclear β-Decay

The foundations of the weak interaction theory were formulated by Fermi [213] in 1933 in order to explain the nuclear β-decay in the reaction

$$(A, Z) \to (A, Z + 1) + e^- + \tilde{\nu} \qquad\qquad (5.1.4a)$$

or

$$(A, Z) \to (A, Z - 1) + e^+ + \nu. \qquad\qquad (5.1.4b)$$

The reaction (5.1.4) proceeds when the initial nuclear mass exceeds the final nuclear mass by more than the electron mass:

$$\Delta_{ZZ'} = (m_{A,Z} - m_{A,Z'})\, c^2 > m_e c^2, \quad Z' = Z \pm 1. \qquad (5.1.5)$$

The signs "+" and "−" correspond to the reactions a) and b), respectively.

According to the perturbation theory [144], the probability of a β-decay is

$$W_\beta = \frac{2\pi}{\hbar} \int |H|_\beta^2 \, \delta \left(\sum_\lambda \epsilon_\lambda \right) dN_e \, dN_{\tilde{\nu}} \, dN_{A,Z'}. \qquad (5.1.6)$$

---

[1] It follows from cosmology that $m_{\nu_\mu} < 40$ eV [88, 100].

Here the initial state of the (A, Z) nucleus is assumed to be given, the δ-function accounts for the conservation of the energy ϵ_λ in β-decay.

The summation is done over the four particles involved in the reaction ($\lambda = 1, 2, 3, 4$). It is assumed in Fermi's theory that β-interaction occurs when all four particles (leptons and nucleons inside the nucleus) are located in the same point of space. This is allowed for in (5.1.6), where the integration over the region of possible interactions (the nuclear volume) is included in the matrix element for the Hamiltonian H_β. The differentials in (5.1.6) imply an integration over only the momentum space. If the β-decay occurs in a medium containing electrons e^- (or e^+) and antineutrinos $\bar{\nu}_e$ (or ν_e) with phase space distribution functions f_e and $f_{\bar{\nu}}$ (see (1.2.2)), then, bearing in mind the Pauli exclusion principle, we obtain

$$dN_e = \frac{1 - f_e}{(2\pi\hbar)^3} \, d^3 p_e, \quad dN_{\bar{\nu}} = \frac{1 - f_{\bar{\nu}}}{(2\pi\hbar)^3} \, d^3 p_{\bar{\nu}}. \tag{5.1.7}$$

In β-decays $\Delta \ll m_{A,Z} c^2$, so we may ignore the nuclear recoil and avoid integration over $dN_{A,Z'}$ since $\int dN_{A,Z'} = 1$ by virtue of the normalization condition. To obtain the total probability of the β-decay, we have to find all possible final (excited) states of the (A, Z') nucleus and perform the summation $\Sigma_i W_{\beta,i}$ over them. With increasing Δ, the number of possible final states grows rapidly.

5.1.2 Matrix Elements in β-Decay

In order to calculate the matrix element H_β for non-polarized nuclei in (5.1.6), one should first specify a special linear combination of products of wave-function components defined below. This linear combination is to be integrated over the nuclear volume V_n with summation performed afterwards over all possible spin states of final particles and averaging over the spin states of the initial nucleus (A, Z). If only free elementary particles, leptons or hadrons, take part in the β-reaction, then the integration is done over the entire space. The summation over spin states is included here in the matrix element, so the phase volume in (5.1.7) is defined regardless of the statistical weight.

In the relativistic field theory a free electron and other free fermions are described by the four-component wave function (bispinor) ψ_e satisfying Dirac's equation

$$\left(i\gamma_i \frac{\partial}{\partial x_i} - \frac{m_e c}{\hbar} \right) \psi_e = 0. \tag{5.1.8}$$

The Dirac matrices γ_i are defined to an arbitrary unitary[2] transformation. In a commonly used representation with two components of the bispinor ψ_e turning to zero in the non-relativistic limit, the matrices γ_i become [69, 171]

---

[2] A transformation in linear space is called unitary when the transformation matrix U has its reciprocal U^{-1} equal to Hermitian conjugates U^+. Here $U^+ = (U^*)^T$,

$$\gamma_1 = \begin{pmatrix} 0 & 0 & 0 & 1 \\ 0 & 0 & 1 & 0 \\ 0 & -1 & 0 & 0 \\ -1 & 0 & 0 & 0 \end{pmatrix}, \quad \gamma_2 = \begin{pmatrix} 0 & 0 & 0 & -i \\ 0 & 0 & i & 0 \\ 0 & i & 0 & 0 \\ -i & 0 & 0 & 0 \end{pmatrix},$$

$$\gamma_3 = \begin{pmatrix} 0 & 0 & 1 & 0 \\ 0 & 0 & 0 & -1 \\ -1 & 0 & 0 & 0 \\ 0 & 1 & 0 & 0 \end{pmatrix}, \quad (5.1.9)$$

$$\gamma_4 = \begin{pmatrix} 1 & 0 & 0 & 0 \\ 0 & 1 & 0 & 0 \\ 0 & 0 & -1 & 0 \\ 0 & 0 & 0 & -1 \end{pmatrix}, \quad \gamma_5 = i\gamma_1\gamma_2\gamma_3\gamma_4 = \begin{pmatrix} 0 & 0 & -1 & 0 \\ 0 & 0 & 0 & -1 \\ -1 & 0 & 0 & 0 \\ 0 & -1 & 0 & 0 \end{pmatrix}.$$

We use in (5.1.8) the scalar product of four vectors

$$uv \equiv u_i v_i = u_4 v_4 - u_1 v_1 - u_2 v_2 - u_3 v_3; \quad x_4 = ct. \quad (5.1.10)$$

The matrix S determining the Lorentz transformation for bispinor

$$\psi' = S\psi \quad (5.1.11)$$

at a fixed representation of γ_i is related to the matrix L_{ik} of the Lorentz transformation for vectors,

$$x' = Lx, \quad x'_i = L_{ik}x_k \quad (5.1.12)$$

by the equality

$$S^{-1}\gamma_i S = L_{ik}\gamma_k, \quad S\gamma_4 S^+ = b\gamma_4, \quad (5.1.13)$$

determined by the invariance condition for Dirac's equation [230]. The sign before γ_4 here is the same as before the matrix element L_{44}, $b = \pm 1$. The first equality in (5.1.13) is equivalent to

$$S\gamma_i S^{-1} = L_{ki}\gamma_k, \quad (5.1.14)$$

coming from the orthogonality of the Lorentz transformations

$$L_{ik}^{-1} = L_{ki}. \quad (5.1.15)$$

The matrix S satisfying the conditions (5.1.13) is generally not unitary and has the structure [214]

---

the asterisk denotes a complex conjugation, and superscript "T" transposition. In a unitary transformation of ψ with matrix U that changes representation in a fixed space–time coordinate system according to $\psi' = U\psi$, the form of the matrices γ_i is also changed according to the relation $\gamma'_i = U\gamma_i U^{-1}$, leaving Dirac's equation (5.1.8) invariant.

$$S = \begin{pmatrix} \alpha & \beta & 0 & 0 \\ \gamma & \delta & 0 & 0 \\ 0 & 0 & \alpha^* & \beta^* \\ 0 & 0 & \gamma^* & \delta^* \end{pmatrix}, \quad \alpha\delta - \beta\gamma = 1. \tag{5.1.16}$$

The matrices γ_i satisfy the anticommutation relations

$$\gamma_i\gamma_k + \gamma_k\gamma_i = 2\delta_{ik}I \quad i,k = 1,2,3,4,5; \tag{5.1.17)}$$

I is the unit matrix,

$$\delta_{ik} = \begin{cases} 1 & \text{for } i = k = 4,5 \\ -1 & \text{for } i = k = 1,2,3 \\ 0 & \text{for } i \neq k. \end{cases}$$

Note that the matrices γ_4 and γ_5 in (5.1.9) are Hermitian while γ_1, γ_2, and γ_3 are anti-Hermitian[3]. 16 bilinear terms can be composed of the two four-component bispinors ψ_e and ψ_ν, and then combined in such a way as to form five covariant quantities. For constructing these terms, the Dirac conjugate spinor

$$\overline{\psi} = \psi^+\gamma_4, \tag{5.1.18}$$

is usually used, which allows us to write the five quantities in the form of [171][4]

the scalar (S) $\overline{\psi}\psi$,

polar vector (V) $\overline{\psi}\gamma_i\psi$,

axial vector (A) $\overline{\psi}\gamma_5\gamma_i\psi$,

pseudoscalar (P) $\overline{\psi}\gamma_5\psi$,   (5.1.19)

antisymmetric tensor (T) $\overline{\psi}\sigma_{ik}\psi$,

$$\sigma_{ik} = \frac{1}{2i}(\gamma_i\gamma_k - \gamma_k\gamma_i).$$

The invariance of the matrix element H relative to Lorentz transformations causes it to include the wave function products in the from of pairwise products of terms from (5.1.19). In the primary version of the β-decay theory [213] Fermi considered only the V interaction in analogy with the electrodynamics. The analysis of experimental data revealed, however, that a variant of V–A interaction takes place in reality, which has been considered by Feynman and Gell-Mann and where the lepton (l) wave functions are included in the H_β matrix element in the form of $j_{l,i}$-like combinations. For the reaction (5.1.4a), this combination has the form [171]

---

[3] Matrix A is Hermitian if $A^+ = A$, and anti-Hermitian if $A^+ = -A$. Hermiticity and anti-Hermiticity are invariant with respect to unitary transformations.

[4] $\overline{\psi}' = \overline{\psi}U^{-1}$ for the unitary transformation, $\overline{\psi}' = b\overline{\psi}S^{-1}$ for the Lorentz transformation (5.1.11).

$$j_{l,i} = \overline{\psi}_e \gamma_i (1 + \gamma_5) \psi_\nu. \tag{5.1.20}$$

The bilinear function $j_{l,i}$ is called the leptonic current[5]. Owing to the properties of the matrix γ_5, the result of this choice of the leptonic current $j_{l,i}$ is that only two (left) combinations of components determine the rate of the weak interaction reaction [171]. This corresponds to the two-component nature of a neutrino which has a left-hand helicity, i.e. its spin and momentum have opposite directions. The antineutrino has the inverse (right-handed) helicity. As a result, the spatial parity (P) is not conserved, while the combined parity (CP) is. The charge conjugation (C) corresponds to the replacement of the particle by its antiparticle. The current constructed by use of the operator $O_i^L = \gamma_i (1 + \gamma_5)$ is called the left-handed current.

The hadronic current $j_{a,i}$ has the form analogous to (5.1.20), but the composite structure of hadrons and production of virtual particles in strong interactions render the matrix element dependent on the transfer momentum $q_i = \frac{1}{c}(p_i' - p_i)$, $q^2 = |q_i q_i|$ so that the form of this current becomes more complicated than that of the leptonic current (5.1.20). We have [72, 469, 86]

$$j_{a,i} = \overline{\psi}_p \left[f_1(q^2)\gamma_i - i f_2(q^2)\sigma_{ik}q_k + g_1(q^2)\gamma_i\gamma_5 + g_3(q^2)q_i\gamma_5 \right] \psi_n. \tag{5.1.21}$$

Weak quasi-elastic[6] nucleon form-factors f_α and g_α are determined from experimental data and from hypothesis of conservation of the vector current (CVC) [87] and partial conservation of the axial current (PCAC) [86]:

$$f_1(q^2) = \left[1 + \frac{q^2}{4m_p^2}(1 + \mu_p - \mu_n) \right] \left(1 + \frac{q^2}{m_v^2} \right)^{-2} \left(1 + \frac{q^2}{4m_p^2} \right)^{-1} \tag{5.1.22}$$

$$f_2(q^2) = \frac{\mu_p - \mu_n}{2m_p} \left(1 + \frac{q^2}{m_v^2} \right)^{-2} \left(1 + \frac{q^2}{4m_p^2} \right)^{-1} \tag{5.1.23}$$

$$g_1(q^2) = 1.25 \left(1 + \frac{q^2}{m_A^2} \right)^{-2}, \tag{5.1.24}$$

$$g_3(q^2) = \frac{2m_p}{m_\pi^2 + q^2} g_1(q^2) \approx \frac{2.5 m_p}{m_\pi^2 + q^2} \left(1 + \frac{q^2}{m_A^2} \right)^{-2}. \tag{5.1.25}$$

In the CVC-hypothesis the "weak charge" of the nucleon is assumed to be constant despite the virtual particle production resulting from strong interactions. This hypothesis is based on analogy with the electrodynamics, where the electric charge of a proton does not change under similar circumstances. So it is assumed that $f_1(0) = 1$, in agreement with experiment. On the other hand, the difference between the magnetic momenta of a proton μ_p and a

---

[5] A conjugate bispinor refers to a particle birth, an ordinary one to a particle annihilation. In (5.1.20) ψ_ν is used since the antineutrino birth in (5.1.4a) is equivalent to annihilation of its antiparticle, i.e. neutrino.

[6] All reactions which yield nucleons without forming heavier hadrons, i.e. all reactions with $\epsilon_{\nu,e} \leq 300$ MeV, are called quasi-elastic.

neutron μ_n yields a form-factor $f_2(q^2)$. According to the CVC-hypothesis, the weak form-factors are taken to be equal to the electromagnetic ones. The PCAC-hypothesis allows us to relate $g_1(q^2)$ to $g_3(q^2)$ in (5.1.25), but the value $g_1(0) = 1.25$ (axial current is not conserved) is determined from experiments. The functions (5.1.22–25) include the following constants

$$\mu_\mathrm{p} = 1.79, \quad \mu_\mathrm{n} = 1.91,$$
$$m_\pi = 273\, m_\mathrm{e} = 139.5 \text{ MeV},$$
$$m_v^2 = 0.71 \text{ GeV}^2, \tag{5.1.26}$$
$$m_A = \begin{cases} 1.34 \pm 0.05 \text{ GeV} & [469], \\ 0.95 \pm 0.12 \text{ GeV} & [86]. \end{cases}$$

Obviously, the accuracy of m_A measurements is not better than 30%, whereas the accuracy for m_v is several per cent [86]. With (5.1.20–21), the matrix element H_β becomes

$$H_\beta = \Sigma_\sigma \frac{g_w}{\sqrt{2}} \int_{V_n} j_{a,i} j_{l,i} d^3 x, \tag{5.1.27}$$

where Σ_σ implies the summation and averaging over spin states, and the integration is performed over the nuclear volume where the nuclear wave functions are localized. Experimental data on neutron β-decay and muon decay give the following values for the weak coupling constant [86] (see below Sects. 5.1.3–4):

$$10^{49} G_W = \begin{cases} 1.4358 \text{ erg cm}^3 & (\mu - \text{decay}) \quad [172] \\ 1.4335 \text{ erg cm}^3 & (\mu - \text{decay}) \quad [86] \\ 1.4132 \text{ erg cm}^3 & (\beta - \text{decay}) \quad [86]. \end{cases} \tag{5.1.28}$$

The normalization of wave functions is different for free particles and nucleons localized in nuclei. For the latter, one uses a standard normalization to unity over the nuclear volume V_n

$$\int_{V_n} (\psi^+ \psi)_{\mathrm{n,p}} d^3 x = 1. \tag{5.1.29}$$

The wave functions of leptons and other free particles are normalized in such a way as to obtain one particle per unit volume (see Sect. 5.1.3–4).

The slight difference between constants for μ- and β-decays in (5.1.28) arises from the fact that the neutron β-decay is determined only by the strangeness-conserving part of the weak hadronic current. Note that in weak interactions involving nuclei a much larger uncertainty in results arises from the fact that the nucleon wave functions inside nuclei are not sufficiently known because of the absence of an accurate strong interaction theory.

5.1.3 Calculation of the Square of Matrix Element and Muon-Decay Probability

The small difference in constants in (5.1.28) supports the universal weak interaction theory which implies that the coupling constant G_W, and $(V-A)$-like interaction are the same for all elementary particles, leptons and quarks[7] [172]. The interaction type of neutral currents discussed in Sect. 5.1.7 is slightly modified. A matrix element with two lepton currents and with no form factors is the simplest to evaluate. An example of this kind of process is given by the muon decay

$$\mu^- \to e^- + \nu_\mu + \tilde{\nu}_e. \tag{5.1.30}$$

The probability for a muon to decay is given by

$$W_\mu = \frac{2\pi}{\hbar} \int |H|_\mu^2 \, \delta \left(\sum_\lambda \epsilon_\lambda \right) dN_e \, dN_{\tilde{\nu}_e} \, dN_{\nu_\mu} \tag{5.1.31}$$

with the matrix element

$$
\begin{aligned}
H_\mu &= \sum_\sigma \frac{G_W}{\sqrt{2}} \int j_{\bar{e}\nu_e, i} \, j_{\bar{\nu}_\mu \mu, i} \, d^3x \\
&= \sum_\sigma \frac{G_W}{\sqrt{2}} \int \left[\bar{\psi}_e \gamma_i (1 + \gamma_5) \psi_{\nu_e} \right] \left[\bar{\psi}_{\nu_\mu} \gamma_i (1 + \gamma_5) \psi_\mu \right] d^3x
\end{aligned}
\tag{5.1.32}
$$

and phase volumes analogous to (5.1.7). According to Dirac's equation (5.1.8), the wave functions for free leptons with given momentum p may be written as

$$\psi(p_i, x_i) = \frac{\sqrt{c}}{\sqrt{2\epsilon}} u(p) e^{-i \frac{p_i x_i}{\hbar}}, \tag{5.1.33}$$

where p_i is the lepton four-momentum, $p \equiv p_i = \left(-\boldsymbol{p}, \frac{\epsilon}{c} \right)$, $u(p)$ is a bispinor independent on space coordinates. Substituting (5.1.33) into (5.1.8), we obtain for $u(p)$

$$(p_i \gamma_i - mc) u(p) = 0, \tag{5.1.34}$$

where the bispinor $u(p)$ satisfies the relations

$$\bar{u} u = 2mc,$$

$$\bar{u} \gamma_i u = 2p_i.$$

Here we use the normalization condition for free particles. As a result, we write for the electron the solution of (5.1.34) in the form

---

[7] The quark structure of nucleons should be taken into account at high energies $E \geq 1000$ MeV but we do not consider these here.

$$u = u_{e^-} = \begin{pmatrix} \sqrt{\frac{\epsilon}{c} + m_e c} \times W_e \\ \sqrt{\frac{\epsilon}{c} - m_e c} \times (\boldsymbol{n} \cdot \boldsymbol{\sigma}) \, W_e \end{pmatrix}, \tag{5.1.35}$$

where $\boldsymbol{n} = \boldsymbol{p}/|\boldsymbol{p}|$ is a unit vector along the momentum direction,

$$\boldsymbol{\sigma} = \left\{ \begin{pmatrix} 0 & 1 \\ 1 & 0 \end{pmatrix} \begin{pmatrix} 0 & -i \\ i & 0 \end{pmatrix} \begin{pmatrix} 1 & 0 \\ 0 & -1 \end{pmatrix} \right\} \tag{5.1.36}$$

is a vector composed of the Pauli matrices σ_k, W_e is an arbitrary two-component spinor satisfying the equality

$$W_e^+ W_e = 1 \tag{5.1.37}$$

and characterizing polarization[8].

A non-polarized electron is characterized by the density matrix Λ_e yielded by summation over spin states and representing the square 4-matrix

$$\Sigma_\sigma u\bar{u} = \Lambda_e = (p_i \gamma_i + m_e c I). \tag{5.1.38}$$

Averaging over spin states of non-polarized muon yields $\frac{1}{2}\Lambda_\mu$ matrix with Λ_μ analogous to Λ_e from (5.1.38). $|H|_\mu^2$ is evaluated with the aid of the Fierz relation [171] of the wave functions

$$\left[\bar{\psi}_e \gamma_i (1 + \gamma_5) \psi_{\nu_e} \right] \left[\bar{\psi}_{\nu_\mu} \gamma_i (1 + \gamma_5) \psi_\mu \right]$$
$$= - \left[\bar{\psi}_e \gamma_i (1 + \gamma_5) \psi_\mu \right] \left[\bar{\psi}_{\nu_\mu} \gamma_i (1 + \gamma_5) \psi_{\nu_e} \right]. \tag{5.1.39}$$

corresponding to the symmetry of the matrix element (5.1.32). To evaluate $|H|_\mu^2$, the expression is also needed for the complex conjugate current which, for the case of (5.1.20), has the form

$$j_{l,i}^* = \left[\bar{\psi}_e \gamma_i (1 + \gamma_5) \psi_{\nu_e} \right]^* = \bar{\psi}_{\nu_e} \gamma_i (1 + \gamma_5) \psi_e. \tag{5.1.40}$$

Define the matrices

$$A_{ik} = \left[\bar{u}_e \gamma_i (1 + \gamma_5) \, u_\mu \right] \left[\bar{u}_\mu \gamma_k (1 + \gamma_5) \, u_e \right],$$
$$B_{ik} = \left[\bar{u}_{\nu_\mu} \gamma_i (1 + \gamma_5) \, u_{\nu_e} \right] \left[\bar{u}_{\nu_e} \gamma_k (1 + \gamma_5) \, u_{\nu_\mu} \right], \tag{5.1.41}$$

where the bispinors u_λ and ψ_λ are related to each other similarly to (5.1.33). Using (5.1.38–41), summing over the permitted spin states of the resulting

---

[8] Besides the bispinor from (5.1.33), (5.1.35) corresponding to a free electron, there is another solution of equation (5.1.8) corresponding to a free positron: $\psi_{e^+} = \frac{\sqrt{c}}{\sqrt{2\epsilon}} \bar{u}_{e^+} \exp\left(\frac{ip_i x_i}{\hbar}\right)$, where $\bar{u}_{e^+} u_{e^+} = -2m_e c$, $\bar{u}_{e^+} \gamma_i u_{e^+} = 2p_i$. The bispinor $u_{e^+} = u(-p)$ may be obtained from (5.1.35) by changing the sign of m_e and substituting $W_e \Rightarrow (\boldsymbol{n}\boldsymbol{\sigma}) W_{e^+}$ with $(\boldsymbol{n}\boldsymbol{\sigma})^2 = 1$.

particles and averaging over the muon spin states in (5.1.32), we obtain for a non-polarized muon[9]

$$|H|_\mu^2 = \frac{G_W^2 c^4 A_{ik} B_{ik}}{32 \epsilon_{\nu_\mu} \epsilon_{\bar{\nu}_e} \epsilon_e \epsilon_\mu} (2\pi\hbar)^3 \delta \left(\Sigma_\lambda \boldsymbol{p}_\lambda\right). \tag{5.1.42}$$

Using (5.1.42) reduces the evaluation of $|H|_\mu^2$ to obtaining the traces of matrices which are products of Dirac's γ-matrices. The evaluation is given in detail in [171]. As a result, we have

$$|H|_\mu^2 = 4G_W^2 c^4 \frac{\left(p_e p_{\nu_\mu}\right)\left(p_\mu p_{\bar{\nu}_e}\right)}{\epsilon_{\nu_\mu} \epsilon_{\bar{\nu}_e} \epsilon_e \epsilon_\mu} (2\pi\hbar)^3 \delta \left(\Sigma_\lambda \boldsymbol{p}_\lambda\right), \tag{5.1.43}$$

with the scalar product defined in (5.1.10). Inserting (5.1.43) into (5.1.31), we integrate over momenta with $f_e = f_{\bar{\nu}_e} = f_{\nu_\mu} = 0$ in (5.1.7), which corresponds to the muon decay in vacuum. We follow here [171] and first integrate over the ν_μ and $\bar{\nu}_e$ momenta to evaluate the integral

$$I_{ik} = \int \frac{p_{\nu_\mu,i} p_{\bar{\nu}_e,k}}{\epsilon_{\nu_\mu} \epsilon_{\bar{\nu}_e}} \delta \left(\boldsymbol{p}_{\nu_\mu} + \boldsymbol{p}_{\bar{\nu}_e} - \boldsymbol{q}\right) \delta \left(\epsilon_{\nu_\mu} + \epsilon_{\bar{\nu}_e} - cq_0\right)$$
$$\times d^3 p_{\nu_\mu} d^3 p_{\bar{\nu}_e}, \quad q_j = p_{\mu,j} - p_{e,j}. \tag{5.1.44}$$

For particles with zero rest-mass and $p_i^2 = 0$ the equality [171]

$$\int \frac{d^3 p_1 d^3 p_2}{\epsilon_1 \epsilon_2} \delta \left(\boldsymbol{p}_1 + \boldsymbol{p}_2 - \boldsymbol{q}\right) \delta \left(\epsilon_1 + \epsilon_2 - cq_0\right) = \frac{2\pi}{c^3} \tag{5.1.45}$$

holds. Taking I_{ik} in the form

$$I_{ik} = Aq^2 \delta_{ik} + Bq_i q_k \tag{5.1.46}$$

and using

$$I_{ik} \delta_{ik} = \pi \frac{q^2}{c^3}, \quad I_{ik} q_i q_k = \frac{\pi q^4}{2c^3}, \tag{5.1.47}$$

gives[10]

$$I_{ik} = \frac{\pi}{6c^3} \left(q^2 \delta_{ik} + 2q_i q_k\right). \tag{5.1.48}$$

Using (5.1.48), (5.1.43), and (5.1.44) we find from (5.1.31)

---

[9] Upon integrating over $d^3 x$ with (5.1.33) and (5.1.32) taken into account, a factor of $(2\pi\hbar)^3 \delta \left(\Sigma_\lambda \boldsymbol{p}_\lambda\right)$ arises from the momentum conservation. The second δ-function results from raising to square and yields the normalization volume which cancels out from the expression for the decay probability [172]. Here we use the equality $\int_{-\infty}^{\infty} e^{-ikx} dx = 2\pi\delta(k)$.

[10] Calculations (5.1.45) and (5.1.47) are done in a coordinate system where $\boldsymbol{p}_{\nu_\mu} = -\boldsymbol{p}_{\bar{\nu}_e}$ with account of $\boldsymbol{p}_{\nu_\mu} + \boldsymbol{p}_{\bar{\nu}_e} = \boldsymbol{q}$, $p_{\nu_\mu}^2 = p_{\bar{\nu}_e}^2 = 0$.

$$W_\mu = \frac{2\pi}{\hbar} \frac{4G_W^2 c^4}{(2\pi\hbar)^6} \frac{\pi}{6c^3} \int \frac{p_{ei}p_{\mu k}}{\epsilon_e \epsilon_\mu} \left(q^2 \delta_{ik} + 2q_i q_k \right) d^3 p_e. \tag{5.1.49}$$

Neglecting the electron mass compared to the muon mass, we have $p_e^2 \approx 0$, and considering the muon in a state of rest ($p_\mu = (0, m_\mu c)$), we have takint into account (5.1.10)

$$q_i = p_{\mu i} - p_{ei} = \left(\mathbf{p}_e, m_\mu c - \frac{\epsilon_e}{c} \right),$$

$$q^2 = (m_\mu c)^2 - 2m_\mu \epsilon_e,$$

$$p_{ei}p_{\mu i} = p_{ei}q_i = m_\mu \epsilon_e,$$

$$p_{\mu i}q_i = m_\mu c \left(m_\mu c - \frac{\epsilon_e}{c} \right). \tag{5.1.50}$$

Inserting (5.1.50) into (5.1.49) yields

$$
\begin{aligned}
W_\mu &= \frac{cG_W^2}{3(2\pi)^4\hbar^7} \int \left[3(m_\mu c)^2 - 4m_\mu \epsilon_e \right] \frac{d^3 p_e}{c^2} \\
&= \frac{G_W^2}{3(2\pi)^4\hbar^7 c^4} 4\pi \int_0^{m_\mu c^2/2} \left[3(m_\mu c)^2 - 4m_\mu \epsilon_e \right] \epsilon_e^2 \, d\epsilon_e \\
&= \frac{G_W^2 m_\mu^5 c^4}{192\pi^3 \hbar^7}.
\end{aligned}
\tag{5.1.51}
$$

Using the muon lifetime $\frac{1}{W_\mu} = \tau_\mu \approx 2.1971 \times 10^{-6}$ s obtained from experiments and recalling (5.1.3), we find from (5.1.51) the weak constant

$$G_W = \left[\frac{192\pi^3\hbar^7}{m_\mu^5 \tau_\mu c^4} \right]^{1/2} = 1.433 \times 10^{-49} \text{ erg cm}^3 \tag{5.1.52}$$

given in (5.1.28).

5.1.4 Calculation of $|H|_b^2$ and the Probability for Neutron and Nuclei to Decay; Influence of Magnetic Field

In β-decays $q \ll m_v$, so we may neglect in (5.1.21) the form factors f_2 and g_3 from (5.1.23) and (5.1.25) and use f_1 and g_1 from (5.1.22) and (5.1.24) at $q = 0$: $f_1(0) = 1$, $g_1(0) = g_A = 1.25$. Consider first the neutron decay. In this case both neutron and proton are free particles, and their wave functions are similar to the leptonic (5.1.33) and (5.1.35).

The matrix element for neutron decay becomes[11]

$$H_n = \sum_\sigma \frac{G_W}{\sqrt{2}} \int \left[\bar{\psi}_p \gamma_i (1 + g_A \gamma_5) \psi_n \right] \left[\bar{\psi}_e \gamma_i (1 + \gamma_5) \psi_\nu \right] d^3 x. \tag{5.1.53}$$

---

[11]The same symbol G_W stands for both β^- and μ-decay constants.

The nucleons may be regarded as non-relativistic to a great accuracy, so the relevant bispinors analogous to (5.1.35) contain only two non-zero components

$$\bar{u}_p = \left(\sqrt{m_p c}\, W_p^+, 0\right), \quad u_n = \begin{pmatrix} \sqrt{m_n c}\, W_n \\ 0 \end{pmatrix}. \tag{5.1.54}$$

Here, W_p and W_n are the spinors similar to W_e in (5.1.37)[12]. With account of (5.1.9), (5.1.36), and (5.1.54), the elements of the hadronic current $j_{a,i}$ become, in the non-relativistic limit

$$\bar{u}_p \gamma_4 u_n = c\sqrt{m_p m_n}\, W_p^+ W_n, \tag{5.1.55}$$
$$\bar{u}_p \gamma u_n = 0,$$

$$\bar{u}_p \gamma_4 \gamma_5 u_n = 0,$$

$$\bar{u}_p \gamma \gamma_5 u_n = c\sqrt{m_p m_n}\, \left(W_p^+, 0\right) \begin{pmatrix} 0 & \sigma \\ -\sigma & 0 \end{pmatrix} \begin{pmatrix} 0 & -I \\ -I & 0 \end{pmatrix} \begin{pmatrix} W_n \\ 0 \end{pmatrix} \tag{5.1.56}$$
$$= -W_p^+ \sigma W_n c\sqrt{m_p m_n}.$$

The square of the matrix element (5.1.53) will be [171]

$$|H|_n^2 = \sum_\sigma \frac{G_W^2 c^2}{32\epsilon_\nu \epsilon_e} \left\{ \left| \left(W_p^+ W_n\right) \left(\bar{u}_e \gamma_4 (1+\gamma_5) u_\nu\right)\right|^2 \right.$$
$$\left. + g_A^2 \left| \left(W_p^+ \sigma W_n\right) \left(\bar{u}_e \gamma (1+\gamma_5) u_\nu\right)\right|^2 \right\} (2\pi\hbar)^3 \delta\left(\Sigma_\lambda p_\lambda\right). \tag{5.1.57}$$

The density matrix for a non-polarized proton is the sum over spin states

$$\sum_\sigma u_p \bar{u}_p = \Sigma_\sigma m c W_p W_p^+ = \Lambda_p = mc\gamma_4 + mcI = 2mc \begin{pmatrix} I & 0 \\ 0 & 0 \end{pmatrix}, \tag{5.1.58}$$

On averaging similarly over spins, we have for neutrons

$$\sum_\sigma{}' u_n \bar{u}_n = \Sigma_\sigma' m c W_n W_n^+ = \frac{1}{2} \Lambda_n = mc \begin{pmatrix} I & 0 \\ 0 & 0 \end{pmatrix}. \tag{5.1.59}$$

Here, Σ' means averaging over spin states. Substituting (5.1.38), (5.1.58), and (5.1.59) into (5.1.57) and evaluating the traces of products of the matrices γ and σ, we obtain [171][13]

$$|M|^2 = G_W^2 \left[\left(1 + 3g_A^2\right) + \frac{c^2 p_e p_{\bar{\nu}}}{\epsilon_e \epsilon_{\bar{\nu}}} \left(1 - g_A^2\right) \right]. \tag{5.1.60}$$

Note that after the replacements $p_\mu \Rightarrow p_n$, $p_{\nu_\mu} \Rightarrow p_p$ in the non-relativistic limit for n and p (5.1.43) corresponds to (5.1.60) at $g_A = 1$ and $|M|^2 = 4G_W^2$.

---

[12]With the normalization condition $W_p^+ W_p = 2$, $W_n^+ W_n = 2$.
[13]$|M|^2$ differs from $|H|^2$ by the absence of the factor $(2\pi\hbar)^3 \delta\left(\sum_\lambda p_\lambda\right)$, see (5.1.57).

Substituting (5.1.60), (5.1.57) into (5.1.6), ignoring the nuclear recoil and integrating over angles gives

$$W_n = \frac{2\pi}{\hbar} \int |M|_n^2 \frac{(4\pi)^2}{2\pi\hbar)^6} \delta(\epsilon_e + \epsilon_{\bar{\nu}} - \Delta_{np}) p_e^2 \, dp_e \, p_{\bar{\nu}}^2 \, dp_{\bar{\nu}}. \tag{5.1.61}$$

The second term in (5.1.60) is not included in (5.1.61), turning into zero after angular integration in momentum space[14]. With $p_{\bar{\nu}} = \epsilon_{\bar{\nu}}/c$, $p_e = \frac{1}{c}\sqrt{\epsilon_e^2 - m_e^2 c^4}$, (5.1.61) gives

$$
\begin{aligned}
W_n &= \frac{G_W^2 \left(1 + 3g_A^2\right)}{2\pi^3 \hbar^7 c^6} \int_{m_e c^2}^{\Delta} \sqrt{\epsilon_e^2 - m_e^2 c^4} \, (\Delta - \epsilon_e)^2 \epsilon_e \, d\epsilon_e \\
&= \frac{G_W^2 \left(1 + 3g_A^2\right) m_e^5 c^4}{2\pi^3 \hbar^7} \int_1^{\delta} \sqrt{u^2 - 1} \, (\delta - u)^2 u \, du,
\end{aligned}
\tag{5.1.62}
$$

$$u = \epsilon_e/m_e c^2, \quad \delta = \Delta/m_e c^2 = 2.531.$$

The integrand in (5.1.62) determines the electron spectrum in neutron β-decay. The integral (5.1.62) can be done analytically and is called the Fermi function F_0 with no Coulomb interaction:

$$
\begin{aligned}
F_0(u) &= \int_1^u \sqrt{x^2 - 1} \, (\delta - x)^2 dx = \frac{1}{60}\sqrt{u^2 - 1}\left[12u^4 - 30u^3\delta \right. \\
&\quad \left. + 4\left(5\delta^2 - 1\right)u^2 + 15u\delta - 4\left(5\delta^2 + 2\right)\right] + \frac{\delta}{4}\ln\left(u + \sqrt{u^2 - 1}\right), \\
F_0(\delta) &= F_0 = \frac{1}{60}\sqrt{\delta^2 - 1}\left(2\delta^4 - 9\delta^2 - 8\right) \\
&\quad + \frac{1}{4}\delta\ln\left(\delta + \sqrt{\delta^2 - 1}\right).
\end{aligned}
\tag{5.1.63}
$$

The neutron half-life $t_{1/2}$ is

$$t_{1/2} = \frac{\ln 2}{W_n}. \tag{5.1.64}$$

The quantity

$$(Ft_{1/2})_n = \frac{2\pi^3 \hbar^7 \ln 2}{G_W^2 \left(1 + 3g_A^2\right) m_e^5 c^4} \approx 1083 \tag{5.1.65}$$

---

[14]If baryons or massive leptons are non-relativistic in beta-interactions, then

$$|M|^2 = G_W^2(\tilde{g}_V^2 + 3\tilde{g}_A^2 + o.t.) g_V^2 (1 + \delta_{mm'})^2 \tag{5.1.60a}$$

at

$$H = \frac{G_W}{\sqrt{2}} \int \left[g_V \overline{\psi}_\nu \gamma_i (1 + \gamma_5) \psi_{\nu'}\right] \cdot \left[\overline{\psi}_m \gamma_i (\tilde{g}_V + \tilde{g}_A \gamma_5) \psi_{m'}\right] (1 + \delta_{mm'}) \, d^3x,$$

where m are the baryons or massive leptons, o.t. are the terms turning into zero upon averaging for non-polarized particles. We may always write H in the form (5.1.60a) by reason of (5.1.39).

depends[15] only on physical constants and the constant part of the matrix element rather than on phase relations, and is determined by experimental measurements of the half-life $t_{1/2}$ and by calculation of the Fermi function F.

In neutron β-decay the neutron and proton are located in the same point so that their wave functions overlap fully and $|H|^2$ is maximized. This corresponds to the equality $\int_{V_n}(\psi_n^+\psi_p)\,d^3x = 1$ in (5.1.27). If a nuclear β-decay occurs where nucleons cannot be treated as free then the neutron and proton wave functions overlap only partially and $(Ft_{1/2})_{A,Z}$ is usually larger than $(Ft_{1/2})_n$. Sometimes, however, $(Ft_{1/2})_{A,Z}$ turns out to be less than $(Ft_{1/2})_n$ (see Table 5.1). The reason for this effect is that several neutrons may take part in the nuclear β-decay, e.g., $(Ft_{1/2})_{6He} < (Ft_{1/2})_n$ though the n and p wave functions overlap only partially; here two neutrons belonging to the nucleons ^{6}He and located on the second neutron shell take part in the β-decay. Measurements of $(Ft_{1/2})_{A,Z}$ provide data on nucleon wave functions inside nuclei.

In the magnetic field, electrons occupy discrete energy (Landau) levels for motion in the plane perpendicular to the field, with energy difference

$$\Delta\epsilon_{Ln} = n\hbar\frac{eB}{m_e c}. \tag{5.1.65a}$$

When this difference is of the order of the energy of beta decay $\Delta \sim m_e c^2$, the decay probability begins to depend on the value of B. This happens because of the change of the electron wave function, matrix element of decay, and due to the change of the electron phase space structure. The critical magnetic field corresponds to

$$\Delta\epsilon_{L1} = m_e c^2, \quad B_{cr} = \frac{m_e^2 c^3}{e\hbar} \simeq 4.4 \times 10^{13}\,Gs. \tag{5.1.65b}$$

The probability of the neutron decay in the strong magnetic field was calculated in [515*]:

$$W_{nB} = W_n\left[1 + 0.17\left(\frac{B}{B_{cr}}\right)^2 + ...\right] \quad \text{for } B \ll B_{cr}, \tag{5.1.65c}$$

and

$$W_{nB} = W_n\,0.77\left(\frac{B}{B_{cr}}\right) \quad \text{for } B > 2.7\,B_{cr} \tag{5.1.65d}$$

with W_n from (5.1.62). In a strongly relativistic plasma with $\epsilon_{Fe} \gg m_e c^2$ or $kT \gg m_e c^2$, or when $\Delta \gg m_e c^2$, the magnetic field begins to influence the beta processes at

$$B_c = \lambda B_{cr} \quad \text{with } \lambda = \frac{\epsilon_{Fe}}{m_e c^2}, \text{ or } \lambda = \frac{kT}{m_e c^2}, \text{ or } \lambda = \frac{\Delta}{m_e c^2}. \tag{5.1.65e}$$

Astrophysical phenomena, where beta processes at large B are important, are considered in Chap. 10, Vol. 2.

---

[15] Here G_W for β-decay from (5.1.28) is used, and the value $g_A = 1.25$, obtained by averaging the results of different experiments. The experimental value for neuron decay $(Ft_{1/2})_n=1187$ determines the lower value $g_A = 1.18$ [232].

Table 5.1. Data on some superallowed transitions

Transition	Spin	$t_{1/2}$, s	E_{max}, keV	$Ft_{1/2}$, s
$^1_0$n $\to$ $^1_1$H	1/2	11.7 min	782	1187
$^3_1$H $\to$ $^3_2$He	1/2	$3.87 \cdot 10^8$	18.65	1132
$^{11}_6$C $\to$ $^{11}_5$B	3/2	1224	968	4030
$^{13}_7$N $\to$ $^{13}_6$C	1/2	603	1202	4700
$^{15}_8$O $\to$ $^{15}_7$N	1/2	124	1739	4475
$^{17}_9$F $\to$ $^{17}_8$O	5/2	66	1748	2380
$^{23}_{12}$Mg $\to$ $^{23}_{11}$Na	3/2	12	3056	4480
$^{25}_{13}$Al $\to$ $^{25}_{12}$Mg	5/2	7.23	3239	4280
$^{27}_{14}$Si $\to$ $^{27}_{13}$Al	5/2	4.19	3793	4500
$^{35}_{18}$Ar $\to$ $^{35}_{17}$Cl	3/2	1.804	4448	5680
$^{41}_{21}$Sc $\to$ $^{41}_{20}$Ca	7/2	0.87	4940	2560
$^6_2$He $\to$ $^6_3$Li	$0 \to 1$	0.813	3500	808
$^{10}_6$C $\to$ $^{10}_5$B*	$0 \to 1$	19.1	2100	1700
$^{18}_{10}$Ne $\to$ $^{18}_9$F	$0 \to 1$	1.6	3200	794.4
$^{50}_{25}$Mn $\to$ $^{50}_{24}$Cr	$0 \to 1$	0.2857	6609	3082
$^{54}_{27}$Co $\to$ $^{54}_{26}$Fe	$0 \to 1$	0.1937	7229	2966

$$E_{max} = \Delta - m_e c^2$$

5.1.5 Classification of β-Decays; Selection Rules

The degree of overlap of nuclear wave functions before and after β-decay depends mostly on the relation between spins and parities of these nuclei. The detailed reasons for such a dependence may be found in special texts [213, 171, 69, 230, 86, 84], here we shall restrict ourselves to describing the results and their qualitative interpretation.

If exponentials in the lepton wave functions are replaced by unity then the lepton pair takes away no orbital angular momentum, and the total spin carried away equals either zero, or unity[16], positive or negative. The vector term (V) in the current $j_{l,i}$ in (5.1.20) is non-zero if both leptons in it have the same spin direction. For β-decay this corresponds to the zero total spin of the electron and antineutrino, so the V-interaction introduced by Fermi implies conservation of nuclear spin I and parity. Thus, the selection rule reads

$$\Delta|I| = 0, \quad |\Delta I| = 0, \quad \text{parity is conserved.} \qquad (5.1.66)$$

The pseudovector A-interaction in (5.1.20) suggested by Gamov and Teller (1936), changes nuclear spins, i.e. the sum of lepton spins in β-decay is equal to unity. The nuclear spin may change here by unity, but may also remain

---

[16]In units of $\hbar$.

a constant factor with a change in direction rendering their vector difference equal to unity. The latter possibility, takes place only for a non-zero spin of the initial nucleus. Thus, the Gamov–Teller selection rules read

$$\left.\begin{array}{l} \Delta|I| = \pm 1, \quad |\Delta I| = 1; \\ \Delta|I| = 0, \quad |\Delta I| = 1, \\ \quad \text{if} \quad I \neq 0. \end{array}\right\} \qquad \text{parity is conserved.} \qquad (5.1.67)$$

If the selection rules (5.1.66–67) are satisfied, the value of $Ft_{1/2}$ depends on the degree of overlap of the wave functions. If the overlap is large, as in neutron decay, the transition is called **superallowed**. The corresponding values $\lg(Ft_{1/2}) = 3.5 \pm 0.7$ [84], see Table 5.1. Such a transition occurs in relatively light nuclei. A good example is a mirror transition in odd-even nuclei, when the neutron-proton number of the initial nucleus equals the proton-neutron number of the final nucleus, say, the positron decay $^{17}_{9}\text{F}_8 \rightarrow ^{17}_{8}\text{O}_9$. Another kind of superallowed transition may occur in even nuclei with $A = 4n + 2$, when two neutrons of the outer shell take part in the decay. The above mentioned transition $^6_2\text{He}_4 \rightarrow ^6_3\text{Li}_3$ belongs to this kind. The positron decays in mirror nuclei are caused by the Coulomb repulsion energy of protons in nucleus.

Transitions with essentially higher values of $lg(Ft_{1/2}) = 5.7 \pm 2$ satisfying the above selection rules are called **allowed** transitions. The allowed and superallowed transitions are characterized by the shape of the electron spectrum determined by the integrand in (5.1.62) (the effect of Coulomb corrections is discussed below).

Nuclear β-decays that do not satisfy the selection rules (5.1.66–67) may occur only via a departure from unity of the lepton wave function exponents inside the nucleus equivalent to taking away the orbital momentum by the lepton pair. A small factor arises now in the expression for the transition probability

$$(R_n/\lambda_1)^2 \approx 10^{-3} \div 10^{-4}, \qquad (5.1.68)$$

where R_n is the nuclear size, $\lambda_1 = \hbar/p_1$ is the characteristic size of periodicity of the lepton wave function. The transitions due to the first term in the exponent expansion are called once-forbidden and have $\lg(Ft_{1/2}) = 7.5 \pm 2$. The lepton pair takes away here not only spins but a unit orbital angular momentum as well. The selection rules for these transitions are

$$
\begin{array}{lll}
\Delta|I| = 0, \pm 1, & \text{parity changes,} & \\
|\Delta I| = 1, & \text{transitions } 0 \rightarrow 0 \text{ are forbidden} & \text{(F)} \quad (5.1.69) \\
\Delta|I| = 0, \pm 1, \pm 2, & \text{parity changes,} & \\
|\Delta I| = 0, 2, & \text{transitions with } |I_1| + |I_2| < 2 & \\
& \text{are forbidden for } |\Delta I| = 2 & \text{(G-T)}. \quad (5.1.70)
\end{array}
$$

If the change in spin exceeds by unity the degree of forbiddance, the transition is then called unique. For a transition with $\Delta I = \pm 2$ in (5.1.70) the equality

$\lg(Ft_{1/2}) = 8.5 \pm 1$ holds. The electron spectra in once-forbidden transitions are determined [23] by the integrand in (5.1.62) times

$$(p_e + p_\nu)^2 \sim p_e^2 + p_\nu^2 \sim (u^2 - 1) + (\delta - u)^2. \tag{5.1.71}$$

Relativistic and the Coulomb field effects lead to a shift of the states $p_{1/2}$ and $s_{1/2}$ with $j = 1/2$ by a value of $\sim \alpha Z$. This causes the contribution to the matrix element to come from not only the second term of the exponent expansion $\sim \frac{1}{\hbar} p_e R_n$, but also from the constant value of the lepton wave function $\sim \alpha Z$. The decay probability becomes small as $\sim \alpha^2 Z^2$ while the electron spectrum at $\Delta|I| = 0, \pm 1$ is still close to allowed.

The matrix elements of n-fold forbidden transitions are determined by the n-th term in the expansion of the lepton exponent and obey the following selection rules:

$\Delta|I| = 0, \pm 1, \ldots \pm n$ (F);

$|\Delta I| = n$, transitions with $|I_1| + |I_2| < n$ are forbidden;

$\Delta|I| = 0, \pm 1, \ldots \pm n \pm (n + 1)$ (G-T);   (5.1.72)

$|\Delta I| = n \pm 1$, where transitions with $|I_1| + |I_2| < n \pm 1$ are forbidden,

parity $\pi_2 = (-1)^n \pi_1$.

The parity is conserved here at $\pi_2 = \pi_1$ and changes at $\pi_2 = -\pi_1$. Each additional degree of forbiddance increases $Ft_{1/2}$ by 3 or 4 orders of magnitude. Data on known nuclear *beta*-decay transitions are given in [99].

5.1.6 Coulomb Corrections

The Coulomb interaction between nuclei and charged leptons results in deviations of the distribution wave function from the free one, enhancing the probability for electron nuclear decay via an increase in the electron wave function inside a nucleus and, accordingly, reducing the probability for positron decay. The correction factor for Coulomb interaction $\Phi(Z, \epsilon)$ may be found similarly to the Coulomb tunnel transition factor in nuclear reactions (see Chap. 4). For nonrelativistic electrons and positrons $e^\mp$ we have

$$\Phi_\mp(Z, \epsilon) = \frac{|\psi_e(0)|^2}{|\psi_e(0)|^2_{\text{free}}} = \frac{2\pi\eta}{|\exp(\mp 2\pi\eta) - 1|},$$

$$\eta = \frac{Ze^2}{\hbar v} = \alpha Z \frac{c}{v}. \tag{5.1.73}$$

Here Z is the charge of the final nucleus. For relativistic $e^\mp$ functions, $\frac{v}{c}\Phi_\mp = \frac{pc}{\epsilon}\Phi_\mp = 2\pi\alpha Z(\Phi_\mp/2\pi\eta)$ are tabulated (see [363]). To take into account the Coulomb interaction in nuclear β-decay, it is sufficient to multiply the integrand in the relation for Fermi function F by $\Phi_\mp(Z, \epsilon)$, so that for allowed transitions we have

$$F_{\mp} = 2\pi\alpha Z \left\langle \frac{\Phi_{\mp}}{2\pi\eta} \right\rangle \left(\frac{\delta^5}{30} - \frac{\delta^2}{3} - \frac{\delta}{2} - \frac{1}{5} \right). \tag{5.1.74}$$

Integration with the aid of tables for $Z = 26$ gives $\langle \Phi_-/2\pi\eta \rangle = 1.6$, $\langle \Phi_+/2\pi\eta \rangle = 0.5$. The electron factor in tables changes by $\sim 10\%$ and the positron by $\sim 50\%$ in the range of $e^{\mp}$ energies essentially determining the β-decay rate [93]. Note that the Coulomb corrections though significantly changing the beta-electron spectrum at low energies have no considerable effect on the β-decay rate and are therefore often ignored in astrophysical calculations where the Fermi function F_0 is used.

5.1.7 Neutral Currents

In leptonic currents similar to (5.1.20), that describe leptonic and nuclear decays, and in the hadronic current (5.1.21) contributing to the β-decay probability the charges of the initial and final particles are different, therefore these currents are called charged. The possibility for the neutral currents with the same charge of the initial and final particles to exist was suggested by Bludman in 1958, but only after the development of the Weinberg–Salam theory in 1967, unifying the weak and electromagnetic interactions, and subsequent experimental discovery of the muon neutrino scattering by electrons in 1973 was the existence of neutral currents universally recognized [231]. Since the electron, muon, τ-lepton and their neutrinos have different conserving leptonic charges, the neutral currents can only convert particles into themselves. For example, the cross-section for the reaction

$$\nu_e + \nu_\mu \to \nu_e + \nu_\mu \tag{5.1.75}$$

is determined only by the product of currents $(\bar{\psi}_{\nu_e} O_i \psi_{\nu_e})(\bar{\psi}_{\nu_\mu} O_i \psi_{\nu_\mu})$, while currents of $(\bar{\psi}_{\nu_e} O_i \psi_{\nu_\mu})$-type are absent. Some processes, such as the scattering

$$\nu_e + e \to \nu_e + e \tag{5.1.76}$$

involve both the charged currents $(\bar{\psi}_{\nu_e} O_i \psi_e)(\bar{\psi}_e O_i \psi_{\nu_e})$ and neutral currents $(\bar{\psi}_{\nu_e} O_i \psi_{\nu_e})(\bar{\psi}_e O_i \psi_e)$. However, the cross-section for the reaction

$$\nu_\mu + e \to \nu_\mu + e \tag{5.1.77}$$

depends on only the neutral currents $(\bar{\psi}_{\nu_\mu} O_i \psi_{\nu_\mu})(\bar{\psi}_e O_i \psi_e)$, since the charged currents $(\bar{\psi}_{\nu_\mu} O_i \psi_e)$ or $(\bar{\psi}_e O_i \psi_{\nu_\mu})$ are absent because of a difference in leptonic charges of a muon neutrino and an electron.

It follows from the Weinberg–Salam theory confirmed by experiments that the neutral currents are produced not only by the left-handed operator $O_i^L = \gamma_i(1 + \gamma_5)$ corresponding to $(V - A)$-interaction, but also by the right-handed operator $O_i^R = \gamma_i(1 - \gamma_5)$ representing the contribution of $(V + A)$-interaction. Another property of the neutral currents consists of their dependence on elementary particle charges, therefore the universal character of the weak interaction should be treated in a more general way here. The neutral currents for leptons and quarks have the form [172]

$$j_i^{(n)} = \sum_\lambda [g_L^\lambda \bar\psi_\lambda \gamma_i (1 + \gamma_5)\psi_\lambda + g_R^\lambda \bar\psi_\lambda \gamma_i (1 - \gamma_5)\psi_\lambda],$$

$$g_L^\lambda = 1/2, \quad g_R^\lambda = 0 \quad \text{for} \quad \lambda = \nu_e, \nu_\mu, \nu_\tau;$$

$$g_L^\lambda = -\frac{1}{2} + \sin^2\theta_W, \quad g_R^\lambda = \sin^2\theta_W \quad \text{for} \quad \lambda = e, \mu, \tau; \qquad (5.1.78)$$

$$g_L^\lambda = \frac{1}{2} - \frac{2}{3}\sin^2\theta_W, \quad g_R^\lambda = -\frac{2}{3}\sin^2\theta_W \quad \text{for} \quad \lambda = u, c, t \quad \text{(quarks)};$$

$$g_L^\lambda = -\frac{1}{2} + \frac{1}{3}\sin^2\theta_W, \quad g_R^\lambda = \frac{1}{3}\sin^2\theta_W \quad \text{for} \quad \lambda = d, s, b \quad \text{(quarks)}.$$

Here θ_W is the Weinberg angle characterizing the mixing of weak and electromagnetic interactions. The term with $\sin^2\theta_W$ in $j_i^{(n)}$ is purely vector similarly to j_i^{em} in the electrodynamics. It has been determined in experiments [120] that

$$\sin^2\theta_W = 0.224 \pm 0.020. \qquad (5.1.79)$$

Neutral currents of nucleons[17] are obtained by summation of quark currents and including form-factors analogous to (5.1.22–25). For small transfer momenta (see, e.g. [73])

$$j_{p,i}^{(n)} = \bar\psi_p \Big[\frac{1}{2}(1 - 4\sin^2\theta_W)\gamma_i + \frac{1}{2}g_A\gamma_i\gamma_5\Big]\psi_p =$$

$$= \bar\psi_p \Big[\frac{1}{2}\gamma_i(1 + g_A\gamma_5) - 2\sin^2\theta_W\gamma_i\Big]\psi_p, \qquad (5.1.80)$$

$$j_{n,i}^{(n)} = -\frac{1}{2}\bar\psi_n\gamma_i(1 + g_A\gamma_5)\psi_n.$$

The matrix element should take into account the identity of the direct and reverse transitions in reactions described by the neutral currents (5.1.75-77). As a result, the products of different currents in the matrix element are increased by a factor of 2 [172, 231]. The product of identical currents is included in the matrix element only once because the identity of the direct and reverse reactions is compensated by a coefficient $1/2$ owing to the identity of the particles involved in the reaction as in the case for the scattering $\nu_e + \nu_e \to \nu_e + \nu_e$. Note that the product of the charged currents of the reaction (5.1.76) enters into the matrix element also once, since the coefficient $1/2$ arises here again owing to the identity of the currents in this product [172]. Thus, the matrix element for a β-reaction associated with neutral currents may be written as

$$H^{(n)} = \sum_\sigma \int \frac{G_W}{\sqrt{2}} \frac{2}{1 + \delta_{\lambda\theta}} j_{\lambda,i}^{(n)} j_{\theta,i}^{(n)} d^3x. \qquad (5.1.81)$$

---

[17]The quark structure of nucleons is $p = uud, n = udd, Q_u = Q_c = Q_t = 2/3, Q_d = Q_s = Q_b = -1/3$ in units of e.

If the reaction with neutral currents involves nuclei, the integration is to be done over the nuclear volume similar to (5.1.27).

For scattering reactions, it is more convenient to use cross-section σ instead of probability W. The differential quantities are related by $dW = jd\sigma$, where j is the particle flux density. In a laboratory frame of reference with particle 0 at rest, and incoming particle 1 moving at velocity v_1, $j = v_1$ according to the adopted normalization condition for wave functions. Expressing laboratory quantities in invariant terms

$$m_0c^2\epsilon_1 v_1 = m_0c^4|\boldsymbol{p}_1| = c^3\sqrt{(p_0p_1)^2 - p_0^2p_1^2} = c^3\sqrt{(p_0p_1)^2 - m_0^2m_1^2c^4},$$

and taking for W one of the formulae (5.1.6), (5.1.31) or (5.1.61) give [172] the scattering cross-section for the particles 0 and 1 (2 and 3 after scattering)

$$\sigma_\beta = \frac{2\pi}{\hbar}\int\frac{\epsilon_0\epsilon_1|H|_{01}^2 d\phi}{c^3\sqrt{(p_0p_1)^2 - m_0^2m_1^2c^4}},$$

$$d\phi = \frac{d^3p_2 d^3p_3}{(2\pi\hbar)^6}\delta(\sum_\lambda \epsilon_\lambda)(1 - f_2)(1 - f_3).$$

(5.1.82)

Here, $d\phi$ is the phase volume element for particles after scattering that includes the δ-function of energy. The absorption cross-section is written analogously to (5.1.82) with appropriate form of $d\phi$. Note that statistical weights here, similarly to (5.1.7), are not included in the definition of $d\phi$ being already included in the summation over spins in the matrix element.

Consider cross-sections for certain processes. For scattering of neutrinos by leptons, the square of the matrix element for charged current is obtained from (5.1.42) by a simple change of variables. The integration in (5.1.82) at $f_2 = f_3 = 0$ gives the following invariant cross-section for charged current [172]

$$\sigma_{e^-\nu_e}^{(ch)} = \sigma_{s0}\frac{s_{\nu_e}^2}{s_{\nu_e} + 1/2} = \begin{cases} \sigma_{s0}s_{\nu_e} & \text{at } s_{\nu_e} \gg 1 \\ 2\sigma_{s0}s_{\nu_e}^2 & \text{at } s_{\nu_e} \ll 1, \end{cases}$$

(5.1.83)

$$\sigma_{e^-\bar{\nu}}^{(ch)} = \frac{1}{3}\sigma_{s0}s_{\bar{\nu}_e}[1 - (1 + 2s_{\bar{\nu}_e})^{-3}] \approx \begin{cases} \frac{1}{3}\sigma_{s0}s_{\bar{\nu}_e} & \text{at } s_{\bar{\nu}_e} \gg 1 \\ 2\sigma_{s0}s_{\bar{\nu}_e}^2 & \text{at } s_{\bar{\nu}_e} \ll 1. \end{cases}$$

(5.1.84)

Here we use the following notation

$$\sigma_{s0} = \frac{2G_W^2 m_e^2}{\pi\hbar^4} = 0.876 \times 10^{-44}\text{cm}^2$$

$$s_{\nu_e} = u_e u_{\nu_e} - \boldsymbol{\pi}_e\boldsymbol{\pi}_{\nu_e}, \quad s_{\bar{\nu}_e} = u_e u_{\bar{\nu}_e} - \boldsymbol{\pi}_e\boldsymbol{\pi}_{\bar{\nu}_e},$$

$$u_e = \epsilon_e/m_ec^2, \quad u_{\nu_e} = \epsilon_{\nu_e}/m_ec^2, \quad u_{\bar{\nu}_e} = \epsilon_{\bar{\nu}_e}/m_ec^2,$$

$$\boldsymbol{\pi}_e = \boldsymbol{p}_e/m_ec, \quad \boldsymbol{\pi}_{\nu_e} = \boldsymbol{p}_{\nu_e}/m_ec, \quad \boldsymbol{\pi}_{\bar{\nu}_e} = \boldsymbol{p}_{\bar{\nu}_e}/m_ec.$$

(5.1.85)

In the rest frame of the electron ($u_e = 1$, $\boldsymbol{\pi}_e = 0$)

$$s_{\nu_e} = u_{\nu_e}, \quad s_{\bar{\nu}_e} = u_{\bar{\nu}_e}.$$

(5.1.86)

Note that at low energies $\sigma_{e^-\nu_e}^{(ch)} = \sigma_{e^-\bar{\nu}_e}^{(ch)}$ and at high energies $\sigma_{e^-\nu_e}^{(ch)} = 3\sigma_{e^-\bar{\nu}_e}^{(ch)}$.

Calculations analogous to those in Sect. 5.1.3, with use of (5.1.78–81) give a matrix element for neutral currents. According to [172, 86], the total cross-section is

$$\sigma_{e\nu}^{tot} \sim |H^{(ch)} + H^{(n)}|^2, \tag{5.1.87}$$

$$\sigma_{e^-\nu_e}^{tot} = d_{\nu_e}\sigma_{e^-\nu_e}^{(ch)}, \quad \sigma_{e^-\tilde{\nu}_e}^{tot} = d_{\tilde{\nu}_e}\sigma_{e^-\tilde{\nu}_e}^{(ch)},$$

$$\left.\begin{array}{l} d_{\nu_e} = \left(\dfrac{1}{2} + \sin^2\theta_W\right)^2 + \dfrac{1}{3}\sin^4\theta_W \approx 0.541 \\[3mm] d_{\tilde{\nu}_e} = \left(\dfrac{1}{2} + \sin^2\theta_W\right)^2 + 3\sin^4\theta_W \approx 0.675 \end{array}\right\} \text{ at } \quad s_{\nu_e}, s_{\tilde{\nu}_e} \gg 1, \tag{5.1.88}$$

$$d_{\nu_e} = d_{\tilde{\nu}_e} = \frac{1}{4}\left[\frac{3}{4} + \left(\frac{1}{2} + 2\sin^2\theta_W\right)^2\right] \approx 0.412 \text{ at } \quad s_{\nu_e}, s_{\tilde{\nu}_e} \ll 1.$$

Here we use (5.1.79); the neutral currents reduce the scattering cross-section by reason of their negative contribution to the matrix element. The scattering $(e^-\nu_\mu)$ in the reaction (5.1.77), is associated only with neutral currents and has the cross-section [172, 86]

$$\sigma_{e^-\nu_\mu} = d_{\nu_\mu}\sigma_{e^-\nu_e}^{(ch)}, \quad \sigma_{e^-\tilde{\nu}_\mu} = d_{\tilde{\nu}_\mu}\sigma_{e^-\tilde{\nu}_e}^{(ch)},$$

$$\left.\begin{array}{l} d_{\nu_\mu} = \left(-\dfrac{1}{2} + \sin^2\theta_W\right)^2 + \dfrac{1}{3}\sin^4\theta_W \approx 0.093 \\[3mm] d_{\tilde{\nu}_\mu} = \left(-\dfrac{1}{2} + \sin^2\theta_W\right)^2 + 3\sin^4\theta_W \approx 0.227 \end{array}\right\} \text{ at } \quad s_{\nu_\mu}, s_{\tilde{\nu}_\mu} \gg 1, \tag{5.1.89}$$

$$d_{\nu_\mu} = d_{\tilde{\nu}_\mu} = \frac{1}{4}\left[\frac{3}{4} + \left(-\frac{1}{2} + 2\sin^2\theta_W\right)^2\right] \approx 0.188 \text{ at } \quad s_{\nu_\mu}, s_{\tilde{\nu}_\mu} \ll 1.$$

In (5.1.89) $\sigma_{e^-\nu_e}^{(ch)}$ depends on u_{ν_μ} instead of u_{ν_e} (see (5.1.85)). We see from (5.1.89) that the cross-section for $(e^-\nu_\mu)$-scattering is about five times smaller than for $(e^-\nu_e)$-scattering.

The coherent elastic scattering of neutrinos off nuclei represents an interesting effect due to neutral currents; its amplitude is linear and its cross-section is quadratic in nuclear mass. Calculations in [73] give for the total scattering cross-section (including the axial current, see (5.1.60a) and (5.1.80))[18]

$$\sigma_{N\nu} = \frac{1}{2}\sigma_{s0}u_\nu^2\left\{\left[A\sin^2\theta_W + \frac{1}{2}(1 - 2\sin^2\theta_W)(N - Z)\right]^2 + \right.$$

---

[18]The sum of contributions from Z protons and $N = A - Z$ neutrons is calculated by use of (5.1.60a).

$$+ \frac{3}{4} g_A^2 (N-Z)^2 \xi_J \Big\} \approx \sigma_{s0} u_\nu^2 \left[0.025 \left(A + \frac{N-Z}{0.81} \right)^2 + \right.$$

$$\left. + \xi_J \times 0.568 (N-Z)^2 \right]. \tag{5.1.90}$$

Here we use (5.1.79), $g_A = 1.25$, $\xi_J = 0.1$ at $J = 1/2$, $\xi_J \ll 1$ at $J \neq 1/2$, $\xi_J = 1$ for nucleons. For non-relativistic nuclei $\sigma_{N\nu} = \sigma_{N\bar{\nu}}$ as for electrons in (5.1.83) and (5.1.84). Note that for the vector current the scattering is determined by the s-wave, therefore the value of the integral in (5.1.27) is maximum, as in the case of a neutron decay. At $J \neq 1/2$, in axial transition the p-wave is the first contribution into the elastic scattering and always $\xi_J \ll 1$. If $J = 1/2$ and in the presence of a single (with no pair) nucleon on the outer nuclear shell a scattering by the s-wave with a nucleon spinflip, when $\xi_J \sim 0.1$ is possible.

5.1.8 On the Weinberg–Salam Theory

The universal weak interaction theory (UWI), though in good agreement with experiments on nuclear β-decay and the decay of many other elementary particles, encounters significant difficulties for high energies. All calculations in the theory are made to the first order in perturbation theory. At high energies the cross-section for weak interactions increases (see (5.1.83)) and for $E \geq 100\,\text{GeV}$, in a centre-of-mass frame, it is necessary to include higher-order terms. However, divergences which arise in high-order terms are unavoidable here by means of renormalization, contrary to the case in quantum electrodynamics. Weinberg and Salam found a solution to this difficulty, developing a renormalizable weak interaction theory in 1967. Their theory incorporates the electromagnetic and weak interactions and is therefore called the **electroweak** interaction theory.

The UWI theory deals with a non-renormalizable four-fermion point interaction. In analogy with the electrodynamics, it has been assumed that the weak interaction is mediated by the vector boson W which must be massive with $M_W > 10\,\text{GeV}$ by reason of a short range of weak interactions [172]. However, an intermediate massive boson theory suffers the same drawback: it is not renormalizable.

The solution to this difficulty comes from assuming that the vector bosons mediating the interaction are initially massless. Their mass that we observe now arises from spontaneous symmetry breaking in coupling of the vector boson field with the independent scalar field φ. The potential of the scalar field is symmetric but has minima at non-zero values $\varphi = \pm\varphi_0$. The spontaneous symmetry breaking occurs when the field settles down in one of the minima with a non-zero expectation value $\varphi = \varphi_0$ in vacuum. One of the terms describing the coupling of the vector field D_i with the scalar field φ in the form $d\varphi^2 D_i D_i$ (d is the coupling constant), near $\varphi = \varphi_0$ exhibits behaviour as though bosons had mass m_D such as $m_D^2 = \frac{1}{2} d\varphi_0^2$. This mass is treated as the rest mass of the boson D_i. This mechanism for spontaneous breakdown of symmetry and acquiring mass was suggested by P. W. Higgs in 1964. If the mass is acquired through the Higgs mechanism, it becomes possible to construct a renormalizable theory with heavy bosons based on locally gauge invariant interactions examined by C. Yang and R. Mills in 1954. The Weinberg–Salam theory uses just this kind of interactions. The same properties are typical for quantum chromodynamics, a rapidly developing strong interaction theory, and diverse versions of unified theories which incorporate three or four (gravity included) interactions. Besides vector bosons, the Higgs field scalar bosons also acquire mass through spontaneous symmetry breaking.

The gauge invariance needed to make the theory renormalizable implies the existence of four intermediate bosons, i.e. besides the photon and $W^{\pm}$-bosons there must be an additional neutral boson which will be responsible for weak neutral currents.

We next discuss how a Lagrangian in the electroweak theory [86] should be constructed. In addition to an ordinary space, a weak isospin space is defined in which matrices τ identically equal to the Pauli spin matrices σ from (5.1.36) will be rotation matrices. Instead of the spinors ν_e and e we consider the isotopic doublet

$$L = \frac{1 + \gamma_5}{2} \begin{pmatrix} \nu_e \\ e \end{pmatrix} \tag{5.1.91}$$

and isotopic singlet[19]

$$R = \frac{1 - \gamma_5}{2} e, \tag{5.1.92}$$

whose isotopic components operate with matrices τ as ordinary spinors and scalars do with the Pauli matrices σ. The weak current j_1 and the purely vector electromagnetic current j_{em} may be represented in terms of L, R, τ as

$$\begin{aligned} j_1^- &= \bar{e}\gamma_i(1 + \gamma_5)\nu_e = 2\bar{L}\gamma_i\tau_- L, \\ j_1^+ &= \bar{\nu}_e\gamma_i(1 + \gamma_5)e = 2\bar{L}\gamma_i\tau_+ L, \\ j_{em} &= \bar{e}\gamma_i e = \frac{1}{2}(\bar{L}\gamma_i L - \bar{L}\gamma_i\tau_3 L) + \bar{R}\gamma_i R. \end{aligned} \tag{5.1.93}$$

Here

$$\bar{L} = \frac{1}{2}(\bar{\nu}_e\bar{e})(1 - \gamma_5), \quad \bar{R} = \frac{1}{2}\bar{e}(1 + \gamma_5) \tag{5.1.94}$$

denote quantities Dirac conjugate with respect to ordinary spinors, similarly to (5.1.18), matrices $\tau_{\pm}$ are

---

[19]In this section e and ν_e denote the respective bispinor wave functions, and units with $\hbar = c = 1$ are used.

$$2\tau_- = \tau_1 - i\tau_2 = \begin{pmatrix} 0 & 0 \\ 2 & 0 \end{pmatrix}, \quad 2\tau_+ = \tau_1 + i\tau_2 = \begin{pmatrix} 0 & 2 \\ 0 & 0 \end{pmatrix}. \tag{5.1.95}$$

The current components in (5.1.93) are incorporated in the

isovector $\quad \bar{L}\gamma_i\tau L$

and $\tag{5.1.96}$

isoscalar $\quad \dfrac{1}{2}(\bar{L}\gamma_i L) + \bar{R}\gamma_i R.$

Four vector bosons also form isovector $\mathbf{C}_i$ and isoscalar B_i in the isospace (i is an ordinary vector subscript). The gauge invariant Lagrangian of interactions between leptons and vector bosons invariant against rotations in the isospace has the form[20]

$$\mathcal{L}_1 = \frac{g}{2}\bar{L}\gamma_i\tau L\mathbf{C}_i + g'\left(\frac{1}{2}\bar{L}\gamma_i L + \bar{R}\gamma_i R\right)B_i, \tag{5.1.97}$$

where g and g' are the independent coupling constants. In a standard notation (with no isospace) $\mathcal{L}_1$ must include terms similar to $j_{1,i}^\pm W_i^\pm$, $j_{1,i}^n Z_i$ and $j_{\mathrm{em},i} A_i$, where A_i is the field vector (photon). Comparing (5.1.93) and (5.1.97) shows that $W^\pm$, Z and A_i form the following linear combinations of $\mathbf{C}_i$ and B_i

$$W_i^\mp = \frac{1}{\sqrt{2}}(C_i^{(1)} \pm iC_i^{(2)}),$$
$$C_i^{(3)} = \cos\theta_{\mathrm{W}} Z_i - \sin\theta_{\mathrm{W}} A_i, \quad B_i = \sin\theta_{\mathrm{W}} Z_i + \cos\theta_{\mathrm{W}} A_i. \tag{5.1.98}$$

Substituting (5.1.93) and (5.1.98) into (5.1.97) gives $\mathcal{L}_1$ in the form

$$\mathcal{L}_1 = \frac{g}{2\sqrt{2}}(j_{1,i}^- W_i^- + j_{1,i}^+ W_i^+)$$
$$+ \left[\cos\theta_{\mathrm{W}}\frac{g}{2}\bar{L}\gamma_i\tau_3 L + \sin\theta_{\mathrm{W}} g'\left(\frac{1}{2}\bar{L}\gamma_i L + \bar{R}\gamma_i R\right)\right]Z_i +$$
$$+ \left[\cos\theta_{\mathrm{W}} g'\left(\frac{1}{2}\bar{L}\gamma_i L + \bar{R}\gamma_i R\right) - \sin\theta_{\mathrm{W}}\frac{g}{2}\bar{L}\gamma_i\tau_3 L\right]A_i. \tag{5.1.99}$$

In order for the coefficient before A_i to be the electromagnetic current j_{em} from (5.1.93) and be a term in the Lagrangian with an appropriate coupling constant e (the electron charge), it must be

$$g\sin\theta_{\mathrm{W}} = g'\cos\theta_{\mathrm{W}} = e\sqrt{4\pi}. \tag{5.1.100}$$

Using (5.1.100), the coefficient before Z_i proportional to the weak neutral currents becomes

$$\frac{g}{4\cos\theta_{\mathrm{W}}}\left[\bar{\nu}\gamma_i(1 + \gamma_5)\nu - \bar{e}\gamma_i(1 - 4\sin^2\theta_{\mathrm{W}} + \gamma_5)e\right]. \tag{5.1.101}$$

---

[20]In the perturbation theory the weak decay probabilities and cross-sections for weak processes are determined by the matrix element for one of the terms of the Lagrangian [144].

Obviously, the structure of neutral leptonic currents given in (5.1.78) may be found from (5.1.101).

To construct a gauge invariant Lagrangian of the Higgs field, its wave function φ must be a scalar isodoublet in which one component may be taken to be zero

$$\varphi = \begin{pmatrix} \varphi^+ \\ \varphi^0 \end{pmatrix} = \frac{1}{\sqrt{2}} \begin{pmatrix} 0 \\ \tilde{\varphi}^0 \end{pmatrix}. \tag{5.1.102}$$

The simplest potential $V(\varphi)$ providing breaking spontaneous symmetry and its contribution to the Lagrangian has the form

$$\mathcal{L}_2 = -V(\varphi) = \mu^2 |\varphi|^2 - h|\varphi|^4, \quad |\varphi|^2 = |\varphi^+|^2 + |\varphi^0|^2. \tag{5.1.103}$$

The minimum in the potential occurs at the point

$$|\varphi|_{\min} = \sqrt{\mu^2/2h} = \lambda/\sqrt{2}. \tag{5.1.104}$$

Writing

$$\tilde{\varphi}^0 = \lambda + \chi, \tag{5.1.105}$$

and omitting the constant terms, we obtain $\mathcal{L}_2$ in the form

$$\mathcal{L}_2 = -h\lambda^2\chi^2 - h\lambda\chi^3 - \frac{h}{4}\chi^4. \tag{5.1.106}$$

For massive scalars, the mass-term contribution in the Lagrangian is equal to $-m^2\chi^2/2$. Comparing with (5.1.106) gives that, as a result of spontaneous symmetry breaking, the Higgs field particles acquire the mass

$$m_\chi = \lambda\sqrt{2h}. \tag{5.1.107}$$

The coupling of the Higgs field with vector bosons is described by the Lagrangian $\mathcal{L}_3$ that in gauge invariant form may be written as

$$\mathcal{L}_3 = \left| \left(\frac{\partial}{\partial x_i} - i\frac{g}{2}\boldsymbol{\tau}\mathbf{C}_i + i\frac{g'}{2}B_i \right)\varphi \right|^2. \tag{5.1.108}$$

Using here (5.1.98) and (5.1.102–105), we see that $\mathcal{L}_3$ does not contain the photon A_i:

$$\mathcal{L}_3 = \frac{1}{2}\left(\frac{\partial\chi}{\partial x_i}\right)^2 + \frac{g^2}{4}\left(W_i^\dagger W_i + \frac{1}{2\cos^2\theta_W}Z_iZ_i\right)(\lambda^2 + 2\lambda\chi + \chi^2). \tag{5.1.109}$$

For the massive vector isosinglet Z_i and isodoublet $W_i^\pm$ the mass-term contribution in the Lagrangian is $\frac{1}{2}m_Z^2 Z_i Z_i + m_W^2 W_i^\dagger W_i$. Comparing with (5.1.109) gives the effective masses of bosons

$$m_{\mathrm{W}} = \frac{g\lambda}{2}, \quad m_{\mathrm{Z}} = \frac{g\lambda}{2\cos\theta_{\mathrm{W}}} = \frac{m_{\mathrm{W}}}{\cos\theta_{\mathrm{W}}}. \tag{5.1.110}$$

In order for spontaneous symmetry breaking to make the electron acquire its mass as well, the Lagrangian should be supplemented by the gauge invariant term[21]

$$\mathcal{L}_4 = \frac{f}{\sqrt{2}}(\bar{R}\varphi^\dagger L + \bar{L}\varphi R), \tag{5.1.111}$$

which, on using (5.1.91–94) and (5.1.102), takes the form

$$\mathcal{L}_4 = \frac{f\lambda}{2}\left(1 + \frac{\chi}{\lambda}\right)\bar{e}e. \tag{5.1.112}$$

The mass term of the electron as a component of the spinor isodoublet in the Lagrangian is $m_e\bar{e}e$, and, on being combined with (5.1.112), yields

$$m_e = \frac{f\lambda}{2}. \tag{5.1.113}$$

At low energies $\epsilon \ll m_{\mathrm{W}}$, the Lagrangian of the interaction (5.1.99) may lead to modification of the UWI theory [86] with the weak interaction constant G_{W} being expressed in terms of constants of the Lagrangian (5.1.99) as

$$\frac{G_{\mathrm{W}}}{\sqrt{2}} = \frac{g^2}{8m_{\mathrm{W}}^2}. \tag{5.1.114}$$

Using (5.1.100) and (5.1.110) gives

$$\frac{G_{\mathrm{W}}}{\sqrt{2}} = \frac{1}{2\lambda^2} = \frac{4\pi e^2}{8m_{\mathrm{W}}^2\sin^2\theta_{\mathrm{W}}}. \tag{5.1.115}$$

Recalling (5.1.79), we estimate from (5.1.115) and (5.1.110) the masses

$$m_{\mathrm{W}} = \left(\frac{4\pi e^2}{4\sqrt{2}G_{\mathrm{W}}\sin^2\theta_{\mathrm{W}}}\right)^{1/2} \approx 78.8 \text{ GeV},$$

$$m_{\mathrm{Z}} = \left(\frac{4\pi e^2}{4\sqrt{2}G_{\mathrm{W}}\sin^2\theta_{\mathrm{W}}\cos^2\theta_{\mathrm{W}}}\right)^{1/2} \approx 89.5 \text{ GeV}. \tag{5.1.116}$$

The photon A_i and neutrinos turn out to be massless by virtue of (5.1.109) and (5.1.102), respectively, but neutrinos may acquire their masses providing the theory is supplemented by a right-handed neutrino singlet $R_\nu = \frac{1-\gamma_5}{2}\nu$ similar to (5.1.92), and a Lagrangian term $\mathcal{L}_5$ for its interaction with the Higgs field (5.1.102) similar to (5.1.111) (see [172]). The total Lagrangian for electroweak interactions of leptons calculated from (5.1.97), (5.1.103),

---

[21]Here † denotes a Hermitian conjugate doublet in the space of weak isospin.

(5.1.108) and (5.1.111) or, in another form, from (5.1.99), (5.1.106), (5.1.109) and (5.1.112), has the form

$$\mathcal{L} = \mathcal{L}_0^{(b)} + \sum_l (\mathcal{L}_0^{(l)} + \mathcal{L}_1^{(l)} + \mathcal{L}_2^{(l)} + \mathcal{L}_3^{(l)} + \mathcal{L}_4^{(l)}), \tag{5.1.117}$$

where $\mathcal{L}_0^{(b)}$ corresponds to free vector bosons [86]

$$\begin{aligned}
\mathcal{L}^{(b)} &= \frac{1}{4}\mathbf{C}_{ik}\mathbf{C}_{ik} + \frac{1}{4}B_{ik}B_{ik}, \\
\mathbf{C}_{ik} &= \frac{\partial \mathbf{C}_i}{\partial x_k} - \frac{\partial \mathbf{C}_k}{\partial x_i} + g[\mathbf{C}_i \times \mathbf{C}_k], \\
B_{ik} &= \frac{\partial B_i}{\partial x_k} - \frac{\partial B_k}{\partial x_i},
\end{aligned} \tag{5.1.118}$$

and $\mathcal{L}_0^{(l)}$ to free leptons [86]

$$\mathcal{L}_0^{(l)} = \mathrm{i}\bar{R}\gamma_i \frac{\partial R}{\partial x_i} + \mathrm{i}\bar{L}\gamma_i \frac{\partial L}{\partial x_i}. \tag{5.1.119}$$

The summation in (5.1.117) is performed over three species of leptons.

Weak coupling with hadrons is included in the Weinberg-Salam theory via quarks. Six quarks are divided into three left-handed doublets similar to (5.1.91) such as $\frac{1+\gamma_5}{2}\begin{pmatrix} u \\ d \end{pmatrix}$, while a right-handed singlet similar to (5.1.92) such as $\frac{1-\gamma_5}{2}u$ and $\frac{1-\gamma_5}{2}d$ corresponds to each quark since all of them are charged and massive. The Lagrangian $\mathcal{L}_6$ of interaction between the weak quark current and the vector bosons $\mathbf{C}_i$, B_i is, analogously to $\mathcal{L}_1$ from (5.1.99), invariant in isospace and may also be divided into charged and neutral weak currents and electromagnetic current. Some complications in comparison with leptonic currents arise from the necessity to allow for the possibility of weak transitions between quarks not only within the same doublet, but between diverse doublets as well. This may be achieved by introduction of three mixing angles; one of them, called the Cabibbo angle, determines the strangeness-changing transitions (see the end of Sect. 5.1.2). The charged current with account of transitions between all three quark doublets is constructed with the help of a Kobayashi–Maskawa (3×3) matrix [331a]. The mixing does not violate the renormalizability of the theory and produces no principal difficulties. A mixing which results in mutual transformations of various species of neutrino may also be introduced in the theory [544a]. The structure of the quark and hadronic weak neutral currents (5.1.78) and (5.1.80) is determined by the structure of the Lagrangian $\mathcal{L}_6$, analogously to the Lagrangian $\mathcal{L}_1$ for leptons.

The Gauge invariance condition imposed on the Lagrangian together with experimental restrictions led to the prediction in 1970 of the fourth quark (c-charm) which forms a doublet with the s-strange quark. This quark was discovered in 1974. Discovery of τ-leptons implies the existence of the third quark doublet: b-beauty, and t-top quarks. Now all 6 quarks have been observed experimentally, the last one (t) was caught in 1994 in Fermilab (see [471*] and references therein).

Note that under the physical conditions in stars with $\epsilon \leq 100$ MeV, the additional contribution of the Weinberg–Salam theory, compared with the UWI theory, consists of introducing the neutral currents and determining their kind since these currents play a significant role in the late evolutionary stages and particularly in supernova explosions.

Problem 1. Show that V_i in (5.1.19) is a vector.

Solution. According to [17, 230], the Lorenz transformation for the Dirac conjugated spinor $\bar{\psi}$ from (5.1.18) is

$$\bar{\psi}' = \bar{\psi}S^{-1}.$$

Then the transformation of V_i, by virtue of (5.1.11) and (5.1.14), takes the form

$$V_i' = (\bar{\psi}\gamma_i\psi)' = \bar{\psi}S^{-1}\gamma_i S\psi = \bar{\psi}L_{ik}\gamma_k\psi = L_{ik}\bar{\psi}\gamma_k\psi = L_{ik}V_k.$$

The matrix for the V_i transformation thus coincides with the Lorentz matrix for the vector (5.1.12) whereby V_i is a vector.

Problem 2. Find the cross-section for the capture of an electron with energy ϵ_e by stable nucleus (A, Z), taking $(Ft_{1/2})_{Z-1}$ to be known and the transition allowed.

Solution. The probability for the nucleus $(A, Z-1)$ to decay is, according to (5.1.61),

$$W_{Z-1}^{\text{dec}} = \frac{2\pi}{\hbar} \int |M|_{Z-1}^2 \frac{(4\pi)^2}{(2\pi\hbar)^6} \delta(\epsilon_e + \epsilon_\nu - \Delta_{Z-1,Z}) p_e^2 dp_e p_\nu^2 dp_\nu =$$
$$= \frac{|M|_{Z-1}^2}{2\pi^3\hbar^7} m_e^5 c^4 F_0, \tag{1}$$

where $|M|_{Z-1}^2 = \text{const.}$, F_0 is defined by (5.1.63). The probability W_Z^{cap} for an electron with energy ϵ_e and concentration n_e to be captured by nucleus (A, Z) is, in analogy with (1),

$$W_Z^{\text{cap}} = \frac{2\pi}{\hbar} n_e \int |M|_Z^2 \frac{4\pi}{(2\pi\hbar)^3} \delta(\epsilon_e - \epsilon_\nu - \Delta_{Z-1,Z}) p_\nu^2 dp_\nu =$$
$$= \frac{n_e |M|_Z^2}{\pi\hbar^4 c^3} (\epsilon_e - \Delta_{Z-1,Z})^2. \tag{2}$$

Here $|M|_Z^2 = \frac{1}{2}|M|_{Z-1}^2 g_{Z-1}/g_Z$ by reason of averaging over electron spins in (2) instead of summing in (1), and a possible difference in statistical weights of the initial and final nuclei. Solving (1) for $(Ft_{1/2})_{Z-1}$ with the aid of (5.1.64) and substituting it into (2) gives

$$
W_Z^{\text{cap}} = \frac{g_{Z-1}}{g_Z} \pi^2 \left(\frac{\hbar}{m_e c}\right)^3 \frac{n_e \ln 2}{(Ft_{1/2})_{Z-1}} \frac{(\epsilon_e - \Delta_{Z-1,Z})^2}{(m_e c^2)^2}. \tag{3}
$$

The cross-section for the electron capture by nucleus is, according to (5.1.82) and (5.1.63),

$$
\begin{aligned}
\sigma_e^{\text{cap}} &= \frac{W_Z^{\text{cap}}}{n_e v_e} = \\
&= \frac{g_{Z-1}}{g_Z} \frac{\pi^2}{c} \left(\frac{\hbar}{m_e c}\right)^3 \frac{\ln 2}{(Ft_{1/2})_{Z-1}} \frac{(\epsilon_e - \Delta_{Z-1,Z})^2}{(m_e c^2)^2 \sqrt{1 - (m_e c^2/\epsilon_e)^2}} = \\
&= \frac{g_{Z-1}}{g_Z} \frac{1.313 \times 10^{-41}}{(Ft_{1/2})_{Z-1}} \frac{(u_e - \delta)^2}{\sqrt{1 - u_e^{-2}}} \quad (\text{cm})^2,
\end{aligned} \tag{4}
$$

$$
u_e = \epsilon/m_e c^2, \quad \delta = \Delta_{Z-1,Z}/m_e c^2, \quad u_e > \delta > 1.
$$

At $0 \le \delta \le 1$ the nucleus $(A, Z - 1)$ is stable against e^--emission, but the nucleus (A, Z) may capture an electron at rest (K-capture) [212]. For this case, (3) and (4) do not apply since no β-decay of the $(A, Z - 1)$ nucleus occurs.

Problem 3. Find the probability for a K-electron to be captured by a nucleus.

Solution. The size of the atomic K-shell is $\frac{a_0}{Z} = \frac{\hbar^2}{m_e e^2 Z}$ (see Sect. 1.4.7), the mean electron density within a filled K-shell with two electrons is $\bar{n}_e = \frac{3Z^3}{2\pi a_0^3}$. In calculations of the K-capture probability the equality $n_e(0) = \frac{4}{3}\bar{n}_e = \frac{2Z^3}{\pi a_0^3}$ [84] must be taken into account. Substituting this into (2) of Problem 2 gives

$$
\begin{aligned}
W_Z^{\text{K}} &= \frac{2|M|_Z^2}{\pi^2} (\alpha Z)^3 \left(\frac{m_e c}{\hbar}\right)^3 \frac{m_e^2 c}{\hbar^4} (u_e - \delta)^2 = \\
&= 2 \frac{|M|_Z^2}{|M|_n^2} \frac{G_W^2 (1 + 3g_A^2)}{\pi^2} (\alpha Z)^3 \left(\frac{m_e c}{\hbar}\right)^3 \frac{m_e^2 c}{\hbar^4} (u_e - \delta)^2 \quad (\text{s}^{-1}),
\end{aligned}
$$

$\delta < u_e \le 1$ for a bound electron.

Problem 4. Find the probability for an electron to be captured by nucleus (A, Z) in degenerate matter with density ρ.

Solution. (Frank-Kamenetsky, 1962 [216]). The probability for an electron to be captured in degenerate matter with $f_e = 1$ at $\epsilon_e < \epsilon_{\text{Fe}} + m_e c^2$ is determined, similarly to (2) of Problem 2, by

$$W_Z^D = \frac{g_{Z-1}}{g_Z} \frac{2\pi}{\hbar} \int |M|_{Z-1}^2 \frac{(4\pi)^2}{(2\pi\hbar)^6} \delta(\epsilon_e - \epsilon_\nu - \Delta_{Z-1,Z}) p_\nu^2 dp_\nu p_e^2 dp_e =$$

$$= \frac{g_{Z-1}}{g_Z} \frac{|M|_{Z-1}^2}{2\pi^3 \hbar^7} m_e^5 c^4 [F_0(u_{Fe}) - F_0(\delta)] =$$

$$= \frac{g_{Z-1}}{g_Z} \ln 2 \frac{F_0(u_{Fe}) - F_0(\delta)}{(Ft_{1/2})_{Z-1}}, \quad u_{Fe} = (\epsilon_{Fe}/m_e c^2) + 1. \tag{1}$$

Here, $F_0(u)$ is defined by (5.1.63). At $u_{Fe} \gg \delta$ we obtain approximately

$$W_Z \approx \frac{g_{Z-1}}{g_Z} \frac{\ln 2}{(Ft_{1/2})_{Z-1}} \frac{u_{Fe}^5 - \delta^5}{5}. \tag{2}$$

Problem 5. Find the total probability for degenerate electrons to be captured by a nucleus in a gaseous model, taking into account the excitation levels of the final nucleus.

Solution [54]. The nucleus should be represented as a uniform sphere of radius

$$R = \frac{e^2}{2m_e c^2} A^{1/3} = 1.4 A^{1/3} \times 10^{-13} \quad \text{cm} \tag{1}$$

The boundary values of the Fermi momentum of protons p_0 and neutrons q_0 in the nucleus are

$$p_0 = 3\left(\frac{\pi}{3}\right)^{1/3} \frac{\hbar c}{e^2} \left(\frac{2Z}{A}\right)^{1/3} m_e c \approx 417 \left(\frac{2Z}{A}\right)^{1/3} m_e c,$$

$$q_0 \approx 417 \left(\frac{2N}{A}\right)^{1/3} m_e c, \quad N = A - Z. \tag{2}$$

The probability for an electron to be captured in a degenerate plasma, on integrating over the nuclear volume V_n and phase volume dN_n will be

$$W_e = \frac{2\pi}{\hbar} |M|^2 V_n \int dN_\nu dN_p dN_e \delta(E_e + E_p - E_\nu - E_n),$$

$$E_n = E_p + \frac{p_{e\nu}^2}{2m} + \frac{p_p p_{e\nu} z_p}{m}, \quad p_{e\nu} = |p_{e\nu}| = |p_e - p_\nu|, \tag{3}$$

$$z_p = \cos(\widehat{p_{e\nu}, p_p}), \quad V_n = \frac{3Z}{8\pi} \left(\frac{2\pi\hbar}{p_0}\right)^3 = \frac{4}{3}\pi R^3.$$

Here $m \approx m_p/2$ is the effective mass of nucleons in the nucleus, V_n is the nuclear volume. $|M|^2 = \text{const.} < |M|_n^2$ from (5.1.60) is the mean square matrix element for (p, n)-transition, resulting from summation over spins e, ν and n and averaging over spins p. Using

$$dN_e = 2\frac{d^3 p_e}{(2\pi\hbar)^3} = \frac{4\pi p_e^2 dp_e dz_e}{(2\pi\hbar)^3}, \qquad dN_\nu = \frac{d^3 p_\nu}{(2\pi\hbar)^3} = \frac{2\pi p_\nu^2 dp_\nu dz_\nu}{(2\pi\hbar)^3},$$

$$dN_p = \frac{d^3 p_p}{(2\pi\hbar)^3} = \frac{2\pi p_p^2 dp_p dz_p}{(2\pi\hbar)^3}, \qquad dN_n = 2\frac{d^3 p_n}{(2\pi\hbar)^3},$$

(4)

we obtain from (3)

$$W_e = \frac{2\pi}{\hbar}|M|^2 2\frac{(2\pi)^3}{(2\pi\hbar)^9}V_n \int p_e^2 dp_e dz_e p_\nu^2 dp_\nu dz_\nu p_p^2 \times$$

$$\times dp_p dz_p \delta\left(E_e - E_\nu - \frac{p_{e\nu}^2}{2m} - \frac{p_p p_{e\nu} z_p}{m}\right).$$

(5)

Find the integration limits in (5). It is obvious that

$$p_e \le p_{Fe}, \qquad p_p \le p_0, \qquad p_n \ge q_0. \tag{6}$$

As an approximation, we may take nucleons as non-relativistic and electrons as ultrarelativistic. To find the low limit for the integration over dp_p, write the energy conservation law for the four-fermion process in the form

$$p_p^2 = p_n^2 - 2mc(p_e - p_\nu). \tag{7}$$

Since $p_n > q_0$, we obtain

$$p_p^2 > q_0^2 - 2mc(p_e - p_\nu) = p_1^2. \tag{8}$$

On the other hand, the δ-function in (5) arising from the energy conservation law implies

$$p_p = \frac{mc(p_e - p_\nu)}{p_{e\nu} z_p} - \frac{1}{2}\frac{p_{e\nu}}{z_p}. \tag{9}$$

Obviously, the minimum in p_p occurs at $z_p = 1$, so

$$p_p > \frac{mc(p_e - p_\nu)}{p_{e\nu}} - \frac{p_{e\nu}}{2} = p_2. \tag{10}$$

Integrating over dp_p, we should use $\max(p_1, p_2)$. Performing the integration in (5) over $dz_p dp_p$ within the limits obtained above gives

$$W_e = \frac{4\pi}{\hbar}|M|^2\frac{(2\pi)^3}{(2\pi\hbar)^9}V_n \int p_e^2 dp_e dz_e p_\nu^2 dp_\nu dz_\nu \frac{m}{2p_{e\nu}} \times$$

$$\times \begin{cases} (p_0^2 - p_1^2) & \text{at} \quad p_1 > p_2 \\ (p_0^2 - p_2^2) & \text{at} \quad p_1 < p_2. \end{cases} \tag{11}$$

Now find the limits for integration over $dz_\nu dp_\nu dp_e$. It follows from (8) and (10) that

$$p_1 < p_0 \quad \text{at} \quad p_\nu < p_{\nu 2}, \tag{12}$$

$$p_2 < p_0 \quad \text{at} \quad z_\nu < z_{\nu 0}, \tag{13}$$

where

$$z_\nu = \cos\left(\widehat{\boldsymbol{p}_e \boldsymbol{p}_\nu}\right),$$
$$p_{\nu 2} = p_e - (q_0^2 - p_0^2)/2mc,$$
$$z_{\nu 0} = (p_e^2 + p_\nu^2 - x_0^2)/2p_e p_\nu,$$
$$x_0 = \sqrt{p_0^2 + 2mc(p_e - p_\nu)} - p_0. \tag{14}$$

We may, further, find the conditions for applying the upper line in (11):

$$p_1 > p_2 \quad \text{at} \quad z_\nu < z_{\nu 1}, \tag{15}$$

where

$$z_{\nu 1} = (p_0^2 + p_\nu^2 - x_1^2)/2p_e p_\nu,$$
$$x_1 = q_0 - \sqrt{q_0^2 - 2mc(p_e - p_\nu)}. \tag{16}$$

The inequalities (13) and (15) lead to

$$z_{\nu 1} < z_{\nu 0} \quad \text{at} \quad p_\nu < p_{\nu 2}, \tag{17}$$

valid under the same condition as (12). The obvious condition $z_{\nu 1} > -1$ gives

$$z_{\nu 1} > -1 \quad \text{at} \quad p_\nu > p_{\nu 1}, \tag{18}$$

where

$$p_{\nu 1} = q_0 - p_e + mc - \sqrt{(q_0 + mc)^2 - 4mcp_e}. \tag{19}$$

Combining (12), (17), and (18) yields

$$p_{\nu 1} < p_{\nu 2} \quad \text{at} \quad p_e < p_{e1}, \tag{20}$$

where

$$p_{e1} = \frac{q_0 + p_0}{2} + \frac{q_0^2 - p_0^2}{4mc}. \tag{21}$$

The condition $z_{\nu 0} > -1$ gives

$$z_{\nu 0} > -1 \quad \text{at} \quad p_\nu > p_{\nu 0}, \tag{22}$$

where

$$p_{\nu 0} = [(mc + p_0)^2 + 4mcp_e]^{1/2} - p_e - p_0 - mc. \tag{23}$$

It follows from (17), (18), and (22) that

$$p_{\nu 1} > p_{\nu 0} \quad \text{at} \quad p_e > p_{e0}, \tag{24}$$

where

$$p_{e0} = \frac{q_0 - p_0}{2} + \frac{q_0^2 - p_0^2}{4mc}. \tag{25}$$

The upper relation in (11) contributes to W_e when integration limits are as follows:

$$p_{e0} < p_e < p_{e1}, \qquad p_{\nu1} < p_\nu < p_{\nu2}, \qquad -1 < z_\nu < z_{\nu1}. \tag{26}$$

Using the lower relation in (11) requires

$$p_1 < p_2 \quad \text{at} \quad z_\nu > z_{\nu1} \tag{27}$$

together with (12) and (13). We have to distinguish two different cases:

$$z_{\nu1} > -1 \quad \text{at} \quad p_\nu > p_{\nu1}, \quad \text{then} \quad z_\nu > z_{\nu1} \tag{28}$$

and

$$z_{\nu1} < -1 \quad \text{at} \quad p_\nu < p_{\nu1}, \quad \text{then} \quad z_\nu > -1. \tag{29}$$

The conditions (17) and (20) follow from (28) and (13), the inequality (24) also holds. The following integration limits are then used if (28) is valid:

$$p_{e0} < p_e < p_{e1}, \quad p_{\nu1} < p_\nu < p_{\nu2}, \quad z_{\nu1} < z_\nu < z_{\nu0}. \tag{30}$$

The condition (29) together with (12) gives two possibilities:

$$p_{\nu1} < p_{\nu2} \quad \text{at} \quad p_e < p_{e1}, \quad \text{then} \quad p_\nu < p_{\nu1}; \tag{31}$$

$$p_{\nu1} > p_{\nu2} \quad \text{at} \quad p_e > p_{e1}, \quad \text{then} \quad p_\nu < p_{\nu2}. \tag{32}$$

In the case of (31) the inequalities (13), (22), (24), and (29) are valid, giving the following integration limits

$$p_{e0} < p_e < p_{e1}, \quad p_{\nu0} < p_\nu < p_{\nu1}, \quad -1 < z_\nu < z_{\nu0}. \tag{33}$$

The condition (32) simultaneously with (22) gives the inequality

$$p_{\nu2} > p_{\nu0} \quad \text{at} \quad p_e > p_{e1}, \tag{34}$$

coinciding with $p_{\nu2} < p_{\nu1}$. The conditions (29), (32), (13), (22), (34), and (6) yield the following integration limits:

$$p_{e1} < p_e < p_{Fe}, \quad p_{\nu0} < p_\nu < p_{\nu2}, \quad -1 < z_\nu < z_{\nu0}. \tag{35}$$

Obviously, $-1 < z_e < 1$. At $p_{Fe} > p_{e1}$ the probability (11) for the capture contains four integrals: one is given by the upper line in (11) and the integration limits (26), three others by the lower line in (11) and the integration limits (30), (33), and (35). At $p_{Fe} < p_{e1}$ the integral with (35) cancels out, and $p_e < p_{Fe}$ should replace $p_e < p_{e1}$ in (26), (30), and (33). W_e is given in general form in [54]. In reality we are interested in the situation where

$$p_{\nu e}^2 \ll mc(p_e - p_\nu) \ll p_0^2, q_0^2 \tag{36}$$

and for which we get

$$W_e = \frac{4\pi}{\hbar} \frac{(2\pi)^3}{(2\pi\hbar)^9} |M|^2 V_n m^2 c \frac{8mcq_0^2}{3(mc+q_0)^3} p_{k,e}^4 \varphi_e, \quad p_{k,e} = p_{Fe} - p_{e0},$$

$$\varphi_e = \frac{1}{6} p_{k,e}^2 + \frac{1}{5} p_{e0} p_{k,e} \left(2 - \frac{q_0}{mc}\right) + \frac{1}{4} p_{e0}^2 \left(1 - \frac{q_0}{mc}\right), \qquad (37)$$

$$p_{e0} = \frac{q_0^2 - p_0^2}{4} \left(\frac{1}{mc} + \frac{1}{q_0}\right).$$

Expressing $|M|^2$ in terms of the probability for a K-muon to be captured calculated from the same gaseous model of nucleus, we have

$$W_e = W_\mu \frac{1}{2\pi} \left(\frac{137}{Z_e m_\mu c}\right)^3 \frac{Z}{Z_e} \frac{8mcq_0^2}{3(mc+q_0)^3} \frac{p_{k,e}^4}{p_{k,\mu}^4} \frac{\varphi_e}{\varphi_\mu}, \qquad (38)$$

where

$$Z_e = Z[1 + (Z/42)^{1.47}]^{-1/1.47},$$

$$\varphi_\mu = \frac{1}{3} p_{k,\mu}^2 + p_{\nu 1\mu} p_{k,\mu} + \frac{1}{2}(p_0^2 - m^2 c^2 + mm_\mu c^2) \frac{p_{\nu 1\mu}^2 - p_{\nu 0\mu}^2}{mcp_{k,\mu}} +$$

$$+ 2mm_\mu c^2 \frac{p_{\nu 1\mu} - p_{\nu 0\mu}}{p_{k,\mu}} - \frac{p_{\nu 1\mu}^3 - p_{\nu 0\mu}^3}{3p_{k,\mu}} - \frac{p_{\nu 1\mu}^4 - p_{\nu 0\mu}^4}{16mcp_{k,\mu}} -$$

$$- \frac{mm_\mu^2 c^3}{p_{k,\mu}} \ln \frac{p_{\nu 1\mu}}{p_{\nu 0\mu}}, \qquad (39)$$

$$p_{k,\mu} = m_\mu c - (q_0^2 - p_0^2)/2mc - p_{\nu 1\mu},$$

$$p_{\nu 1\mu} = mc + q_0 - [(mc + q)^2 - 2mm_\mu c^2]^{1/2},$$

$$p_{\nu 0\mu} = -mc - p_0 + [(mc + p_0)^2 + 2mm_\mu c^2]^{1/2}.$$

It follows from (37) that including the excited states enhances the dependence of W_e on p_{Fe}: $\sim p_{Fe}^6$ in (37) in lieu of $\sim p_{Fe}^5$ in Problem 4, at $p_{Fe} \gg (\delta m_e c, p_{e0})$. Experimental values of W_μ for various nuclei are given in [74]. Comparing them with results calculated from a more complicated shell model of nucleus [73, 316] exhibits a satisfactory accuracy of (38).

5.2 Neutrino Cooling of Stars

Neutrinos escape freely from the star and thus cool its interior at all evolutionary stages up to formation of a dense neutron core during the collapse. We now consider the most important processes of neutrino production in stars.

5.2.1 Annihilation of e^+e^- Pairs

At high temperatures and low densities the pair annihilation

$$e^+ + e^- \rightarrow \nu + \tilde{\nu} \tag{5.2.1}$$

gives rise to neutrino production processes. The cross-section for the reaction (5.2.1) is determined by the product of charged $(\bar{\nu}^- O_i^L e^-)(\bar{\nu}^+ O_i^L e^+)$ and neutral $(\bar{e}^- O_i^{n,e} e^-)(\bar{\nu} O_i^{n,\nu} \nu)$ currents. Here O_i^L is the left-handed operator (5.1.20), $O^{n,e}$ and $O^{n,\nu}$ are given in (5.1.78). The latter product contributes to the reaction (5.2.1) because the production of e^- is equivalent to annihilation of e^+, and the annihilation of ν to the production of $\tilde{\nu}$. The cross-section for (5.2.1) due to charged currents has been calculated in [268]. The $(\bar{\nu}_e O_i \nu_e)$-like neutral currents have been included in calculations in [336]. Neutral currents related to all neutrino species have been taken into account in [496]. With (5.1.82), the differential cross-section obtained in [496] becomes

$$\sigma = \frac{G_W^2 c (m_e c/\hbar)^4}{6\pi v \epsilon \epsilon'} \left\{ \frac{r_+ + r'_+}{2} \left[1 + 3\frac{p_i p'_i}{m_e^2 c^2} + 2\frac{(p_i p'_i)^2}{m_e^4 c^4} \right] + \right.$$
$$\left. + 3\frac{r_- + r'_-}{2} \left(1 + \frac{p_i p'_i}{m_e^2 c^2} \right) \right\} \quad \text{cm}^2. \tag{5.2.2}$$

Here, p_i, p'_i, ϵ, ϵ' are the 4-momenta and energies of electrons and positrons, respectively, v is the relative speed of e^+, e^-.

$$r_\pm = \frac{c_V^2 \pm c_A^2}{2}, \quad r'_\pm = n\frac{c_V'^2 \pm c_A'^2}{2},$$
$$c_V = \frac{1}{2} + 2\sin^2\theta_W, \quad c_A = 1/2, \tag{5.2.3}$$
$$c_V' = -\frac{1}{2} + 2\sin^2\theta_W, \quad c_A' = -1/2.$$

The quantities c_V and c_A correspond to the sum of charged and neutral currents due to ν_e, while c_V' and c_A' represent the contribution of neutral currents of other neutrino species; n is the number of neutrino species except ν_e known to be $n = 2$ (see (5.1.3), [553a]). For the charged current [268] the cross-section is obtained from (5.2.2) with $r_+ = 1$, $r_- = 0$ and $n = 0$. The energy loss rate due to the reaction (5.2.1) is

$$Q_{\text{pair}} = \int (\epsilon + \epsilon') \sigma v \, dn_{e^-} \, dn_{e^+}. \tag{5.2.4}$$

Here

$$dn_{e^\pm} = \frac{2}{(2\pi\hbar)^3} \frac{p^2 dp \, d\cos\theta \, d\varphi}{1 + \exp\left(\frac{\epsilon \pm \mu_{te^-}}{kT}\right)} \tag{5.2.5}$$

is determined according to (1.2.2), (1.2.4), and (1.2.11), but, contrary to (1.2.4), the angular differentials in the momentum space are taken into account. The factor 2 in (5.2.5) accounts for statistical weight because averaging has been made in (5.2.2) over spins of e^+, e^-. Integrating (5.2.4) and taking into account (5.2.5) gives [496]

$$Q_{\text{pair}} = \frac{G_W^2}{18\pi^5 m_e c} \left(\frac{m_e c}{\hbar}\right)^{10} \{(r_+ + r'_+)[5(G_0^- G_{-1/2}^+ + G_{-1/2}^- G_0^+) +$$
$$+ 7(G_0^- G_{1/2}^+ + G_{1/2}^- G_0^+) - 2(G_1^- G_{-1/2}^+ + G_{-1/2}^- G_1^+) +$$
$$+ 8(G_1^- G_{1/2}^+ + G_{1/2}^- G_1^+)] + (r_- + r'_-)9(G_{1/2}^- G_0^+ +$$
$$+ G_0^- G_{1/2}^+ + G_0^- G_{-1/2}^+ + G_{-1/2}^- G_0^+)\} \quad (\text{erg cm}^{-3}\text{s}^{-1}). \tag{5.2.6}$$

The functions $G_n^\pm(\alpha, \beta)$ are similar to the integrals (1.2.10) and have the form

$$G_n^\pm(\alpha, \beta) = \alpha^{-3-2n} \int\limits_\alpha^\infty \frac{x^{2n+1}(x^2 - \alpha^2)^{1/2}}{1 + \exp(x \pm \beta)} dx. \tag{5.2.7}$$

Here, $\alpha = m_e c^2/kT$ and $\beta = \mu_{te}/kT$ defined in (1.2.8), $x = \epsilon/kT = \sqrt{p^2 c^2 + m_e^2 c^4}/kT$. Asymptotic values of Q_{pair} are given in [268] for charged current, and in [336] neutral currents are also taken into account at $n = 0$. At $n \neq 0$ we have for Q_{pair} in various ranges

$$(c_V^2 + n c_V'^2)\frac{G_W^2}{\pi^4 m_e c}\left(\frac{m_e c}{\hbar}\right)^{10}\frac{1}{\alpha^3}e^{-2\alpha}, \quad \alpha \gg 1, \quad \beta \ll \alpha, \tag{NRND}$$

$$(c_V^2 + n c_V'^2)\frac{\sqrt{2\pi}}{\pi^3}\frac{G_W^2}{m_e c}\left(\frac{m_e c}{\hbar}\right)^7\frac{1}{\alpha^{3/2}}\frac{\rho}{\mu_Z m_u}e^{-(\alpha+\beta)},$$
$$\alpha \gg 1, \quad 1 + \alpha \ll \beta \ll 2\alpha, \tag{NRD}$$

$$(r_+ + r'_+)\frac{\sqrt{2\pi}}{10\pi^3}\frac{G_W^2}{m_e c}\left(\frac{m_e c}{\hbar}\right)^7\frac{1}{\alpha^{3/2}}\left(\frac{\beta}{\alpha}\right)^2\frac{\rho}{\mu_Z m_u}e^{-(\alpha+\beta)},$$
$$\alpha \gg 1, \quad \beta/\alpha \gg 1, \tag{RD}$$

$$(r_+ + r'_+)\frac{127.8}{\pi^5}\frac{G_W^2}{m_e c}\left(\frac{m_e c}{\hbar}\right)^{10}\frac{1}{\alpha^9}, \quad \alpha \ll 1, \quad \beta \ll 1, \tag{RND}$$

$$(r_+ + r'_+)\frac{8}{5\pi^3}\frac{G_W^2}{m_e c}\left(\frac{m_e c}{\hbar}\right)^7\frac{1}{\alpha^4}\left(\frac{\beta}{\alpha}\right)^2\frac{\rho}{\mu_Z m_u}e^{-\beta}, \quad \alpha \ll 1, \quad \beta \gg 1. \tag{RD}$$

(5.2.8)

The first factors in (5.2.8) differ from unity owing to neutral currents. In deriving (5.2.8) the expansions in (5.2.7) are obtained by use of formulae analogous to (1.2.50), (1.2.51), (1.2.24), (1.2.40), and (1.2.43), μ_Z is the number of nucleons per electrons given by (1.2.17), (N)R stands for (non)relativistic, (N)D for (non)degenerate.

5.2.2 Neutrino Photoproduction

At low density and not very high temperature $T \leq 4 \times 10^8\,\mathrm{K}$, the reaction

$$\gamma + e^- \rightarrow e^- + \nu + \tilde{\nu}, \tag{5.2.9}$$

is an important neutrino source, and is called the photoneutrino reaction. Its cross-section is calculated in a much more complicated way and reduces to a five-dimensional integral evaluated by a Monte-Carlo method [336, 496, 543, 562]. Interpolation formulae calculated with the aid of these techniques in [268, 562] are given below. In the limiting cases the photoneutrino losses due solely to the charged currents are [543, 185]

$$Q_{\mathrm{photo}}^{\mathrm{ch}} \quad (\mathrm{erg\ cm}^{-3}\mathrm{s}^{-1}) =$$

$$= \begin{cases} 0.976 \times 10^8 T_9^8 \rho/\mu_Z, & \alpha \gg 1, \quad \beta \ll \alpha & \text{(NRND)} \\ 4.851 \times 10^{11} T_9^9 (\rho/\mu_Z)^{1/3}, & \alpha \gg 1, \quad 1+\alpha \ll \beta \ll 2\alpha & \text{(NRD)} \\ 1.477 \times 10^{13} T_9^9 (\lg T_9 - 0.536), & \alpha \ll 1, \quad \beta \ll 1 & \text{(RND)} \\ 1.514 \times 10^{13} T_9^9, & \beta \gg 1, \quad \beta/\alpha \gg 1. & \text{(RD)} \end{cases}$$

$$\tag{5.2.10}$$

The neutral currents may be included here with the aid of (5.2.28) and (5.2.29).

5.2.3 Plasma Neutrino

Electromagnetic waves in a plasma undergo dispersion and absorption and are called plasmons Γ. The dispersion relation for plasmons contains a term $\sim \hbar w_{\mathrm{p}}$ equivalent to the rest energy, hence, neutrino production becomes possible in the plasmon decay [185]

$$\Gamma \rightarrow \nu + \tilde{\nu}. \tag{5.2.11}$$

For the case of non-degenerate electrons, the electron plasma frequency is $\omega_{\mathrm{p}}^2 = 4\pi n_e e^2/m_e$. For a general case, we may write the following dispersion relations for transversal (t) and longitudinal (l) plasmons [268]:

$$\omega^2 = \omega_0^2 + k^2 c^2 + \frac{1}{5}(\omega_1/\omega)^2 k^2 c^2, \qquad \text{(t)}$$

$$\omega^2 = \omega_0^2 + \frac{3}{5}(\omega_1/\omega)^2 k^2 c^2. \qquad \text{(l)} \tag{5.2.12}$$

Here, k is the wave-vector. The relations (5.2.12) do not include the heat motion of electrons, the terms with ω_1, represent relativistic corrections to the plasma frequency. The frequencies ω_0 and ω_1 are

$$\left(\frac{\hbar\omega_0}{m_e c^2}\right)^2 = \frac{4e^2}{3\pi\hbar c}[2G^+_{-1/2} + 2G^-_{-1/2} + G^+_{-3/2} + G^-_{-3/2}],$$

$$\left(\frac{\hbar\omega_1}{m_e c^2}\right)^2 = \frac{4e^2}{3\pi\hbar c}[2G^+_{-1/2} + 2G^-_{-1/2} + G^+_{-3/2} + G^-_{-3/2} -$$

$$- 3G^+_{-5/2} - 3G^-_{-5/2}],$$

$$(5.2.13)$$

where $G^\pm_n$ are defined by (5.2.7). For the charged currents, the energy losses due to the reaction (5.2.11) are [268]:

$$Q^{ch}_t = A_0\gamma^6\alpha^{-9}\int_\gamma^\infty \frac{x(x^2 - \gamma^2)^{1/2}}{e^x - 1}dx,$$

$$Q^{ch}_l = A_0\gamma^9\alpha^{-9}\left(\frac{5}{3}\right)^{7/2}\frac{1}{2}\left(\frac{\omega_0}{\omega_1}\right)^7\int_1^a \frac{y^{10}(y^2 - a^2)^2(y^2 - 1)^{1/2}}{e^{\gamma y} - 1}dy,$$

$$(5.2.14)$$

where

$$A_0 = \frac{G^2_W c}{48\pi^4 e^2}\left(\frac{m_e c}{\hbar}\right)^9 \approx 3 \times 10^{21} \quad (\text{erg cm}^{-3}\text{s}^{-1})$$

$$\gamma = \frac{\hbar\omega_0}{kT}, \quad a^2 = 1 + \frac{3}{5}\left(\frac{\omega_1}{\omega_0}\right)^2.$$

$$(5.2.15)$$

For the limiting cases [22] we have [268]:

$$(\gamma/\alpha)^2 \equiv \left(\frac{\hbar\omega_0}{m_e c^2}\right)^2 \approx \frac{4}{3\pi}\frac{e^2}{\hbar c}\left(\frac{\beta}{\alpha}\right)^2 \approx 0.0031\left(\frac{\epsilon_{Fe}}{m_e c^2}\right)^2 = \left(\frac{\rho/\mu_Z}{6.05 \cdot 10^9}\right)^{2/3}$$

at $\quad \beta/\alpha \gg 1, \quad \alpha \ll 1 \quad$ (RD) $\qquad\qquad (5.2.16)$

$$\gamma^2 \equiv \left(\frac{\hbar\omega_0}{kT}\right)^2 = \frac{8\sqrt{2}}{3\pi}\frac{e^2}{\hbar c}\sqrt{\alpha}\left(\frac{\epsilon_{Fe}}{kT}\right)^{3/2} = \frac{\rho}{\mu_Z m_u}4\pi\frac{e^2}{\hbar c}\left(\frac{\hbar}{m_e c}\right)^3\alpha^2$$

at $\quad \alpha \ll \beta \ll 2\alpha, \quad \alpha \gg 1 \quad$ (NRD) $\qquad (5.2.17)$

$$Q^{ch}_t = \begin{cases} A_0\gamma^6\alpha^{-9}2\zeta(3) & \text{at } \gamma \ll 1, \\ A_0\gamma^6\alpha^{-9}e^{-\gamma}\sqrt{\frac{\pi}{2}}\gamma^{3/2} & \text{at } \gamma \gg 1, \end{cases} \qquad (5.2.18)$$

$$Q^{ch}_l = \begin{cases} A_0\gamma^8\alpha^{-9}\frac{4}{105} & \text{at } \gamma \ll 1, \\ A_0\gamma^9\alpha^{-9}(\frac{5}{3})^{3/2}e^{-\gamma}\gamma^{-3/2}\sqrt{\frac{\pi}{8}} & \text{at } \gamma \gg 1. \end{cases} \qquad (5.2.19)$$

---

[22]Here we use the equality $\int_0^\infty \frac{x^{\nu-1}dx}{e^x-1} = \Gamma(\nu)\zeta(\nu)$, see [145].

Taking the neutral currents (5.2.3) into account, we obtain for the energy losses due to plasma neutrinos [496]:

$$Q_{\text{plasm}} = (c_V^2 + nc_V'^2)(Q_t^{\text{ch}} + Q_l^{\text{ch}}). \tag{5.2.20}$$

Consideration of plasma neutrino emission valid in the relativistic limit was given in [311a]. Other emission mechanisms are usually more effective in these conditions.

5.2.4 Interpolation Formulae

The interpolation formulae corresponding to the above three mechanisms of energy losses due to neutrino production have been obtained for charged currents in [268], the neutral currents have been included in [496]. The more detailed calculations in [562] may be approximated by interpolation formulae where

$$Q_{\text{tot}} = Q_{\text{pair}} + Q_{\text{photo}} + Q_{\text{plasm}}. \tag{5.2.21}$$

For all kinds of losses these formulae may be written as

$$Q_d = K(\rho, \alpha)e^{-c\xi} \frac{a_0 + a_1\xi + a_2\xi}{\xi^2 + b_1\alpha + b_2\alpha^2 + b_3\alpha^3}. \tag{5.2.22}$$

Here

$$K(\rho, \alpha) = g(\alpha)e^{-2\alpha} \quad \text{for } d = \text{pair}, \tag{5.2.23}$$

with

$$g(\alpha) = 1 - \frac{13.04}{\alpha^2} + \frac{133.5}{\alpha^4} + \frac{1534}{\alpha^6} + \frac{918.6}{\alpha^8}; \tag{5.2.24}$$

$$K(\rho, \alpha) = (\rho/\mu_Z)\alpha^{-5} \quad \text{for } d = \text{photo}; \tag{5.2.25}$$

$$K(\rho, \alpha) = (\rho/\mu_Z)^3 \quad \text{for } d = \text{plasma}; \tag{5.2.26}$$

$$\xi = \left(\frac{\rho/\mu_Z}{10^9}\right)^{1/3}\alpha. \tag{5.2.27}$$

The coefficients c, a_i, b_i for various d are given in Table 5.2 from [562]. The formulae (5.2.22–27) are similar to those from [268] for charged currents. For the limiting cases, the neutral currents are included in calculations by merely multiplying the results [268] by a constant factor. For the non-relativistic case these coefficients D_d^{NR} are

$$D_{\text{pair}}^{\text{NR}} = c_V^2 + nc_V'^2 = 0.925,$$

$$D_{\text{photo}}^{\text{NR}} = \frac{1}{6}(c_V^2 + nc_V'^2) + \frac{5}{6}(c_A^2 + nc_A'^2) = 0.7791, \tag{5.2.28}$$

$$D_{\text{plasm}}^{\text{NR}} = c_V^2 + nc_V'^2 = 0.925.$$

Table 5.2. Coefficients for (5.2.22) from [562]

Process	a_0	a_1	a_2	b_1	b_2	b_3	c
			10^8 K $\leq T \leq 10^{10}$ K				
pair	5.026(19)	1.745(20)	1.568(21)	9.383(-1)	-4.141(-1)	5.829(-2)	5.5924
photo	3.897(10)	5.906(10)	4.693(10)	6.290(-3)	7.483(-3)	3.061(-4)	1.5654
plasm	2.146(-7)	7.814(-8)	1.653(-8)	2.581(-2)	1.734(-2)	6.990(-4)	0.56457
			10^{10} K $\leq T \leq 10^{11}$ K				
pair	5.026(19)	1.745(20)	1.568(21)	1.2383	-8.1141(-1)	0.0	4.9924
photo	3.897(10)	5.906(10)	4.693(10)	6.290(-3)	7.483(-3)	3.061(-4)	1.5654
plasm	2.146(-7)	7.814(-8)	1.653(-8)	2.581(-2)	1.734(-2)	6.990(-4)	0.56457

For an ultrarelativistic case the coefficients $D_{\mathrm{d}}^{\mathrm{R}}$ are

$$D_{\mathrm{pair}}^{\mathrm{R}} = \frac{1}{2}(c_V^2 + nc_V'^2 + c_A^2 + nc_A'^2) = 0.8375,$$

$$D_{\mathrm{photo}}^{\mathrm{R}} = \frac{1}{2}(c_V^2 + nc_V'^2 + c_A^2 + nc_A'^2) = 0.8375, \qquad (5.2.29)$$

$$D_{\mathrm{plasm}}^{\mathrm{R}} = c_V^2 + nc_V'^2 = 0.925.$$

For the coefficients form (5.2.3), we set here $\sin^2 \theta_W = 0.23$, $n = 2$. The coefficients (5.2.28) and (5.2.29) may be used for including neutral currents in (5.2.10). The relations (5.2.21–27) determine neutrino losses due to the above mechanisms in the range $10 \leq \rho \leq 10^{14}$ g cm^{-3}, $10^{8.2} \leq T \leq 10^{11}$ K with an accuracy of $\leq 20\%$ for all kinds of losses [562]. Interpolation formulae for Q_{pair} yielding a high accuracy at high temperatures have been obtained in [67].

5.2.5 Energy Losses Due to URCA-Processes

At temperatures $T \geq 10^9$ K, an important role in the cooling process belongs to nuclear $e^{\pm}$-captures followed by $e^{\pm}$-decays or corresponding $e^{\mp}$-captures:

$$
\begin{array}{ll}
(A, Z) + e^{\pm} \rightarrow (A, Z \pm 1) + \nu^{\pm} & \text{(a)} \\
(A, Z \pm 1) \rightarrow (A, Z) + e^{\pm} + \nu^{\mp} & \text{(b)} \\
(A, Z \pm 1) + e^{\mp} \rightarrow (A, Z) + \nu^{\mp} & \text{(c)}
\end{array}
\qquad
\begin{pmatrix} \nu^+ \equiv \tilde{\nu}_e \\ \nu^- \equiv \nu_e \end{pmatrix}. \qquad (5.2.30)
$$

The reactions (5.2.30) with electrons were first examined by G. Gamov and M. Shenberg (1941) as a possible mechanism for cooling of stars. They had this idea at the casino de Urca in Rio de Janeiro to which this process owes its present name. The reactions (5.2.30) with positrons were considered later by V.S. Pinaev (1963). The $e^{\pm}$-capture rates are determined by the integral (1) of Problem 4, Sect. 5.1, with the additional factor f_e from (1.2.2) in the integrand. Similarly, the $e^{\pm}$-decay rates in matter are determined by the integral (5.1.61) with the additional factor $(1 - f_e)$.

The energy losses through URCA-processes depend drastically on the nuclear composition of matter which, in turn, depends on the reactions (5.2.30). In a general self-consistent formulation of the problem under conditions of nuclear equilibrium the composition is determined by solving equations (1.3.3) and (1.3.6). The rates of neutrino energy losses differ from corresponding reaction rates by the presence of additional factors in the integrand (see (5.1.62)): $(\epsilon_e - \Delta_{Z'Z})$ for the reaction (a), $(\Delta_{Z'Z} - \epsilon_e)$ for (b) and $(\Delta_{Z'Z} + \epsilon_e)$ for (c) from (5.2.30) [23]. Taking the values of $(Ft_{1/2})_{Z'}$ to be known (see (5.1.62–65) and Problems 2–4 from Sect. 5.1), we may represent the reaction probabilities W and energy loss rates per nucleus Q in the form [85]

$$W^{(a)} = \frac{g_{Z'}}{g_Z} \frac{\ln 2}{(Ft_{1/2})_{Z'}} \left(\frac{kT}{m_e c^2} \right)^5 I_2,$$

$$Q^{(a)} = \frac{g_{Z'}}{g_Z} \frac{\ln 2}{(Ft_{1/2})_{Z'}} \left(\frac{kT}{m_e c^2} \right)^6 m_e c^2 I_3,$$

$$W^{(b)} = \frac{\ln 2}{(Ft_{1/2})_{Z'}} \left(\frac{kT}{m_e c^2} \right)^5 I_2',$$

$$Q^{(b)} = \frac{\ln 2}{(Ft_{1/2})_{Z'}} \left(\frac{kT}{m_e c^2} \right)^6 m_e c^2 I_3', \qquad (5.2.31)$$

$$W^{(c)} = \frac{\ln 2}{(Ft_{1/2})_{Z'}} \left(\frac{kT}{m_e c^2} \right)^5 J_2,$$

$$Q^{(c)} = \frac{\ln 2}{(Ft_{1/2})_{Z'}} \left(\frac{kT}{m_e c^2} \right)^6 m_e c^2 J_3.$$

When integrating over the $e^{\pm}$ spectrum, the energies ϵ_e range within the limits

$$\Delta_{Z'Z} < \epsilon_e < \infty \qquad \text{(a)},$$
$$m_e c^2 < \epsilon_e < \Delta_{Z'Z} \qquad \text{(b)}, \qquad (5.2.32)$$
$$m_e c^2 < \epsilon_e < \infty \qquad \text{(c)}.$$

Using (5.2.32), the integrals in (5.2.31) become [24]

---

[23]The nucleus (A, Z) is assumed to be stable so that $\Delta_{Z'Z} > m_e c^2$.
[24]Coulomb corrections are not included.

$$I_k = \int\limits_0^\infty \frac{x^k(x+x_0)\sqrt{(x+x_0)^2 - \alpha^2}\ dx}{1 + \exp(x + x_0 - \beta_\pm)}, \qquad k = 2,3$$

$$J_k = \int\limits_0^\infty \frac{(x+x_0+\alpha)^k(x+\alpha)\sqrt{x^2 + 2\alpha x}\ dx}{1 + \exp(x + \alpha - \beta_\pm)}, \qquad k = 2,3 \qquad (5.2.33)$$

$$I_k' = \int\limits_0^{x_0-\alpha} \frac{(x_0-\alpha-x)^k(x+\alpha)\sqrt{x^2 + 2\alpha x}\ dx}{1 + \exp(\beta_\pm - x - \alpha)}, \qquad k = 2,3.$$

Here

$$x_0 = \Delta_{Z'Z}/kT, \quad \beta_- \equiv \beta, \quad \beta_+ = -\beta, \quad \text{see } (1.2.11). \qquad (5.2.34)$$

Analytical expressions for I_k and J_k obtained by expanding the integrands for $\alpha \ll x$ are given in [85]. I_k, J_k and I_k' may, in the general case, be evaluated numerically with the aid of (1.2.68).

An approximate evaluation of neutrino URCA losses has been made in [111], where the matter is assumed to be a mixture of nucleons and iron with concentrations determined by equations (1.3.3) and (1.3.5) with $\mu_\nu = 0$. The resulting interpolation formula has the form

$$Q_{\text{URCA}} = 1.3 \times 10^9 \rho\eta(T)\Phi(x)T_9^6 \quad (\text{erg cm}^{-3}\text{s}^{-1}) \qquad (5.2.35)$$

where

$$\eta(T) = \begin{cases} 1 & \text{at} \quad T_9 < 7 \\ 664.31 + 51.024(T_9 - 20) & \text{at} \quad 7 < T_9 < 20 \\ 664.31 & \text{at} \quad T_9 > 20, \end{cases}$$

$$\Phi(x) = \begin{cases} 1 & \text{at } \lg x \le -0.2 \\ -0.128334(\lg x)^2 - 0.036\lg x + 0.997933 & \text{at } -0.2 \le \lg x \le 0.8 \\ -0.348\lg x + 1.1654 & \text{at } 0.8 \le \lg x \le 2.3 \\ 0.0601682(\lg x)^2 - 0.543177\lg x + 1.29602 & \text{at } 2.3 \le \lg x \le 4.2, \end{cases}$$

$$(5.2.36)$$

$$x = \frac{\pi^2}{m_p}\left(\frac{\hbar}{m_e c}\right)^3\left(\frac{m_e c^2}{kT}\right)^3 \rho = 7.08595 \times 10^{-5}\frac{\rho}{T_9^3}.$$

Analogous calculations have been made in [168] with the inclusion of a larger number of species of nuclei in accordance with [224]. The results differ from (5.2.36) at $\rho \sim 10^9$ g/cm^3 and $T_9 \le 5$, where the calculations in [168] reveal an increase in Q_{URCA} arising from the β-decay of nuclei ^{55}Cr and ^{53}V with $\epsilon_\beta = 2.6$ MeV and 3.4 MeV, respectively (see (1.4.1)). Detailed tables for the rates of the reactions (5.2.30) and neutrino losses of some nuclei, taking account of the excited states of the initial and final nuclei are given in [367] for the transitions ^{26}Al$\leftrightarrow^{26}$Mg, ^{30}P$\leftrightarrow^{30}$Si, ^{31}S $\leftrightarrow^{31}$P, ^{32}S$\leftrightarrow^{32}$P, ^{33}S$\leftrightarrow^{33}$P, ^{35}Cl$\leftrightarrow^{35}$S. Both $e^\mp$-decays and $e^\pm$-captures are included here. Calculations have been made by use of $Ft_{1/2}$ values between various nuclear levels, with introduction of Coulomb corrections. Analogous calculations have been made in [368–370] for 226 nuclei with $A = 21 \div 60$. URCA processes for the most abundant nuclei ^{12}C, ^{16}O, ^{20}Ne etc. have been studied in [269].

An interesting feature of the URCA processes and their role in the cooling of very dense matter with $\epsilon_{Fe} \gg kT$ have been examined in [217]. At a low temperature the concentration of positrons is small (see Sect. 1.2.1), while the electron phase space is almost filled so that the electron capture in (5.2.30a) is not accompanied at the same density by the reactions (b) and (c). The situation is different in the presence of convection or oscillatory motion of matter in the star. A e^--capture occurs at relatively high densities around the boundary $\epsilon_{Fe} = \Delta_{Z'Z} - m_e c^2$, and an e^--decay at low densities. Hence, the neutrino losses result from the reactions (5.2.30) with e^- proceeding on both sides of the boundary $\epsilon_{Fe} = \Delta_{Z'Z} - m_e c^2$. In [217] the star shell near $\epsilon_{Fe} = \Delta_{Z'Z} - m_e c^2$ has been called the URCA shell. The neutrino losses in URCA shells may become important at the stage of degenerate presupernova prior to thermonuclear explosion [526, 349]. The URCA shells consisting of elements with odd A and a low threshold for electron capture have been examined in [217]:

^{35}Cl$\leftrightarrow$^{35}S with $\Delta = m_e c^2 + 0.168$ MeV, ^{31}P$\leftrightarrow$^{31}Si with $\Delta = m_e c^2 + 1.48$ MeV and others. The competition between additional neutrinos losses, and non-equilibrium heating in β-reactions (see Sect. 5.3) made it difficult to do a simple estimation of the influence of convective URCA shells on presupernova evolution. Recent analyses of this problem have been done in [586*] and [282e].

5.2.6 Other Mechanisms of Cooling by Neutrino Emission

Other mechanisms of cooling by neutrino emission are pointed out in [363]:

bremsstrahlung

$$e^{\pm} + (A, Z) \rightarrow e^{\pm} + (A, Z) + \nu + \tilde{\nu}, \tag{5.2.37}$$

photon–photon interaction

$$\gamma + \gamma \rightarrow \nu + \tilde{\nu} \qquad \text{and}$$
$$\gamma + \gamma \rightarrow \gamma + \nu + \tilde{\nu}, \tag{5.2.38}$$

neutrino photoproduction in nuclear field

$$\gamma + (A, Z) \rightarrow (A, Z) + \nu + \tilde{\nu}. \tag{5.2.39}$$

Calculations have shown [363, 268] that cooling by the reactions (5.2.37)–(5.2.39) is always less intensive than in the case of the reactions considered above.

5.3 Matter Heating in Non-equilibrium β-Processes

When the weak interaction reactions proceed under conditions far from thermodynamic equilibrium, they may result in a matter heating despite the neutrino energy losses. The decay of β-radioactive nuclei at low temperatures $kT \ll \epsilon_\beta$ (see (1.4.1)) may serve as the most simple example. The temperature increase is due to thermalization of the fast electrons produced in β-decays. Under astrophysical conditions, β-captures turn out often to be non-equilibrium at high densities, when $\epsilon_{Fe} > \epsilon_\beta$.

Consider matter consisting of nuclei (A, Z) at zero temperature with a step-like electron distribution function f_e from (1.2.2) (Fig. 5.1a). Let ϵ_{Fe} exceed ϵ_β owing to compression. If compression proceeds slowly (adiabatically), the electrons on the edge of the step have time to be captured by nuclei. The energy of outflowing neutrinos is then $\epsilon_\nu \approx 0$, no changes occur in the shape of the step, the entropy and temperature remain zero. At a fast contraction a finite difference $\epsilon_{Fe} - \epsilon_\beta$ arises, and all electrons with $\epsilon_\beta < \epsilon_e < \epsilon_{Fe}$ may be captured by the nucleus (A, Z). These captures produce holes in the distribution function f_e (Fig. 5.1b) smoothed down by the further thermal relaxation and subsequently taking a form different from the initial step, with non-zero temperature and entropy [282] (Fig. 5.1c). The fast compression is thus non-equilibrium and is accompanied by an irreversible increase in entropy.

5.3.1 Non-equilibrium Heating Rate

In order to derive the heating rate in non-equilibrium β-capture of electrons in the reaction (5.2.30a), we apply the second law of thermodynamics for a system of particles of variable composition [145, 56, 503] to a unit mass:

$$dE - \frac{P}{\rho^2}d\rho = TdS + \sum_i \mu_{ti}d\left(\frac{n_i}{\rho}\right) = dQ = -\frac{Q_\nu}{\rho}dt. \qquad (5.3.1)$$

Neglecting the temperature effect on the reaction rate and the rate of losses, we get from (5.2.31) and (1) of Problem 4, Sect. 5.1

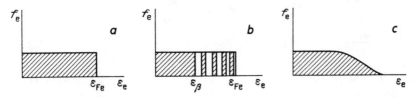

Fig. 5.1a–c. Cold matter heating in non-equilibrium beta-capture.

$$W^{(a)} = \frac{g_{Z'}}{g_Z} \frac{\ln 2}{(Ft_{1/2})_{Z'}} \int_{\delta}^{u_{Fe}} (u^2 - 1)^{1/2} u(u - \delta)^2 du =$$

$$= \frac{g_{Z'}}{g_Z} \frac{\ln 2}{(Ft_{1/2})_{Z'}} [F_0(u_{Fe}) - F_0(\delta)] \text{ (react. per nucl. per sec.)},$$

(5.3.2)

$$Q^{(a)} = \frac{g_{Z'}}{g_Z} \frac{\ln 2}{(Ft_{1/2})_{Z'}} m_e c^2 \int_{\delta}^{u_{Fe}} (u^2 - 1)^{1/2} u(u - \delta)^3 du =$$

$$= \frac{g_{Z'}}{g_Z} \frac{\ln 2}{(Ft_{1/2})_{Z'}} m_e c^2 [F_\epsilon(u_{Fe}) - F_\epsilon(\delta)] \text{ (erg} \cdot \text{s}^{-1} \text{per nucleus)},$$

(5.3.3)

with $F_0(u)$ defined by (5.1.63) and

$$F_\epsilon(u) = \frac{\sqrt{u^2 - 1}}{240} [40u^5 - 144\delta u^4 + 10u^3(18\delta^2 - 1) +$$
$$+ 16\delta u^2(3 - 5\delta^2) - 15u(1 + 6\delta^2) + 16\delta(6 + 5\delta^2)] -$$
$$- \frac{1 + 6\delta^2}{16} \ln(\sqrt{u^2 - 1} + u).$$

(5.3.4)

Equations (5.3.2) and (5.3.3) give

$$Q_\nu = n_{A,Z} Q^{(a)}, \quad \rho d(n_{A,Z}/\rho) = -n_{A,Z} W^{(a)} dt.$$

(5.3.5)

The chemical potentials of non-relativistic nuclei and cold electrons are

$$\mu_{t,Z} \approx m_{A,Z} c^2, \quad \mu_{t,Z-1} \approx m_{A,Z-1} c^2, \quad \mu_{t,e} \approx \epsilon_{Fe} + m_e c^2.$$

(5.3.6)

Using (5.3.2–6), we find from (5.3.1)

$$\rho T \frac{dS}{dt} = [(\epsilon_{Fe} + m_e c^2 - \Delta_{Z'Z}) W^{(a)} - Q^{(a)}] n_{A,Z}$$
$$= \frac{g_{Z'}}{g_Z} \frac{\ln 2}{(Ft_{1/2})_{Z'}} n_{A,Z} m_e c^2 \{(u_{Fe} - \delta)[F_0(u_{Fe}) - F_0(\delta)]$$
$$- F_\epsilon(u_{Fe}) + F_\epsilon(\delta)\}.$$

(5.3.7)

For $u_{Fe} \gg \delta \gg 1$,

$$F_0(u_{Fe}) \approx \frac{u_{Fe}^5}{5}, \quad F_0(\delta) \approx \frac{\delta^5}{5}, \quad F_\epsilon(u_{Fe}) \approx \frac{u_{Fe}^6}{6}, \quad F_\epsilon(\delta) \approx \frac{\delta^6}{6},$$

$$\rho T \frac{dS}{dt} \approx \frac{g_{Z'}}{g_Z} \frac{\ln 2}{(Ft_{1/2})_{Z'}} n_{A,Z} m_e c^2 \frac{u_{Fe}^6 - \delta^6}{30}.$$

(5.3.8)

The matter heating thus takes at $u_{Fe} \gg \delta \gg 1$ five times less energy than the neutrino emission. It has been pointed out in [108] that at $u_{Fe} \gg \delta$ a neutrino takes away the energy of $5/6\epsilon_{Fe}$ per electron capture. For $u_{Fe} \approx \delta$ we have

$$F_0(u) \approx \delta\sqrt{\delta^2 - 1}\frac{(u - \delta)^3}{3},$$

$$F_e(u) \approx \delta\sqrt{\delta^2 - 1}\frac{(u - \delta)^4}{4}, \tag{5.3.9}$$

$$\rho T\frac{dS}{dt} = \frac{g_{Z'}}{g_Z}\frac{\ln 2}{(Ft_{1/2})_{Z'}}n_{A,Z}m_e c^2\frac{(u_{Fe} - \delta)^4}{12}.$$

This shows that the heating takes 3 times less energy than the neutrino emission. Including temperature corrections gives, for a more general case, in lieu of (5.3.7),

$$\rho T\frac{dS}{dt} = [(\mu_{te} - \mu_{t,Z-1} + \mu_{t,Z})W^{(a)} - Q^{(a)}]n_{A,Z} \tag{5.3.7a}$$

with the general relations (1.2.9) and (1.3.2) for μ_t. The temperature dependence of non-equilibrium heating has been examined in [186] where it is shown that for the reaction (5.3.14d) a neutrino cooling takes place instead of heating at $kT > 0.24 m_e c^2$.

5.3.2 The Effect of Capture onto Excitation Levels of the Final Nucleus

The capture onto excitation levels causes the excitation energy to convert into heat and the efficiency of non-equilibrium heating to increase. This role of excitation levels has been discussed in [56, 187]. The non-equilibrium heating has been calculated in [54] with inclusion of excitation levels for a gaseous nuclear model (see Problem 5, Sect. 5.1). In order to calculate the neutrino emission power, we have to evaluate the integral yielded by multiplying the integrand in (5) of the given problem by cp_ν. Performing calculations to the approximation (36) of Problem 5, Sect. 5.1, we have [54]

$$Q^{(a)} = \frac{W_\mu}{2\pi}\left(\frac{137}{Z_e m_\mu c}\right)^3\frac{Z}{Z_e}\frac{8m^2c^3q_0^2}{3(mc + q_0)^4}\frac{p_{k,e}^4}{p_{k,\mu}}\frac{\varphi_L}{\varphi_\mu}, \tag{5.3.10}$$

where Z_e, φ_μ, $p_{k,e}$, $p_{k,\mu}$ are defined in (37) and (39) of the Problem 5, Sect. 5.1 and

$$\varphi_L = \left(\frac{1}{7} + \frac{x^2}{35}\right)p_{k,e}^3 + \left(\frac{1}{2} - \frac{x}{3} + \frac{x^2}{30}\right)p_{e0}p_{k,e}^2 +$$

$$+ \left(\frac{3}{5} - \frac{4}{5}x + \frac{x^2}{5}\right)p_{e0}^2 p_{k,e} + \frac{1}{4}(1 - x)^2 p_{e0}^3; \quad x = \frac{q_0}{mc}. \tag{5.3.11}$$

For $p_{k,e} \gg p_{e0}$, the mean energy of the emitted neutrino according to (38) of Problem 5, Sect. 5.1 and (5.3.10) will be

$$\bar{\epsilon}_\nu = \frac{Q^{(a)}}{W_e} = \frac{6m^2c^2 + 0.2q_0^2}{7mc(mc + q_0)}\epsilon_{Fe} \approx 0.6\epsilon_{Fe} \tag{5.3.12}$$

instead of $\bar{\epsilon}_\nu = \frac{5}{6}\epsilon_{Fe}$ for the similar case with no excitation levels included in the calculation. The equation for changes in entropy including the excitation levels is determined by the first equality (5.3.7) or (5.3.7a) at $W^{(a)} = W_e$ from (38) of Problem 5, Sect. 5.1 and $Q^{(a)}$ from (5.3.10). The value of $\Delta_{Z'Z}$ is still determined by the energy difference (5.1.5) of the nuclear ground states.

5.3.3 Two-Step Non-equilibrium Electron Capture in White Dwarfs near the Chandrasekhar Limit

In the central regions of white dwarfs with mass close to the limit mass M_{Ch} the neutronization reaction takes place, and if the core of the new phase is small enough but finite, the white dwarf remains stable ([191], see Chap. 12, Vol. 2). As even-even nuclei are more stable than odd-odd ones, the threshold energy for the electron to be captured by even-even nucleus (A, Z) exceeds the corresponding energy for an odd-odd nucleus $(A, Z - 1)$[25]:

$$\Delta_{Z-1,Z} > \Delta_{Z-2,Z-1}. \tag{5.3.13}$$

The electron capture by the nucleus $(A, Z - 1)$ is then non-equilibrium at $\epsilon_{Fe} = \epsilon_\beta(A, Z) > \epsilon_\beta(A, Z - 1)$ (see (1.4.1)) and accompanied by heating. The heat effect of this reaction has been considered in [56] for certain reaction chains:

(a) $^{32}_{16}S(0^+) \rightarrow ^{32}_{15}P(1^+) \rightarrow ^{32}_{14}Si(0^+)$, (1708; 213)

(b) $^{38}_{18}Ar(2^-) \rightarrow ^{38}_{17}Cl(2^-) \rightarrow ^{38}_{16}S(0^+)$, (4810; 3000)

(c) $^{42}_{20}Ca(0^+) \rightarrow ^{42}_{19}K(2^-) \rightarrow ^{42}_{18}Ar(0^+)$, (3524; 583) $\hspace{1cm}$ (5.3.14)

(d) $^{56}_{26}Fe(0^+) \rightarrow ^{56}_{25}Mn^*(1^+) \rightarrow ^{56}_{25}Mn(3^+) \rightarrow ^{56}_{24}Cr(0^+)$, (3809; 109; 1610)

(e) $^{60}_{28}Ni(0^+) \rightarrow ^{60}_{27}Co^*(2^+) \rightarrow ^{60}_{27}Co(5^+) \rightarrow ^{60}_{26}Fe(0^+)$, (2890; 60; 260)

(f) $^{60}_{30}Zn(0^+) \rightarrow ^{60}_{29}Cu(1^+) \rightarrow ^{60}_{28}Ni(0^+)$, (2630; 200).

Here, the energies ϵ_β (in keV) for the two successive transitions are indicated in parentheses on the right. When the excited nuclei $^{56}Mn^*$ and $^{56}Co^*$ form, the second number in parentheses is the excitation energy of an intermediate nucleus. The chemical symbol is followed by indication of the nuclear spin and parity in the ground or relevant excited state. Taking approximately $\sim \frac{1}{6}\epsilon_{Fe}$ from (5.3.8) for heat losses per electron capture gives for heat release in the reactions (5.3.14) 249, 302, 490, 476, 498, 405 keV, respectively. For the reactions (d) and (e), the heating due to the excitation level of the intermediate nucleus with excitation energy conversion into heat is included in the calculations. The effect of this phenomenon on the white dwarf cooling is considered in Sect. Chap. 11, Vol. 2.

---

[25] We mean here of nuclei with odd Z and odd $N = A - Z$.

5.3.4 Non-equilibrium Matter Heating During Collapse

The increase in density in the collapsing stellar core after losing hydrodynamical stability (see Chap. 10, Vol. 2) results in a strong non-equilibrium with $\epsilon_{\text{Fe}} \gg \epsilon_\beta$ and a rapid heating. For the case of a free collapse, the equation (5.3.7) is solved simultaneously with the chemical kinetics equation similar to (4.1.6) in [187, 501]. A self-consistent model for stellar collapse with equations of motion for parameters averaged over the star has being solved in [19, 20] by use of techniques from [26]. This problem has been solved in exact formulation in [66]. It may be seen from these papers that the initially cold iron stellar core starts collapsing at $\rho = 1.3 \times 10^9 \, \text{g cm}^{-3}$, then heats up to $T = 4 \times 10^9 \, \text{K}$ at $\rho = 2 \times 10^{10} \, \text{g cm}^{-3}$, when the nuclear equilibrium sets in.

5.4 Neutrino Transfer and Neutrino Heat Conduction

In the central regions of a hot neutron star resulting from collapse the temperature and density are so high that neutrinos can not escape freely and remain in a state close to thermodynamic equilibrium [28, 111]. If we take neutrinos as massless, the transfer equation analogous to (2.1.9)

$$
\frac{1}{c}\frac{\partial I_{\nu\epsilon}}{\partial t} + \frac{\partial I_{\nu\epsilon}}{\partial s} = -\alpha_{\nu\epsilon}^* \rho (I_{\nu\epsilon} - I_{\nu\epsilon}^{\text{p}}) +
$$
$$
+ \rho \int \left[\frac{\epsilon_\nu}{\epsilon_\nu'} K_\nu(l_i \cdot l_i', \epsilon_\nu' \to \epsilon_\nu) I_{\nu\epsilon'}' \left(1 - \frac{c^2 h^3}{\epsilon_\nu^3} I_{\nu\epsilon}\right) - \right.
$$
$$
\left. - K_\nu(l_i \cdot l_i', \epsilon_\nu \to \epsilon_\nu') I_{\nu\epsilon} \left(1 - \frac{c^2 h^3}{\epsilon_\nu'^3} I_{\nu\epsilon'}'\right) \right] d\epsilon_\nu' d\Omega'/4\pi \qquad (5.4.1)
$$

will be valid for them under LTE conditions [115, 430, 301]. Here the subscript ν denotes neutrino instead of frequency in (2.1.9). The intensity of neutrino emission $I_{\nu\epsilon} \equiv I_\nu(l_i, \epsilon_\nu) \, \text{erg cm}^{-2} \, \text{s}^{-1} \, \text{ster}^{-1} \, \text{erg}^{-1}$ is normalized to unit energy range rather than frequency range as in (2.1.4). The equation (5.4.1) holds as well at $m_{\nu 0} \neq 0$ if $\epsilon_\nu \gg m_{\nu 0} c^2$. The left hand side of (5.4.1) equals (2.1.11) and (2.1.12) for the case of plane and spherical symmetry, respectively. The equilibrium neutrino intensity $I_{\nu\epsilon}^{\text{p}}$ is, similarly to (2.1.14),

$$
I_{\nu\epsilon}^{\text{p}} = \frac{\epsilon_\nu^3}{c^2 h^3} \frac{1}{1 + \exp\left(\frac{\epsilon_\nu - \mu_\nu}{kT}\right)}. \qquad (5.4.2)
$$

The variables l_i, l_i', μ_ν stand for the same quantities as in Sect. 2.1. The neutrino absorption coefficient allowing for induced processes is [430]

$$
\alpha_{\nu e}^* = \frac{\sigma_\nu^a N_a}{\rho} \left[1 + \exp\left(\frac{\mu_\nu - \epsilon_\nu}{kT}\right)\right], \qquad \alpha_{\nu e} = \frac{\sigma_\nu^a N_a}{\rho}, \qquad (5.4.3)
$$

where σ_ν^a is the absorption cross-section, N_a is the concentration of absorbing particles. Hence, the induced processes enhance the absorption by fermions contrary to absorption by photons (bosons) in (2.1.13a).

We now derive the relations (2.1.13a) and (5.4.3) for the absorption coefficients allowing for induced processes. The radiative and absorption terms in the right-hand sides of (2.1.9), (5.4.1) are

$$\rho\left[j_\nu\left(1 + \frac{c^2}{2h\nu^3}I_\nu\right) - \alpha_\nu I_\nu\right] \quad \text{and}$$

$$\rho\left[j_{\nu\epsilon}\left(1 - \frac{c^2 h^3}{\epsilon_\nu^3}I_{\nu\epsilon}\right) - \alpha_{\nu\epsilon}I_{\nu\epsilon}\right], \tag{5.4.4}$$

respectively. Under LTE conditions j_ν and $j_{\nu\epsilon}$ coincide with their equilibrium values, and the quantities in square brackets become zero at $I_\nu = B_\nu$, $I_{\nu\epsilon} = I_{\nu\epsilon}^P$ from (2.1.14) and (5.4.2). We then have for equilibrium j_ν and $j_{\nu\epsilon}$

$$j_\nu = \alpha_\nu\left(1 - e^{-\frac{h\nu}{kT}}\right)B_\nu, \quad j_{\nu\epsilon} = \alpha_{\nu\epsilon}\left(1 + e^{\frac{\mu_\nu - \epsilon_\nu}{kT}}\right)I_{\nu\epsilon}^P. \tag{5.4.5}$$

Using (5.4.5) allows us to represent (5.4.4) in the desirable form

$$\rho\alpha_\nu\left(1 - e^{-\frac{h\nu}{kT}}\right)(B_\nu - I_\nu) = \rho\alpha_\nu^*(B_\nu - I_\nu),$$

$$\rho\alpha_{\nu\epsilon}\left(1 + e^{\frac{\mu_{t\nu} - \epsilon_\nu}{kT}}\right)(I_{\nu\epsilon}^P - I_{\nu\epsilon}) = \rho\alpha_{\nu\epsilon}^*(I_{\nu\epsilon}^P - I_{\nu\epsilon}). \tag{5.4.6}$$

The last scattering term in (5.4.1) involves induced fermion processes, similarly to (2.1.16) for bosons, and

$$\int K_\nu(l_i \cdot l_i', \epsilon_\nu \to \epsilon_\nu')d\epsilon_\nu'\frac{d\Omega'}{4\pi} = \sigma_{\nu\epsilon} = \frac{\sigma_\nu^s N_s}{\rho}, \tag{5.4.7}$$

where σ_ν^s is the scattering cross-section, N_s is the number of scattering particles, $\sigma_{\nu\epsilon}$ is the angle-averaged coefficient for neutrino scattering in analogy with (2.1.8). The principle of detailed balance yields [430, 301]

$$K_\nu(l_i \cdot l_i', \epsilon_\nu \to \epsilon_\nu')\epsilon_\nu^2 e^{-\frac{\epsilon_\nu}{kT}} = K_\nu(l_i \cdot l_i', \epsilon_\nu' \to \epsilon_\nu)\epsilon_\nu'^2 e^{-\epsilon_\nu'/kT}. \tag{5.4.8}$$

The equilibrium intensity (5.4.2) combined with (5.4.8) reduces the last term in (5.4.1) to zero. We note that a relation similar to (5.4.8) with the substitution $\epsilon \to \nu$ holds also for photon scattering, and (2.1.16) is therefore reduced to zero by the Planck intensity (2.1.14).

Unlike the photon, the neutrino has its antiparticle $\tilde{\nu}$. A kinetic equation similar to (5.4.1) may be written for the antineutrino as well. At a large neutrino depth, when the distribution function is close to the equilibrium given by (5.4.2), the equality $\mu_{\tilde{\nu}} = -\mu_\nu$ holds (see (1.3.5)), and the first relation

(1.2.56) with $2(n_\nu - n_{\bar\nu})$ in the left side, and β_ν instead of β should be used for finding μ_ν. The neutrino diffusion is characterized by the transfer of leptonic charge Q_{ve}, in addition to the energy transfer, which is the only describing the photon transport, thus being the analog of an ordinary diffusion (see (2.4.13)). It is convenient to write the set of transfer equations for energy and leptonic charge derived in [115] for a spherically symmetric star as [498]:

$$\Lambda = \frac{Q_{ve}}{\rho} = \frac{n_\nu + n_{e^-} - n_{\bar\nu} - n_{e^+}}{\rho},$$

$$\frac{\partial E}{\partial t} + P\frac{\partial(1/\rho)}{\partial t} = -4\pi\frac{\partial}{\partial m}(r^2 H), \tag{5.4.9}$$

$$\frac{\partial \Lambda}{\partial t} + 4\pi\frac{\partial}{\partial m}(r^2 F) = 0,$$

$$\frac{\partial r}{\partial m} = \frac{1}{4\pi\rho r^2}. \tag{5.4.10}$$

Here, E and P are specific energy and pressure containing the neutrino contribution (see Chap. 1). The thermal H and diffusion F neutrino fluxes are [115]

$$H = -\frac{7}{8}\frac{4acT^3}{3}4\pi r^2\rho\left(l_{\rm T}\frac{\partial T}{\partial m} + Tl_\psi\frac{\partial\beta_\nu}{\partial m}\right),$$

$$F = -\frac{7}{8}\frac{4acT^3}{3k}4\pi r^2\rho\left(\frac{\lambda_{\rm T}}{T}\frac{\partial T}{\partial m} + \lambda_\psi\frac{\partial\beta_\nu}{\partial m}\right), \tag{5.4.11}$$

$$\beta_\nu = \mu_\nu/kT.$$

Calculating the Rosseland mean similarly to (2.2.8), gives for neutrino mean free paths

$$l_{\rm T} = l_{{\rm T}\nu} + l_{{\rm T}\bar\nu}, \quad l_\psi = l_{\psi\nu} - l_{\psi\bar\nu},$$

$$\lambda_{\rm T} = \lambda_{{\rm T}\nu} - \lambda_{{\rm T}\bar\nu}, \quad \lambda_\psi = \lambda_{\psi\nu} + \lambda_{\psi\bar\nu}, \tag{5.4.12}$$

where

$$l_{{\rm T}\nu} = \frac{15}{7\pi^4}\int_0^\infty l_{\nu 0}\frac{x^4\exp[2(x - \beta_\nu)]}{[1 + \exp(x - \beta_\nu)]^3}\,dx,$$

$$l_{\psi\nu} = \lambda_{{\rm T}\nu} = \frac{15}{7\pi^4}\int_0^\infty l_{\nu 0}\frac{x^3\exp[2(x - \beta_\nu)]}{[1 + \exp(x - \beta_\nu)]^3}\,dx, \tag{5.4.13}$$

$$\lambda_{\psi\nu} = \frac{15}{7\pi^4}\int_0^\infty l_{\nu 0}\frac{x^2\exp[2(x - \beta_\nu)]}{[1 + \exp(x - \beta_\nu)]^3}\,dx, \quad x = \frac{\epsilon_\nu}{kT}.$$

The quantities $l_{{\rm T}\bar\nu}, l_{\psi\bar\nu}, \lambda_{{\rm T}\bar\nu}, \lambda_{\psi\bar\nu}$ for the antineutrino may be obtained from (5.4.13) by substituting $\beta_\nu \Rightarrow (-\beta_\nu)$, $l_{\nu 0} \Rightarrow l_{\bar\nu 0}$. The neutrino free path before interaction with nucleus $(A, Z - 1)$ is

$$l_{\nu 0}^{A,Z-1} = \frac{1}{\sigma_\nu^{A,Z-1} n_{A,Z-1}}, \quad l_{\bar\nu 0}^{A,Z} = \frac{1}{\sigma_{\bar\nu}^{A,Z} n_{A,Z}}. \tag{5.4.14}$$

The cross-section for neutrino capture, analogous to that for electron capture given by (4), Problem 2, Sect. 5.1, is

$$\sigma_\nu^{A,Z-1} = \frac{W_\nu^{A,Z-1}}{n_\nu c}, \tag{5.4.15}$$

with $W_\nu^{A,Z-1}$ obtained from the electron capture probability W_Z^{cap} from (2), Problem 2, Sect. 5.1, by the substitutions[26]

$$n_e \Rightarrow n_\nu, \quad \delta(\epsilon_e - \epsilon_\nu - \Delta_{Z-1,Z}) \Rightarrow \delta(\epsilon_\nu + \Delta_{Z-1,Z} - \epsilon_e),$$

$$p_\nu^2 dp_\nu \Rightarrow (1 - f_e) p_e^2 dp_e = \frac{1}{c^3}(1 - f_e)\sqrt{\epsilon_e^2 - m_e^2 c^4}\, \epsilon_e d\epsilon_e, \tag{5.4.16}$$

$$|M|_Z^2 \Rightarrow |M|_{Z-1}^2.$$

Recalling f_e from (1.2.2), we then have

$$\sigma_\nu^{A,Z-1} = \frac{1}{4\pi c}\left(\frac{h}{m_e c}\right)^3 \frac{\ln 2}{(Ft_{1/2})_{Z-1}} \frac{\epsilon_e \sqrt{\epsilon_e^2 - m_e^2 c^4}}{(m_e c^2)^2} \exp\left(\frac{\epsilon_e - \mu_{te}}{kT}\right) \times$$

$$\times \left[1 + \exp\left(\frac{\epsilon_e - \mu_{te}}{kT}\right)\right]^{-1} \quad \text{at} \quad \epsilon_e = \epsilon_\nu + \Delta_{Z-1,Z}. \tag{5.4.17}$$

Substituting (5.4.17) into (5.4.14) gives

$$l_{\nu 0}^{A,Z-1} = \frac{(m_e c^2)^2}{\epsilon_e \sqrt{\epsilon_e^2 - m_e^2 c^4}\, \sigma_0 n_{A,Z-1}} \frac{1}{} \frac{1 + \exp\left(\frac{\epsilon_e - \mu_{te}}{kT}\right)}{\exp\left(\frac{\epsilon_e - \mu_{te}}{kT}\right)}, \tag{5.4.18}$$

$$\epsilon_e = \epsilon_\nu + \Delta_{Z-1,Z},$$

$$\sigma_0 = \frac{1}{4\pi c}\left(\frac{h}{m_e c}\right)^3 \frac{\ln 2}{(Ft_{1/2})_{Z-1}} \quad (\approx 2 \times 10^{-44} \mathrm{cm}^2 \text{ for } Ft_{1/2} \text{ determined}$$

from the neutron decay).

In the presence of several nuclear species we have

$$l_{\nu 0} = \left(\sum \frac{1}{l_{\nu 0}^{A,Z-1}}\right)^{-1}. \tag{5.4.19}$$

If there is nothing but nucleons in the matter, we introduce, using (1.3.2), (1.3.5)

---

[26] $W_{\bar\nu}^{A,Z}$ is obtained from $W_\nu^{A,Z}$ by replacing $\nu \Rightarrow \bar\nu$, $e^- \Rightarrow e^+$ and changing sign before the term with $\Delta_{Z-1,Z}((A,Z) + \bar\nu \to (A, Z-1) + e^+)$.

$$\theta = \frac{n_n}{n_p} = \exp[(\mu_{te} - \mu_\nu - \Delta_n)/kT]. \tag{5.4.20}$$

For ultrarelativistic electrons and neutrinos, when $m_e c^2$ and Δ_n are negligible in comparison with $\epsilon_{e,\nu}$ the integrals (5.4.13) may be evaluated analytically [115, 498]:

$$l_{T\nu} = \frac{15}{7\pi^4} \left(\frac{m_e c^2}{kT}\right)^2 \frac{1}{\sigma_0 n_n} \times$$

$$\times \int_0^\infty \frac{x^2 \{\exp[2(x - \beta_\nu)] + \exp(\beta - \beta_\nu)\exp(x - \beta_\nu)\}}{[1 + \exp(x - \beta_\nu)]^3} dx =$$

$$= \frac{15}{4\pi^4} \left(\frac{m_e c^2}{kT}\right) \frac{1}{\sigma_0} \left[\frac{F_0(\beta_\nu) + F_1(\beta_\nu)}{n_n} - \frac{F_0(\beta_\nu) - F_1(\beta_\nu)}{n_p}\right]. \tag{5.4.21}$$

Similarly,

$$l_{T\tilde\nu} = \frac{15}{7\pi^4} \left(\frac{m_e c^2}{kT}\right)^2 \frac{1}{\sigma_0} \left[\frac{F_0(-\beta_\nu) + F_1(-\beta_\nu)}{n_p} - \frac{F_0(-\beta_\nu) - F_1(-\beta_\nu)}{n_n}\right]. \tag{5.4.22}$$

Thus, using (5.4.12) and (1.2.55), we obtain

$$l_T = B(1 + \theta)\left[\frac{1 + \theta}{2\theta}\left(\beta_\nu^2 + \frac{\pi^2}{3}\right) - \frac{\theta - 1}{\theta}\beta_\nu\right], \tag{5.4.23}$$

$$B = \frac{15}{7\pi^4}\left(\frac{m_e c^2}{kT}\right)^2 \frac{m_p}{\sigma_0 \rho}.$$

Other coefficients in (5.4.12) and (5.4.13) are evaluated in a similar way [115]:

$$l_\psi = \lambda_T = \frac{1}{2}B\frac{(1 + \theta)^2}{\theta}\left(\beta_\nu - \frac{\theta - 1}{\theta + 1}\right)$$

$$\lambda_\psi = \frac{1}{2}B\frac{(1 + \theta)^2}{\theta}. \tag{5.4.24}$$

The scattering of neutrinos by electrons is allowed for by adding the term $(1/l_{\nu 0}^s)$ under the sign of summation in (5.4.19), where

$$l_{\nu 0}^s = \frac{1}{\langle\sigma_\nu^s\rangle n_e},$$

and the scattering cross-sections are averaged, with use of (5.1.82), over the distribution of scattering electrons. At this point it is necessary, according to (5.1.82), to allow for filling in the phase space of scattered neutrinos by a factor of $\left(1 - \frac{c^2 h^2 I_{\nu e}^p}{\epsilon_\nu^3}\right)$, together with $(1 - f_e)$ in (5.4.16). The integration over the phase space becomes more complicated compared to $W_\nu^{A,Z-1}$ because of the necessity to take into account relativistic effects in terms involving e and ν. Analytical expressions like (5.4.23) have not been obtained here. As the scattering cross-section including neutral currents is several times smaller than the absorption cross-section of nucleons, the scattering becomes important for electron neutrinos only at such high temperatures that the number of pairs e^+e^- is far above the number of nucleons. For other neutrino species that do not undergo absorption on nucleons, the scattering due to neutral currents must always be taken into account. For $\tau_\nu > 1$, the $\nu\nu$-scattering is as important as νe-scattering. Approximate techniques for calculation of neutrino scattering at $\tau_\nu \leq 1$ have been developed in [430].

6 Equations of Equilibrium
and Stellar Evolution
and Methods for Their Solution

6.1 Spherically Symmetric Stars

Most evolutionary calculations to date deal with non-rotating, spherically symmetric stars. This model has allowed us to obtain evolutionary tracks on the Hertzsprung–Russell diagram $\lg L$–$\lg T_{\text{ef}}$ (see Sect. 2.2) giving a satisfactory explanation for observational stellar types: the main sequence representing the longest phase in a star's life with hydrogen conversion into helium; giant and supergiant branch corresponding to stars with a hot dense core consisting of helium or carbon and with an extended, hydrogen rich envelope; white dwarfs as stars with completed evolution and radiation due to cooling; T Tauri stars as young contracting stars with luminosity accounted for by gravitational energy released by contraction, and others.

The surface temperature of a star is characterized by a spectral class related to the effective temperature T_{ef} as is shown in Table 6.1 (from [562a]). Stellar evolution equations in a spherically symmetric approximation become [229, 165]

$$\frac{dP}{dr} = -\rho\frac{Gm}{r^2} \quad \text{the equilibrium equation,} \tag{6.1.1}$$

$$\frac{dm}{dr} = 4\pi\rho r^2 \quad \text{the continuity equation,} \tag{6.1.2}$$

$$\frac{dL_r}{dr} = 4\pi\rho r^2\left(\epsilon_n - \epsilon_\nu - \frac{\partial E}{\partial t} + \frac{P}{\rho^2}\frac{\partial\rho}{\partial t}\right) \quad \text{the energy equation,} \tag{6.1.3}$$

$$L_r = L_r^{\text{rad}} = -\frac{4acT^3}{3\kappa\rho}4\pi r^2\frac{dT}{dr} \quad \text{the radiation heat transfer equation,}$$

$$\nabla_{\text{rad}} < \nabla_a + \nabla_\mu \quad \text{(see (3.1.11)),} \tag{6.1.4a}$$

$$L_r = 4\pi r^2 F_{\text{conv}} + L_r^{\text{rad}} = 4\pi r^2 c_p\rho\left[-\frac{Gm}{\rho r^2}\left(\frac{\partial\rho}{\partial T}\right)_P\right]^{1/2}$$

$$\times \frac{l^2}{4}(\Delta\nabla T)^{3/2} + L_r^{\text{rad}} \quad \text{the convection heat transfer equation derived}$$

$$\text{from the mixing-length theory,} \tag{6.1.4b}$$

Table 6.1. Relationship between spectral class and surface effective temperature $T_{ef}(K)$ for the main sequence stars (ms), giants (g) and supergiants (sg), from [562a]

Spectral Class	T_{ef} ms	T_{ef} g	T_{ef} sg
O3	52 500	50 000	47 300
O6	41 000	39 500	39 000
O9	33 000	32 000	32 600
B0	30 000	29 000	26 000
B2	22 000	20 300	18 500
B5	15 400	15 000	13 600
B7	13 000	13 200	12 200
B9	10 500	11 000	10 300
A0	9 520	10 100	9 730
A3	8 720	8 600	8 770
A8	7 580	7 450	7 950
F0	7 200	7 150	7 700
F2	6 890	6 870	7 350
F8	6 200	6 150	6 100
G0	6 030	5 850	5 550
G2	5 860	5 450	5 200
G8	5 570	4 900	4 600
K0	5 250	4 750	4 420
K2	4 900	4 420	4 250
K5	4 350	3 950	3 850
K7	4 060	3 850	3 700
M0	3 850	3 800	3 650
M3	3 470	3 530	3 200
M6	3 050	3 240	2 600
M8	2 640	-	-

$$\frac{dT}{dr} = \left(\frac{\partial T}{\partial P}\right)_{S,\mu} \frac{dP}{dr} \qquad \text{the adiabatic convection equation,} \qquad (6.1.4c)$$

$$\frac{\partial x_H}{\partial t} = -4m_p \left(\frac{\epsilon_{CNO}}{Q_{CNO}} + \frac{\epsilon_{pp}}{Q_{pp}}\right) \qquad \text{the hydrogen burning equation,} \quad (6.1.5)$$

$$\left.\begin{aligned}
\frac{\partial x_\alpha}{\partial t} &= 4m_p \left(\frac{\epsilon_{CNO}}{Q_{CNO}} + \frac{\epsilon_{pp}}{Q_{pp}}\right) - \\
&\quad - \frac{3m_\alpha \epsilon_{3\alpha}}{Q_{3\alpha}} - m_\alpha \epsilon_{12C\alpha}/Q_{12C\alpha} \\
\frac{\partial x_{12C}}{\partial t} &= \frac{3m_\alpha \epsilon_{3\alpha}}{Q_{3\alpha}} - \frac{m_{12C}\epsilon_{12C\alpha}}{Q_{12C\alpha}}
\end{aligned}\right\} \quad \begin{aligned} &\text{the helium and carbon} \\ &\text{burning equations,} \end{aligned} \quad (6.1.6)$$

$$\epsilon_n = \epsilon_{pp} + \epsilon_{3\alpha} + \epsilon_{12C\alpha}.$$

Analogous equations may be written for other nuclear reactions. Here, m is the mass inside radius r, $L_r \equiv L(r)$, ϵ_n and ϵ_ν are the rates of nuclear energy release and neutrino losses, respectively. The two last terms in (6.1.3) represent the heat release or absorption at stellar contraction or expansion (a gravitational energy source). The convective flux from (6.1.4b) for a mixture of a perfect, totally ionized gas with radiation is given by (3.1.20). Expressions for c_p, $\left(\frac{\partial T}{\partial P}\right)_S = \frac{T}{P}\gamma_2$ at $\mu = $ const. are given in (1.1.12), (1.1.15), and (1.1.20); $\Delta \nabla \tilde{T}$ is determined by (3.1.6).

Equations (6.1.1–4) serve for calculating the structure of each evolutionary model. Equations (6.1.5) and (6.1.6) are needed for finding the chemical composition at the transition to the next evolutionary model. In convective regions where the chemical composition is constant owing to the mixing, equations (6.1.5) and (6.1.6) should be replaced by those averaged over the convective region and including terms that account for additional matter gains caused by the expansion of the convective zone mass.

For a mixture of a perfect gas with radiation at total ionization (1.1.1–3,7), the logarithmic gradients in (6.1.4) read

$$\nabla_a = \gamma_2 = \left(\frac{\partial \ln T}{\partial \ln P}\right)_S = \frac{2(4 - 3\beta_g)}{32 - 24\beta_g - 3\beta_g^2} \qquad \text{(see (1.1.20)),} \qquad (6.1.7)$$

$$\nabla_\mu = \left(\frac{\partial \ln T}{\partial \ln \mu}\right)_{P,\rho} \frac{d \ln \mu}{d \ln P} = \frac{\beta_g}{4 - 3\beta_g} \frac{d \ln \mu}{d \ln P} \qquad \text{(see (3.1.11)),} \qquad (6.1.8)$$

$$\nabla_{\text{rad}} = \frac{d \ln T}{d \ln P} = \frac{\kappa L_r}{16\pi c G m (1 - \beta_g)} \qquad \text{(see (6.1.1) and (6.1.4a)).} \quad (6.1.9)$$

The rates of energy release ϵ_{pp}, ϵ_{CNO}, $\epsilon_{3\alpha}$, $\epsilon_{12C\alpha}$, and released energies per reaction Q_{pp}, Q_{CNO}, $Q_{3\alpha}$, $Q_{12C\alpha}$ are given in (4.2.4–6), (4.2.12–14), and (4.2.38–41). The thermodynamic functions of matter $P(\rho, T, x_i)$, $E(\rho, T, x_i)$ can be found in Chap. 1, calculation of the opacity κ in Sect. 2.3, and neutrino loss rates in Sect. 5.2.

An evolutionary calculation usually starts from a homogeneous model with a uniform chemical composition. If we deal with an evolutionary stage before the main sequence, then the initial distribution of mass is taken from calculations of the gas cloud collapse. If the calculation starts from the main sequence, a chemically homogeneous model in thermal equilibrium and without gravitational contraction sources is taken to be initial. The model calculations require specifying boundary conditions. Obviously, in the centre we have

$$L_r = 0, \qquad m = 0 \qquad \text{at } r = 0. \tag{6.1.10}$$

An outer boundary condition is specified on the photosphere radius identified with the star's radius R

$$L = \pi a c R^2 T_{\text{ef}}^4, \qquad T = T_{\text{ef}} \text{ at } m = M. \tag{6.1.11}$$

The other outer boundary condition is suggested by an approximate solution of the equilibrium equation in the region almost transparent for radiation. For a mixture of a perfect gas with radiation we find from (6.1.1), (6.1.4a), and (2.2.17) with $L_r = L = $ const. and $m = M$ the equilibrium equation in the form

$$\frac{dP_g}{d\tau} = \frac{GM}{\kappa R^2}\left(1 - \frac{L}{L_{cr}}\right), \quad L_{cr} = \frac{4\pi cGM}{\kappa}, \quad \tau = \int_r^\infty \kappa \rho \, dr. \quad (6.1.12)$$

Here L_{cr} is the critical Eddington luminosity. We see from (6.1.12) that at $L > L_{cr}$ the static equilibrium is impossible[1]. Substituting approximately

$$\kappa = \kappa_1 + \kappa_0 P_g \quad (6.1.13)$$

recovers the solution (6.1.12) in the form

$$P_g = g\frac{\tau_0}{\kappa}\frac{L}{L_{cr}}\left(\frac{L_{cr}}{L}\frac{\kappa}{\kappa - \kappa_1}\ln\left(\frac{1 - \frac{L\kappa_1}{L_{cr}\kappa}}{1 - \frac{L}{L_{cr}}}\right) - 1\right)^{-1},$$

$$P = \frac{P_g}{\beta_g}, \quad (6.1.14)$$

where $g = GM/R^2$, $\tau_0 = 2/3$ (2.2.21) or $\tau_0 = 0.756$ (2.2.23). At $L \ll L_{cr}$ the parenthesized expression in (6.1.14) equals $\frac{\kappa + \kappa_1}{2\kappa}\frac{L}{L_{cr}}$, yielding the boundary condition

$$P_g = \frac{2g\tau_0}{\kappa + \kappa_1}, \quad P = P_g/\beta_g, \quad (6.1.15a)$$

obtained in [165]. At $(\kappa - \kappa_1)/\kappa \ll L_{cr}/L - 1$ the above expression in (6.1.14) equals $\left(\frac{\kappa + \kappa_1}{2\kappa} - L/L_{cr}\right)\frac{L/L_{cr}}{(1 - L/L_{cr})^2}$ yielding now the boundary condition

$$P_g = g\frac{\tau_0}{\kappa}\frac{(1 - L/L_{cr})^2}{\frac{\kappa + \kappa_1}{2\kappa} - L/L_{cr}}, \quad P = \frac{P_g}{\beta_g}. \quad (6.1.15b)$$

At $\kappa = \kappa_1 = $ const., $\beta_g = 1$ and $L/L_{cr} \ll 1$ we have from (6.1.15a, b) the boundary condition $P = g\tau_0/\kappa$ given in [229].

On the red giant branch a star has an extended envelope where sphericity effects must be taken into account. In this situation the boundary condition is

$$T = T_0, \quad \rho = \rho_0 (\approx 0) \text{ at } \tau = 0 \quad (6.1.16)$$

and at $\tau > 0$ the equation (2.2.34) is solved. At $\tau \geq 2/3$ (2.2.34) transforms smoothly into the radiative heat conductivity equation (6.1.4a). Instead of (2.2.33), the outer boundary condition in [520] is taken to be

---

[1] At $\kappa = $ const.

$$T_0 = \left(\frac{L}{2\pi a c R^2} \right)^{1/4} \tag{6.1.17}$$

as for the plane atmosphere (2.2.16–21) where $T^4(\tau = 0) = \frac{1}{2}T_{\mathrm{ef}}^4$. A boundary condition for P_g should also be specified at $\tau = 0$, and at $\tau > 0$ the equation

$$\frac{dP}{d\tau} = \frac{Gm}{\kappa r^2} \left(1 + f \frac{2aT^3 r^{1/2} T_0 R^{1/2}}{3Gm\rho} \right) \quad \text{with specified } \rho(\tau = 0) \tag{6.1.18}$$

is to be integrated. This equation applies in transparent regions with $\tau < 2/3$ if (6.1.12) with (m, r) instead of (M, R), and (2.2.34) are valid, and allows for mass variability $m(\tau)$ at $\tau > 0$. Equations (2.2.34) and (6.1.18) should be combined and solved for $\tau \leq 2/3$ simultaneously with (6.1.2) and the relation for τ from (6.1.12). At $\tau > 2/3$ these equations coincide with equilibrium equations (6.1.1–4). In order to make the computations easier, the density on the outer boundary is adopted to be non-zero. In [520] $\rho(\tau = 0) = 10^{-12}$ (g cm^{-3}). The stellar radius R and luminosity L are eigenvalues of the problem and are obtained from constructing the model. To calculate the convection from the mixing-length theory (6.1.4b), it is necessary to solve a cubic equation in order to obtain an explicit form of the differential equation with respect to dT/dr. This approximation applies in convective envelopes. In convective cores the approximation of adiabatic convection (6.1.4b) is usually used. In evolutionary computations by the Henyey method (see below) the envelope models are often calculated beforehand for a wide range of L and R up to the specified values of m or τ (see e.g. [241]), so that the outer boundary of the core model turns out to be shifted towards higher temperatures and densities.

To solve spherically symmetric stellar evolution equations, Schwarzschild [229] and Henyey (quasi-static relaxation) [194] methods are mainly applied. In the Schwarzschild method the differential equations (6.1.1–4) are integrated from the centre outward and from the surface inward, and fitting conditions for functions yield the model parameters R and L together with central values T_c and P_c. The Henyey method is based on writing the differential equations in a finite difference form and solving the obtained algebraic set of equations by an iterative procedure. The last method is better suited to automatic evolutionary computations and more widespread than the first one. On critical evolutionary stages (dense core formation, simultaneous burning of several layer sources) using the plain Henyey method may encounter difficulties owing their origin to a small radius of iteration convergence. Different modifications of the relaxation method have been developed allowing us to make these calculations.

6.1.1 The Schwarzschild Method

This method is based on solving the set of differential equations (6.1.1–4) by conventional techniques for each evolutionary model. Integration from the centre inward requires specifying, in addition to (5.3.10), the central values P_c and T_c. Integration from the surface inward with the boundary conditions (6.1.11) and (6.1.15) (or (6.1.16) and (6.1.17) for (2.2.34) and (6.1.18)) requires specifying the radius R and luminosity L. Instead of the radius r, the inner mass m usually serves as an independent variable.

The simplest procedure for constructing a model might involve specifying P_c and T_c, performing trial integrations from the centre and finding the unique P_c and T_c which would satisfy the boundary conditions at the surface. An analogous integration from the surface inward might supply R and L and satisfy the boundary conditions in the centre. Unfortunately, it is impossible to apply these simple procedures because of solution divergences in the centre in (6.1.1) and at the surface in (6.1.4) due to the zero value of r and a small T, respectively. The procedure used by Schwarzschild involves finding solutions, after specifying the trial values of

$$P_c, T_c, L, R, \tag{6.1.19}$$

by integrations in opposite senses up to an intermediate mass m_f. The functions P, T, L and r must be continuous in this point:

$$P_{if} = P_{ef}, \quad T_{if} = T_{ef}, \quad L_{if} = L_{ef}, \quad r_{if} = r_{ef}, \tag{6.1.20}$$

where the subscripts i and e refer to intrinsic and extrinsic quantities, respectively. After six trial integrations, three of which are from the centre, and three others from the surface to the fitting point, we have to find the partial derivatives of discrepancies with respect to parameters

$$\frac{\partial \Delta_q}{\partial P_c}, \quad \frac{\partial \Delta_q}{\partial T_c}, \quad \frac{\partial \Delta_q}{\partial R}, \quad \frac{\partial \Delta_q}{\partial L}, \quad q = P, T, L, r,$$

$$\Delta_P = P_{if} - P_{ef}, \quad \Delta_T = T_{if} - T_{ef},$$

$$\Delta_L = L_{if} - L_{ef}, \quad \Delta_r = r_{if} - r_{ef}. \tag{6.1.21}$$

The corrections to the parameters (6.1.19)

$$\Delta P_c, \Delta T_c, \Delta L, \Delta R, \tag{6.1.22}$$

should be found with the aid of the Newton iteration scheme by equating the discrepancies to zero:

$$\Delta_q + \frac{\partial \Delta_q}{\partial P_c} \Delta P_c + \frac{\partial \Delta_q}{\partial T_c} \Delta T_c + \frac{\partial \Delta_q}{\partial R} \Delta R + \frac{\partial \Delta_q}{\partial L} \Delta L = 0,$$

$$q = P, T, L, r. \tag{6.1.23}$$

A solution to the linear non-uniform set of equations (6.1.23) enables us to find corrections to the previous values (6.1.19) and to proceed to the next iteration. We have to stop iterating when a predetermined accuracy ϵ is achieved, or

$$\delta = \sqrt{\sum_q \left(\frac{\Delta_q}{\bar{q}}\right)^2} < \epsilon, \quad q = P, T, L, r, \quad \bar{q} = \frac{1}{2}(q_i + q_e). \tag{6.1.24}$$

This procedure is given in [154] and is well suited with its dimensionless variables to the use of computers.

Schwarzschild first took [229] ϵ_n, κ, $P(\rho, T)$ in the form of a power function, and $\epsilon_\nu = 0$. It allowed the number of unknown parameters to be reduced by a transformation of similarity from four to two[2], significantly facilitating integration of equation without computers. This leads, however, to the use of different variables in the envelope and core, which complicates the fitting procedure. As pointed out in [229], in the case of automatic computations a complicated fitting brings to nothing the advantages of this scheme. There is no transformation of similarity for real $\epsilon_n, \kappa, P(\rho, T), \epsilon_\nu$ which are not power functions, and the number of unknown functions cannot be reduced in this case.

The iteration convergence may be improved by using logarithmic instead of ordinary variables, varying by many orders of magnitude over the star. Schwarzschild took the variables

$$l_P = \lg P, \quad l_T = \lg T, \quad l_L = \lg L, \quad l_r = \lg r, \quad l_m = \lg m. \tag{6.1.25}$$

Four logarithmic variables from (6.1.25) (except l_m) have been used in [80] in the techniques [154] described above. This has allowed us to extend the calculations up to the onset of helium burning, leaving far behind all calculations performed in ordinary variables. Note that integration from the centre outward implies the first step to correspond to the expansion [229]

$$m = \frac{4}{3}\pi\rho_c r^3,$$

$$P = P_c - \frac{2}{3}\pi G\rho_c^2 r^2,$$

$$L_r = \frac{4}{3}\pi\rho_c\epsilon_c r^3,$$

$$T = T_c - \frac{2}{3}\gamma_{2c}\pi G\frac{T_c\rho_c^2}{P_c}r^2 \quad \text{(adiabatic convection)}, \tag{6.1.26}$$

$$T = T_c - \frac{\kappa_c\epsilon_c\rho_c^2}{8acT_c^3}r^2 \quad \text{(radiative heat transfer)},$$

$$\epsilon_c = \epsilon_{nc} - \epsilon_{\nu c} - \frac{\partial E_c}{\partial t} + \frac{P_c}{\rho_c^2}\frac{\partial\rho_c}{\partial t}.$$

---

[2] Fitted are the dimensionless quantities [229] $U = \frac{r}{m}\frac{dm}{dr} = \frac{4\pi\rho r^2}{m}$, $V = -\frac{r}{P}\frac{dP}{dr} = \frac{\rho Gm}{Pr}$ and $n+1 = \frac{T}{P}\frac{dP}{dT}$.

The value of the derivatives $\partial E/\partial t$, $\partial \rho/\partial t$ and chemical composition of the given evolutionary model are calculated from the previous model after a time step Δt. In (6.1.3)

$$\frac{\partial E}{\partial t} = \frac{E(m,t) - E(m,t - \Delta t)}{\Delta t},$$

$$\frac{\partial \rho}{\partial t} = \frac{\rho(m,t) - \rho(m,t - \Delta t)}{\Delta t}, \qquad (6.1.27)$$

while the chemical composition in any point of star at a given time t is

$$x_i(m,t) = x_i(m,t - \Delta t) + \frac{\partial x_i}{\partial t}\Delta t, \qquad i = H, \alpha, {}^{12}\mathrm{C}. \qquad (6.1.28)$$

Here the functions $f(m, t - \Delta t)$ are taken to be known, and $f(m, t)$ unknown. The derivatives $\partial x_i/\partial t$ are found from (6.1.5) and (6.1.6). In an explicit calculation scheme these derivatives refer to the time $t - \Delta t$ and are assumed to be known. In an implicit scheme their values refer to the time t, or to some intermediate time (see (6.1.41) and (6.1.42)). It makes the numerical scheme more stable but somewhat complicates the solution. The values of $E(m)$, $P(m)$, $x_H(m)$, $x_\alpha(m)$, $x_{12C}(m)$ and $r(m)$ referring to the previous time step $t - \Delta t$ have to be recorded in a set of points such that it would be possible to make a fairly accurate interpolation between them. Note that the necessity to record the values referring to the previous evolutionary step is common for all methods. In the Henyey method the points at which the previous step values have to be recorded are determined by the structure of the method itself.

The Schwarzschild method is applicable to evolutionary calculations, if the determinant of the system (6.1.23) has a non-zero value. A zero determinant is an indication of the model instability (bifurcation point) and requires a special investigation.

6.1.2 The Henyey Method

In the method for solving stellar evolution equations proposed by Henyey the star is conceptually divided into J discrete zones, and the differential equations (6.1.1–4) are written in a finite difference form. For solving of such type system of linearized difference equations a technique called the "back substitution method" ("progonka") was developed by Gelfand and Lokuzievskii (see [20a], [92]) which allows us to obtain the solution effectively. Modifications of the Henyey method applied by diverse authors [112, 406, 522] do not differ greatly.

The mass is taken as an independent variable and the mass interval from 0 to M (the stellar mass) is broken up into J generally non-equal zones. The mass is specified at integer points M_j with $M \equiv M_J$. The mass difference is determined both at integer and semi-integer points:

$$\Delta M_{j+1/2} = M_{j+1} - M_j, \qquad j = 0, 1 \ldots J - 1;$$

$$\Delta M_j = \frac{1}{2}(\Delta M_{j+1/2} + \Delta M_{j-1/2}) = \frac{1}{2}(M_{j+1} - M_{j-1}),$$

$$j = 1, 2 \ldots J - 1;$$

$$\Delta M_J = \frac{1}{2}\Delta M_{J-1/2}. \qquad\qquad\qquad\qquad (6.1.29)$$

Equations (6.1.1–4) in a difference form become

$$P_{j+1/2} - P_{j-1/2} + \frac{GM_j \Delta M_j}{4\pi r_j^4} = 0 \qquad j = 1, 2 \ldots J; \qquad (6.1.30)$$

$$\rho_{j+1/2} - \frac{\Delta M_{j+1/2}}{\pi(r_{j+1} - r_j)(r_{j+1} + r_j)^2} = 0; \qquad j = 0, 1 \ldots J - 1; \qquad (6.1.31)$$

$$L_{j+1} - L_j - \Delta M_{j+1/2} \cdot \epsilon_{j+1/2} = 0; \qquad j = 0, 1 \ldots J - 1; \qquad (6.1.32)$$

$$T_{j+1/2} - T_{j-1/2} + \frac{G}{4\pi} \frac{\Delta M_j M_j \cdot T_j}{r_j^4 P_j}(\nabla_{\mathrm{ad}})_j = 0;$$

$$j = 1, 2 \ldots J \ \text{(adiabatic convection)}; \qquad\qquad (6.1.33a)$$

$$T_{j+1/2} - T_{j-1/2} + \frac{3}{64\pi^2 ac} \frac{\Delta M_j \kappa_j L_j}{r_j^4 T_j^3} = 0;$$

$$j = 1, 2 \ldots J \ \text{(radiative heat transfer)}. \qquad\qquad (6.1.33b)$$

Here

$$\epsilon = \epsilon_n - \epsilon_\nu - \frac{\partial E}{\partial t} + \frac{P}{\rho^2}\frac{\partial \rho}{\partial t}. \qquad (6.1.34)$$

The equation of state, opacity and rate of energy release are written as

$$P_{j+1/2}, \ \epsilon_{j+1/2} = P, \ \epsilon(T_{j+1/2}, \rho_{j+1/2}, x_{k,j+1/2}),$$

$$j = 0, 1 \ldots J - 1; \quad k = H, \alpha, {}^{12}\mathrm{C}; \qquad (6.1.35)$$

$$\kappa_j, \ P_j = \kappa, \ P(T_j, \rho_j, x_{k,j}), \quad j = 1, 2 \ldots J,$$

where

$$(T_j, \rho_j, x_{k,j}) = \left[\frac{1}{2}(T_{j+1/2} + T_{j-1/2}), \ldots, \frac{1}{2}(x_{k,j+1/2} + x_{k,j-1/2})\right]$$

$$j = 1, 2 \ldots J; \qquad\qquad\qquad\qquad (6.1.36)$$

$$(\nabla_{\mathrm{ad}})_j = \nabla_{\mathrm{ad}}(\beta_{gj}), \quad \beta_{gj} = \frac{(P_g)_j}{P_j}. \qquad (6.1.37)$$

The central boundary condition at the point $j = 0$: $r_0 = 0$, $L_0 = 0$ has to be included in (6.1.31) and (6.1.32). The outer boundary conditions at the point $J + 1/2$ have the form

$$\frac{L_J}{4\pi \left(\frac{3}{2}r_J - \frac{1}{2}r_{J-1}\right)^2} = \sigma T_{J+1/2}^4 \quad \text{(from (6.1.11))},$$ (6.1.38)

$$P_{J+1/2} = \frac{2/3\,GM_J}{\beta_{g,J+1/2} \left(\frac{3}{2}r_J - \frac{1}{2}r_{J-1}\right)^2 \kappa_{J+1/2}}.$$ (6.1.39)

Here $\kappa + \kappa_1 = 2\kappa$, $\tau_0 = 2/3$ is adopted in (6.1.15a), the total radius should be found by the linear extrapolation $r_{J+1/2} = \frac{3}{2}r_J - \frac{1}{2}r_{J-1}$ and $L_J = L_{J+1/2}$. The relation (6.1.38) is used for finding $T_{J+1/2} = T_{\text{ef}}$, relation (6.1.39) for $P_{J+1/2}$, $\rho_{J+1/2}$ is determined as an implicit function from the equation of state. The values of $T_{J+1/2}$ and $P_{J+1/2}$ are used in (6.1.30) and (6.1.33). Upon including the boundary conditions (6.1.38) and (6.1.39) and tabulating M_j, ΔM_j and $\Delta M_{j+1/2}$, the set of $4J$ equations (6.1.30–33) contains $4J$ unknowns $\rho_{j+1/2}$, $T_{j+1/2}$ ($j = 0, 1 \ldots J - 1$) and r_j, L_j ($j = 1, 2 \ldots J$). In order to solve this set of equations it must be linearized, and the Newton iteration scheme should be used in analogy with (6.1.23). For solving this set of $4J$ linear equations, the "progonka" technique is used at each iteration [92]. The evolutionary changes in chemical composition and gravitational energy release are calculated at each semi-integer point where parameters are recorded after each evolutionary step. In order for the calculations to be accurate and for the iterations to converge quickly, changes in any function from point to point should not be large, the limiting fractional change being determined by the number of discrete intervals. In [112] at $J = 150$ it is required that

$$\frac{L_{j+1} - L_{j-1}}{L_j} \leq 0.05, \qquad \frac{P_{j+3/2} - P_{j-1/2}}{P_{j+1/2}} \leq 0.02.$$ (6.1.40)

The conditions (6.1.40) are checked after each evolutionary step, and if they no longer hold, a permutation of points is performed in such a way as to make them valid again. With increasing J the conditions (6.1.40) may be specified more rigidly. In the same way as in the Schwarzschild method, the evolutionary changes in chemical composition may be calculated in an explicit or implicit way. A time-explicit calculation scheme has been used in [522, 112]. In [406] the chemical composition is evaluated at all points in the radiative zone at a time t_{n+1} with the aid of the relation

$$x_i^{(n+1)}(j + 1/2) = x_i^{(n)}(j + 1/2) + \dot{x}_i^{(n+1/2)}(j + 1/2)\Delta t,$$ (6.1.41)

$$\dot{x}_i^{(n+1/2)}(j + 1/2) = \frac{1}{2}\left[\dot{x}^{(n)}(j + 1/2) + \dot{x}^{(n+1)}(j + 1/2)\right].$$ (6.1.42)

The points stand here for the time derivatives (6.1.5) and (6.1.6). In calculating the convective zone its boundary is assumed to be fixed and is corrected upon completing the iterations.

6.1.3 The Henyey Method with Matched Envelope

The experience of computations has shown that the convergence and accuracy of the method improve if the outer stellar envelope containing $(3 \div 10)\%$ of the mass with a constant luminosity L is calculated separately by solving differential equations. Consider a modification of the Henyey method with a fitted envelope. Integrating the equilibrium equations (6.1.1–4) for the envelope with $L = \mathrm{const.}$, $\epsilon = 0$, and the boundary conditions (6.1.1) and (6.1.15) at $\tau = \tau_0$, or (6.1.17) and (6.1.18) at $\tau = 0$ for (6.1.18) and (2.2.34) allows us to obtain the functions (R is the outer radius)

$$T = f_1(m, L, R), \tag{6.1.43}$$
$$P = f_2(m, L, R), \tag{6.1.44}$$
$$r = f_3(m, L, R). \tag{6.1.45}$$

Three requirements to the point $m = M_1 < M, M_1 = M_{J+1/2}$ taken as the core-envelope boundary, are

$$
\begin{aligned}
S_1 &= T_{J+1/2} - f_1(M_1, L_{J+1/2}, R) = 0, \\
S_2 &= P_{J+1/2} - f_2(M_1, L_{J+1/2}, R) = 0, \\
S_3 &= r_{J+1/2} - f_3(M_1, L_{J+1/2}, R) = 0.
\end{aligned}
\tag{6.1.46}
$$

Here the subscript $J + 1/2$ denotes the fitting point where the luminosity $L_{J+1/2}$ and radius $r_{J+1/2}$ are to be expressed in terms of their values at the integer point J. The constancy of L gives

$$L_{J+1/2} = L_J = L. \tag{6.1.47}$$

We write the mass conservation law in the form

$$4\pi \frac{r_{J+1/2}^3 - r_J^3}{3\Delta M_{J+1/2}} = \frac{1}{\rho_{J+1/4}}, \tag{6.1.48}$$

where $\Delta M_{J+1/2}$ is the mass inside the semi-zone $[J, J + 1/2]$. We may write for small $r_{J+1/2} - r_J$

$$r_{J+1/2}^3 = r_J^3 + 3r_J^2(r_{J+1/2} - r_J). \tag{6.1.49}$$

Obtaining $\rho_{J+1/4}$ from a linear interpolation by mass we finally obtain for the fitting radius

$$
r_{J+1/2} = r_J + \frac{\Delta M_{J+1/2}}{4\pi r_J^2}
$$
$$
\times \frac{1}{\rho_{J+1/2} + \frac{\Delta M_{J+1/2}}{\Delta M_{J-1/2} + 2\Delta M_{J+1/2}}(\rho_{J-1/2} - \rho_{J+1/2})}. \tag{6.1.50}
$$

Inserting (6.1.47) and (6.1.50) into (6.1.46) yields three relations with six unknowns

$$T_{J+1/2}, \quad \rho_{J+1/2}, \quad r_J, \quad L_J, \quad \rho_{J-1/2}, \quad R.$$

All these unknowns except R enter also into the $4J$ difference equations (6.1.30–33) for the inner part of the star which incorporate $4J + 2$ unknowns (see above, Sect. 6.1.2). Adding three equations (6.1.46), we get $4J + 3$ equations in all with the same number of unknowns, or a closed system. This system is solved by the same Newton iteration scheme as in the above mentioned case. Equations for the envelope take most of the computer time. To save this time, the envelope models may be calculated and relations (6.1.46) tabulated beforehand.

6.1.4 Atmosphere Fitting in Self-Consistent Models with Mass Loss [52]

The above method of exact fitting may be applied to a stationary outflowing envelope (atmosphere). It is convenient to describe an outflowing atmosphere by the density at the critical point ρ_{cr} instead of the photosphere radius R.

The second parameter is the total (including kinetic) energy flux L_t from the star. To obtain relations for the outflowing atmosphere, we have to substitute in (6.1.43–50) ρ_{cr} and L_t for R and L. The inner boundary of the outflowing envelope is chosen as the place where the kinetic energy flux is negligible and the heat flux $L = L_t$. Solution of the equations in the envelope demand integration outward from the critical point at $r = r_{cr}$ in order to satisfy boundary conditions far from the star ($\rho \to 0$, $T_g \to 0$ at $r \to \infty$). This includes a transition between optically thick and optically thin regions which must be continuously described by equations. On performing iterations and finding exact values of L_t and r_{cr} the conditions on the photosphere, the values of T_{ef}, L, R and star location on the Hertzsprung–Russell diagram follow from the solution. This calculation scheme applies to a star with an outflowing atmosphere optically thick at $r < r_{cr}$ and having $R > r_{cr}$; it has been used in [290]. A set of equations for outflowing atmosphere and an algorithm for its solution are presented in [52] and will be given in Chap. 9, Vol. 2.

Techniques for self-consistent calculations of the evolution of stars with an outflowing atmosphere which is optically thin at $r < r_{cr}$ and stellar radius $R < r_{cr}$ have not been worked out yet but can be constructed analogously.

The global description of the outflowing spherically symmetric envelope at an arbitrary optical depth was done in [288a], where the radiation field was described by an approach equivalent to the Eddington approximation with a variable Eddington factor (see Sect. 2.2.3).

6.1.5 Relaxation Method Stable in the Presence of Various Time Scales

The numerical computations based on the methods described above turn out to be unstable and evolutionary calculations become impossible at some evolutionary stages (red giants with dense core and extended envelope, phases of unstable nuclear burning in degenerate cores and shell sources (see Chap. 9, Vol. 2), pre-supernovae in transition from the static to dynamic evolution (see Chap. 10, Vol. 2), accreting stars or stars losing mass, and so on). The instability comes from a large difference in time scales

$$\tau_h(Z) = \frac{3c_p \kappa \rho^2 Z^2}{4acT^3} \quad (Z \text{ is a characteristic size}) \tag{6.1.51}$$

related to the heat transfer in diverse parts of the star (core and envelope). In order for the implicit-time difference scheme in the Schwarzschild and Henyey methods to be stable, the time step must be large enough

$$\Delta t > \tau_h(r). \tag{6.1.52}$$

At a fast evolution with core time scale τ_n determined by nuclear burning or rapid contraction, the inequality $\tau_n < \tau_h(r)$ in the envelope may invalidate (6.1.52), and evolutionary calculations using these methods are no longer possible. A method suggested in [592] remains stable at a time step $\Delta t < \tau_h$ and allows us to perform calculations at a large difference in time scales $\tau_h(r)$ over the star. We write out the equilibrium equations (6.1.1–4) in the form

$$\frac{d\ln P}{d\ln m} = -\frac{Gm^2}{4\pi r^4 P}, \tag{6.1.53}$$

$$\frac{d\ln r}{d\ln m} = \frac{m}{4\pi r^3 \rho}, \tag{6.1.54}$$

$$\frac{dL_r}{d\ln m} = m\left(-T\frac{\partial S}{\partial t} + \epsilon_n - \epsilon_\nu\right), \quad \frac{\partial S}{\partial t} = \frac{S(t) - S(t - \Delta t)}{\Delta t}, \tag{6.1.55}$$

$$\frac{d\ln T}{d\ln m} = -\frac{3\kappa L_r m}{64\pi^2 acT^4 r^4} \quad \text{(radiation heat transfer)}, \tag{6.1.56a}$$

$$\frac{d\ln T}{d\ln m} = \gamma_2 \frac{d\ln P}{d\ln m} \quad \text{(convective core)}. \tag{6.1.56b}$$

Defining the four-dimensional vector $y_i = (\ln P, \ln r, L_r, \ln T)$, independent variable $x = \ln m$, and corresponding vector for the right-hand sides allows us to rewrite the system (6.1.53–56) in the form

$$\frac{dy_i}{dx} = \Phi_i(x, y_i). \tag{6.1.57}$$

We divide the star into J discrete zones and take all functions only at integer points. Consider the following difference form [592] of the system (6.1.57)

$$y_i^{(j+1)} - y_i^{(j)} = \Delta x^{(j)} \left[\beta_i \Phi_i(x^{(j)}, y_i^{(j)}) + (1 - \beta_i) \Phi_i (x^{(j+1)}, y_i^{(j+1)}) \right],$$

$$j = 1, 2 \ldots J, \qquad \Delta x^{(j)} = x^{(j+1)} - x^{(j)}. \tag{6.1.58}$$

Here, β_i are four arbitrary numbers $0 \leq \beta \leq 1$. Note that the difference form (6.1.30–33) is equivalent to fixing all $\beta_i = 1/2$. This method implies $\beta_1 = \beta_2 = 1/2$ in the difference analogies of the hydrostatic equilibrium equations (6.1.53) and (6.1.54). It is shown in [592] that a difference scheme for solving the system (6.1.58) by linearization and with the aid of the Newton iteration scheme, similar to the Henyey method, is stable at two sets of β_3 and β_4 values:

(1) $\beta_3 = 1, \quad \beta_4 = 0;$

(2) $\beta_3 = 0, \quad \beta_4 = 1.$

$$\tag{6.1.59}$$

One of the functions $y_3 = L_r$ or $y_4 = \ln T$ may then be calculated as explicit and the other as implicit functions over the space

(1) $y_3^{(j+1)} - y_3^{(j)} = \Phi_3 \left(x^{(j)}, y^{(j)} \right) \Delta x^{(j)},$

$\quad y_4^{(j+1)} - y_4^{(j)} = \Phi_4 \left(x^{(j+1)}, y^{(j+1)} \right) \Delta x^{(j)};$

(2) $y_3^{(j+1)} - y_3^{(j)} = \Phi_3 \left(x^{(j+1)}, y^{(j+1)} \right) \Delta x^{(j)},$

$\quad y_4^{(j+1)} - y_4^{(j)} = \Phi_4 \left(x^{(j)}, y^{(j)} \right) \Delta x^{(j)}.$

$$\tag{6.1.60}$$

A time step may be carried out with the aid of relations (6.1.27), (6.1.41), and (6.1.42). Simplified boundary conditions

$$r = L_r = 0 \text{ for } m = 0,$$
$$P = T = 0 \text{ for } m = M, \tag{6.1.61}$$

have been used in [592]. In [594] the stable numerical method is extended to the dynamic stages of stellar evolution. On these stages the dynamic term

$$-\frac{m}{4\pi r^2 P} \frac{\partial^2 r}{\partial t^2}$$

$$= -\frac{m}{4\pi r(t) P(t)} \left[\frac{\ln r(t) - 2 \ln r(t - \Delta t) + \ln r(t - 2\Delta t)}{\Delta t^2} \right.$$

$$\left. + \left(\frac{\ln r(t) - \ln r(t - \Delta t)}{\Delta t} \right)^2 \right] \tag{6.1.62}$$

must be added to the right-hand side of (6.1.53), while the difference system still has the functional form (6.1.58). The solution to the system by the Henyey method is stable at the coefficients (6.1.59) even if the condition (6.1.52) does not hold. This gives a possibility, for example, to go from the static to dynamic evolution smoothly in studying supernova explosions (see Chap. 10, Vol. 2).

Similar account for hydrodynamical effects by relaxation type numerical method was carried out in [626, 346*]. On quasi-static evolutionary phases calculations with time step Δt much larger then characteristic equilibration times permit us to obtain a solution for a sequence of models in static and thermal equilibrium. Another variant of the Henyey prescription applied to integration of five hydrodynamical equations of the stellar structure, including the one for Lagrangian velocity u, has been used in [358a].

6.1.6 Relaxation Method with Adaptive Grid

As was mentioned in Sect. 6.1.2, after each evolutionary step a reconstruction of the grid is made for mesh points, where the difference of parameters within one interval exceeds some given value. The adaptive method suggested in [344a] makes this reconstruction automatically.

For finding a distribution of the mesh points, which will ensure that within each interval no quantity of physical importance varies by a large amount, a new variable q is introduced, which varies from 0 to Q through the star, with equal increments between mesh points. The equation, determining a mass distribution $m(q)$ is derived from a minimum of appropriate functional Φ, which is written from physical reasons with lots of freedom of choice. Writing

$$\Phi = \sum_{k=2}^{N}(f_k - f_{k-1})^2 \approx \int_0^M \left(\frac{df}{dm}\right)^2 \frac{dm}{dq} dm \tag{6.1.63}$$

we arrive, after taking the variation $\delta\Phi = 0$, to the equation

$$\frac{dq}{dm} = A\frac{df}{dm} \tag{6.1.64}$$

where the constant A is a kind of "eigenvalue" whose value is not known until the equations are solved. The equation (6.1.64) has boundary conditions

$$q = 0 \quad \text{at} \quad m = 0; \qquad q = Q \quad \text{at} \quad m = M \tag{6.1.65}$$

and A is determined by the fact that a first-order differential equation has two boundary conditions (6.1.65). The second condition in (6.1.65) may be formulated for a mass-losing or mass-accreting star [344c]

$$\frac{\partial M}{\partial t} = -\dot{M}_{wind} + \dot{M}_{acc} \quad \text{at} \quad q = Q. \tag{6.1.66}$$

The choice of the function $f(m)$ is made for implying a larger number of mesh points in the region with large gradients of P, L and g. The best choice of the function f may be different at different evolutionary stages and is done empirically. The following functions have been used up to now

$$f(m) = -a_1 \log P + a_2 L_H + a_3 m \quad \text{in [344a],} \tag{6.1.67a}$$

$$f(m) = -0.02 \log P + 0.1 x_H + (m/M)^{2/3} \quad \text{in [344b],} \tag{6.1.67b}$$

$$f(m) = -a_1 \log P + 1.5 a_2 m^{2/3} \quad \text{in [344d],} \tag{6.1.67c}$$

where L_H is the luminosity from the hydrogen burning. Variations of a_1, a_2, a_3 by factors 2 or 3 produce changes in the model of 2 or 3 per cent for 100 mesh points [344a]. With the choice of (6.1.67c), (6.1.64) reads as [344d]

$$\frac{dm}{dq} = \frac{1}{A} \left[\frac{a_1}{4\pi r^2 \rho H} + \frac{a^2}{m^{1/3}} \right]^{-1}, \tag{6.1.68}$$

where

$$H = \frac{P r^2}{\rho G m}. \tag{6.1.69}$$

The equations (6.1.1)–(6.1.6) may be written using an independent variable q instead of r (or m) multiplying them by dm/dq from (6.1.68). The Lagrangian time derivatives are expressed through time derivatives at $q = $ const. (with a dot above) as

$$\frac{\partial F}{\partial t} = \dot{F} - \frac{dF}{dm} \dot{m} \tag{6.1.70}$$

with $F = E, \rho, x_H, x_\alpha, x_{12C}$. To account for a convective mixing an additional diffusion-type term $\frac{d}{dm}\left(\sigma \frac{dX}{dm}\right)$ is added to the right side of (6.1.5)–(6.1.6) with $X = x_H, x_\alpha, x_{12C}$ and

$$\sigma \approx v l \left(\frac{dm}{dr}\right)^2. \tag{6.1.71}$$

Here, convective velocity v and mixing length l are determined in Sect. 3.1.3; $\sigma = 0$ in the convectively stable regions. Second-order equations (6.1.5–6) with terms of the type (6.1.71) are solved with the boundary conditions [334d]

$$\frac{\partial X}{\partial q} = 0 \quad \text{at} \quad q = 0 \quad \text{and} \quad q = Q. \tag{6.1.72}$$

In this method (6.1.1–4) are solved together with s equations for chemical compositions and with (6.1.68). The $5 + s$ equations are solved by the same type of relaxation method as Henyey in the Sect. 6.1.2. Each equation $\frac{\partial X}{\partial t} =$

$-XR$ from the modified system (6.1.5–6) for composition X is written in the form

$$\frac{d}{dm}(\sigma\xi) = \dot{X} - \xi\dot{m} + XR, \quad \xi = \frac{dX}{dm}. \tag{6.1.74}$$

In order to obtain stable calculations on the stage of thin shell burning, the difference analog of (6.1.74) must be written in the form

$$X_j - X_{j-1} = \xi_{j-1}\,\delta m_{j-1}$$

$$\sigma_j\xi_j - \sigma_{j-1}\xi_{j-1} = (\dot{X}_j + X_j R_j - \xi_j\dot{m}_j)\delta m_j. \tag{6.1.75}$$

Here

$$\dot{X}_j = \frac{X_j - X_j^{(0)}}{\Delta t}, \quad \dot{m}_j = \frac{m_j - m_j^{(0)}}{\Delta t}, \quad \delta m_j = h\left(\frac{dm}{dq}\right)_j, \tag{6.1.76}$$

where dm/dq is from (6.1.68), and the step $h = q_j - q_{j-1}$ is equal for all j.

Similar account of mixing in convective and semiconvective zones, but without grid adaptation, has been used in [462a].

6.2 Equilibrium of Rotating Magnetized Stars

Rotation is observed in many stars and is their intrinsic property [199]. In the theory of equilibrium the role of rotation reduces to breaking the spherical symmetry and thus essentially complicating methods used for constructing equilibrium solutions. In a fairly wide region of angular momenta the rotating stars have an axial symmetry. At a low compressibility and a large angular momentum only triaxial figures similar to the Jacobi ellipsoids of a compressible fluid are stable [220]. On the contrary, at a large compressibility in rigidly rotating star the matter outflow starts before the formation of triaxial figures has become energetically allowable. For rigidly rotating polytropes[3] described by the equation of state $P = K\rho^{1+1/n}$, a triaxial configuration forms earlier at $n < 0.808$, while at $n > 0.808$ it is preceded by the onset of outflow from the equator [437]. The theory of evolution of rapidly rotating stars is far from being completed. In reality, we have a theory of equilibrium configurations of barotropic stars described by the equation of state $P(\rho)$. Most evolutionary calculations use simplifying assumptions not always quite justified in the case of rapidly rotating stars.

Stars usually have a magnetic field. Dynamic effects of the field are not strong, but they may be important for finding the rotation law, meridional circulation, chemical mixing which, in turn, affect the evolution significantly. More important still is the role of the magnetic field in various types of

---

[3] The theory of equilibrium of non-rotating polytropes is developed by Emden (1907) and described in detail in [218] (see also Chap. 10, Vol. 2).

stellar activity: outbursts, non-thermal heating, chromosphere, corona and stellar wind formation, significant ultraviolet excess arising in stellar spectra. The magnetic field is closely related to rotation, for the latter is predominant in its enhancement via dynamo mechanisms. In the case of axial symmetry constructing models for rotating barotropic stars has much in common from a mathematical standpoint with constructing models for magnetized stars with infinite conductivity.

6.2.1 Equilibrium Equations for Rotating Magnetized Stars

The steady state of a star with a magnetic field and infinite conductivity is determined by the set of equations [96, 124, 6, 75]

$$\nabla P + \rho \nabla \Phi + \frac{1}{4\pi} \boldsymbol{B} \times (\nabla \times \boldsymbol{B}) + \rho (\boldsymbol{v} \cdot \nabla) \boldsymbol{v} = 0; \tag{6.2.1}$$

$$\nabla (\rho \boldsymbol{v}) = 0, \tag{6.2.2}$$

$$\nabla^2 \Phi = 4\pi G \rho, \tag{6.2.3}$$

$$\Phi = -G \int \frac{\rho(\boldsymbol{r}) dV'}{|\boldsymbol{r} - \boldsymbol{r}'|}, \qquad (dV \text{ is the volume element});$$

$$\nabla \times (\boldsymbol{v} \times \boldsymbol{B}) = 0; \tag{6.2.4a}$$

or

$$(\boldsymbol{v} \cdot \nabla) \left(\frac{\boldsymbol{B}}{\rho} \right) - \left(\frac{\boldsymbol{B}}{\rho} \cdot \nabla \right) \boldsymbol{v} = 0, \tag{6.2.4b}$$

$$\nabla \cdot \boldsymbol{B} = 0. \tag{6.2.5}$$

Here $\boldsymbol{B}$ is the magnetic field strength, $\nabla = \left(\frac{\partial}{\partial x}, \frac{\partial}{\partial y}, \frac{\partial}{\partial z} \right)$ is the Hamiltonian operator. Using vector calculus gives

$$\boldsymbol{B} \times (\nabla \times \boldsymbol{B}) = \frac{1}{2} \nabla (B^2) - (\boldsymbol{B} \cdot \nabla) \boldsymbol{B},$$
$$\nabla \times (\boldsymbol{v} \times \boldsymbol{B}) = \boldsymbol{v}(\nabla \boldsymbol{B}) + (\boldsymbol{B} \cdot \nabla) \boldsymbol{v} - \boldsymbol{B}(\nabla \boldsymbol{v}) - (\boldsymbol{v} \cdot \nabla) \boldsymbol{B}. \tag{6.2.6}$$

For an axially symmetric ($\partial/\partial\phi = 0$) star described by a barotropic equation of state and having purely rotational motion ($v_r = v_z = 0, v_\varphi = \Omega r$), the set of equations (6.2.1–5) in cylindrical coordinates (r, φ, z) becomes

$$\frac{\partial P}{\partial r} + \rho \left(\frac{\partial \Phi}{\partial r} - \Omega^2 r \right)$$
$$+ \frac{1}{8\pi} \left(\frac{\partial B_\varphi^2}{\partial r} + \frac{\partial B_z^2}{\partial r} - 2B_z \frac{\partial B_r}{\partial z} + 2\frac{B_\varphi^2}{r} \right) = 0, \tag{6.2.7}$$

$$B_r \frac{\partial B_\varphi}{\partial r} + B_z \frac{\partial B_\varphi}{\partial z} + \frac{B_r B_\varphi}{r} = 0, \tag{6.2.8}$$

$$\frac{\partial P}{\partial z} + \rho \frac{\partial \Phi}{\partial z} + \frac{1}{8\pi} \left(\frac{\partial B_\varphi^2}{\partial z} + \frac{\partial B_r^2}{\partial z} - 2B_r \frac{\partial B_z}{\partial r} \right) = 0, \tag{6.2.9}$$

$$\frac{1}{r} \frac{\partial}{\partial r} \left(r \frac{\partial \Phi}{\partial r} \right) + \frac{\partial^2 \Phi}{\partial z^2} = 4\pi G\rho, \tag{6.2.10}$$

$$B_r \left(\frac{\partial v_\varphi}{\partial r} - \frac{v_\varphi}{r} \right) + B_z \frac{\partial v_\varphi}{\partial z} = 0 \quad \text{or} \quad B_r \frac{\partial \Omega}{\partial r} + B_z \frac{\partial \Omega}{\partial z} = 0, \tag{6.2.11}$$

$$\frac{1}{r} \frac{\partial}{\partial r} (rB_r) + \frac{\partial B_z}{\partial z} = 0. \tag{6.2.12}$$

The continuity equation (6.2.2) turns identically into zero, thereby leaving some freedom in specifying the angular velocity $\Omega(r,z)$. The equation of motion along the angle coordinate reduces to the relation (6.2.8) that accounts for the absence of azimuthal magnetic force. It follows from (6.2.12) that

$$B_r = -\frac{1}{r} \frac{\partial \psi}{\partial z}, \qquad B_z = \frac{1}{r} \frac{\partial \psi}{\partial r}. \tag{6.2.13}$$

Substituting (6.2.13) into (6.2.8) and (6.2.11) gives the solution

$$B_\varphi = \frac{F_B(\psi)}{r}, \qquad \Omega = F_\Omega(\psi). \tag{6.2.13a}$$

Besides, (6.2.8), (6.2.11), and (6.2.2) have particular solutions

$$B_r = B_z = 0, \quad B_\varphi, \Omega \neq 0, \tag{6.2.14a}$$

$$B_r = 0, \quad B_z = B_z(r), \quad \Omega = \Omega(r) \quad B_\varphi = B_\varphi(r). \tag{6.2.14b}$$

The solution (6.2.13) with $\Omega = $ const. has been considered in [492]. For (6.2.14a), the equations of motion (6.2.7), (6.2.9) become

$$\frac{1}{\rho} \nabla \left(P + \frac{B_\varphi^2}{8\pi} \right) + \nabla\Phi + \left(\frac{B_\varphi^2}{4\pi\rho r^2} - \Omega^2 \right) \frac{\nabla r^2}{2} = 0. \tag{6.2.15}$$

For (6.2.14b), these equations give

$$\frac{1}{\rho} \nabla P + \nabla\Phi + \left(\frac{B_\varphi^2}{4\pi\rho r^2} - \Omega^2 \right) \frac{\nabla r^2}{2} + \frac{1}{8\pi\rho} (\nabla B^2 \cdot \nabla r) \nabla r = 0, \tag{6.2.16}$$

$$B^2 = B_\varphi^2 + B_z^2.$$

In this solution magnetic field lines wind around a circular cylinder coaxial with the rotation axis. For a field determined by the relation (6.2.13), the equilibrium equation may be derived in a similar way. We write out the equilibrium equation (6.2.1) in spherical coordinates (R, θ, ϕ) for axially symmetric case with no field

$$\frac{1}{\rho} \nabla P + \nabla\Phi - \frac{\Omega^2}{2} \nabla (R^2 \sin^2 \theta) = 0, \tag{6.2.17}$$

which is to be solved simultaneously with Poisson's equation

$$\nabla^2 \Phi = \frac{1}{R^2} \frac{\partial}{\partial R} \left(R^2 \frac{\partial \Phi}{\partial R} \right) + \frac{1}{R^2 \sin \theta} \frac{\partial}{\partial \theta} \left(\sin \theta \frac{\partial \Phi}{\partial \theta} \right) = 4\pi G \rho. \qquad (6.2.18)$$

6.2.2 Conditions for Stationary Solutions

For a barotropic medium we may define the enthalpy

$$H(\rho) = \int \frac{dP(\rho)}{\rho}. \qquad (6.2.19)$$

Equation (6.2.17) may then be written as

$$\nabla(H + \Phi) - \frac{\Omega^2}{2} \nabla(R^2 \sin^2 \theta) = 0. \qquad (6.2.19)$$

Equation (6.2.19) does not necessarily have a solution for any value of $\Omega(R, \theta)$. To derive restrictions to the function $\Omega(R, \theta)$ take a curl of the equation (6.2.19). While $\nabla \times (\nabla S) = 0$ for scalar S, we obtain

$$\nabla(R^2 \sin^2 \theta) \times \nabla \Omega^2 = 0. \qquad (6.2.20)$$

The condition (6.2.20) reduces to the law $\Omega = \Omega(R \sin \theta)$. The angular velocity is thus constant on cylinders coaxial with the rotation axis (Poincaré's theorem) [199]. In a cylindrical system $\Omega = \Omega(r)$.

For the toroidal field (6.2.14a), the compatibility condition may be derived analogously from (6.2.15) [284]:

$$\nabla \frac{1}{\rho} \times \nabla B_\varphi^2 - \nabla \left(4\pi \Omega^2 - \frac{B_\varphi^2}{\rho r^2} \right) \times \nabla r^2 = 0. \qquad (6.2.21)$$

The condition (6.2.21) restricts the functions $\Omega(r, z)$ and $B_\varphi(r, z)$. Restrictions to the functions $B_\varphi(r)$, $B_z(r)$, $\Omega(r)$ from (6.2.16) and to the functions ψ, $F_B(\psi)$, $F_\Omega(\psi)$ from (6.2.13) are derived in a similar way.

We look for solution of the equilibrium equations (6.2.17), (6.2.15) or (6.2.16) at a fixed $P(\rho)$ and allowed functions $\Omega(r, z)$, $B(r, z)$ for a star of a given mass with zero density on the boundary. The boundary values of fields may vary depending on specified values of surface currents i_s. Outside the star $B_\varphi = 0$. At $i_s = 0$ the toroidal component $B_\varphi = 0$ at the surface inside the star, while the boundary values of the poloidal field components B_r and B_z must satisfy the compatibility conditions. The solution of equilibrium equations for magnetic stars is rather complicated. It may be obtained by expanding in series in the Legendre polynomials and associated functions. For odd multipoles, for example, the quantity ψ from (6.2.13) in spherical coordinates takes the form [493] (i is odd)

$$\psi = \sum_{i=1}^{\infty} b_i(R) \sin \theta P_i^1 (\cos \theta), \qquad B_R = \sum i(i+1) \frac{b_i}{R^2} P_i(\cos \theta),$$

$$B_R = -\frac{1}{R^2 \sin\theta}\frac{\partial\psi}{\partial\theta}, \qquad B_\theta = \frac{1}{R\sin\theta}\frac{\partial\psi}{\partial R}.$$

We then retain only a small number of terms in this series. Magnetic field configurations may be fairly complicated, but the strength of stellar magnetic fields is usually not so large as to produce a considerable distortion of the stellar shape.

6.2.3 The Self-Consistent Field Method

In the absence of a magnetic field the uncertainty in equilibrium solutions decreases and remains only in the function $\Omega(r)$. For constructing the rotating star models with a barotropic equation of state various methods have been suggested which may be applied up to the limiting rotation case [199]. One of the best methods is the self-consistent field method suggested by Ostriker and Mark [519]. We consider the modified version of this method developed by Blinnikov [64].

The barotropic equation of state $P(\rho)$, equilibrium equation in the form (6.2.19), and integral expression (6.2.3) for the gravitational potential are the equations to be solved. The equilibrium state should be found for a star with a given mass M and specified $\Omega(r)$ or angular momentum distribution:

$$j = \Omega^2 r = j(p), \quad p = \frac{m_r}{M}, \tag{6.2.22}$$

m_r is the mass inside a cylinder of radius r.

The condition (6.2.20) implies the existence of the centrifugal potential

$$\chi(r) = \int_0^r \Omega^2 r'\, dr', \quad \nabla\chi = \Omega^2 r. \tag{6.2.23}$$

The model is constructed by an iteration scheme. At the beginning, an initial density distribution $\rho = \rho_0(\mathbf{r})$ is specified, and the gravitational potential $\Phi(r)$ is calculated from the integral expression (6.2.3). With $\rho_0(\mathbf{r})$, we find the function $m_r(r)$, solve (6.2.22) for $\Omega(r)$ in terms of the specified $j(p)$, and then obtain the centrifugal potential $\chi(r)$ from (6.2.23). With a centrifugal potential we can integrate the equilibrium equation to obtain

$$H + \Phi - \chi = B_s \tag{6.2.24}$$

where $B_s = \Phi_s - \chi_s$ is the total potential at the surface ($H_s = 0$). Finding the distribution $H(r)$ from (6.2.24) gives, on reversing the enthalpy $H(\rho)$, a new density distribution $\rho_1(\mathbf{r})$. If $\rho_1(\mathbf{r})$ differs from $\rho_0(\mathbf{r})$, by more than some predetermined magnitude, $\rho_1(\mathbf{r})$ is then taken for the initial iteration, and the procedure is iterated until a self-consistency is obtained with an accuracy smaller than the predetermined value.

Implementation of this scheme requires calculating the potential $\Phi(\mathbf{r})$ and mass m_r from the determined density distribution $\rho(\mathbf{r})$. Calculations are performed in the dimensionless variables

$$\rho' = \rho/\rho_c, \qquad x = R/R_0, \qquad s = r/R_0,$$
$$\varphi = \Phi/4\pi G\rho_c R_0^2, \qquad \chi' = \chi/4\pi G\rho_c R_0^2 \qquad B'_s = B_s/4\pi G\rho_c R_0^2. \tag{6.2.25}$$

Here, R_0 is the maximum stellar radius, unknown at the beginning, R is the current spherical radius, ρ_c is the central density. In order to calculate ϕ we make use of the expansion in the Legendre polynomials $P_n(\mu)$, $\mu = \cos\theta$. We define the quantity $\rho_l(x)^4$ which may be evaluated with the Gauss formula [1, 137]

$$\rho_l(x) \equiv \int_0^1 \rho(x,\mu)P_{2l}(\mu)\,d\mu = \sum_{j=1}^{N_\mu} \rho(x,\mu_j)w_j P_{2l}(\mu_j), \tag{6.2.26}$$

where μ_j are the abscissae, w_j are the weight functions of the gaussian quadrature in the interval $[0,1]$. With the above definition of $\rho_l(x)$ we have

$$\rho(x,\mu) = \sum_l (4l+1)\rho_l(x)P_{2l}(\mu). \tag{6.2.27}$$

If $\rho_l(x)$ is taken in the form (6.2.26) and (6.2.27), the expression for the potential $\varphi(x,\mu)$ becomes, on using the integral expression (6.2.3),

$$\varphi(x,\mu) = \sum_l \varphi_l(x)P_{2l}(\mu), \tag{6.2.29}$$

$$\varphi_l(x) = -\frac{1}{x^{2l+1}} \int_0^x \rho_l(y)y^{2l+2}dy - x^{2l} \int_x^1 \rho_l(y)y^{-2l+1}dy. \tag{6.2.28}$$

The expressions (6.2.26–28) contain only even Legendre polynomials because of the symmetry with respect to the equatorial plane. It is necessary to know the functions $\varphi_l(x)$ in the same points where $\rho_l(x)$ are given. Specifying the function $\rho_l(x)$ in the interval $0 \le x \le 1$ at a step $h = 1/N_x$, we are in a position to evaluate the integrals in (6.2.29) by use of Simpson's formula [1]. On evaluating $\varphi_l(x)$, we find $\varphi(x,\mu)$ using (6.2.28). In terms of the given density distribution $\rho(x,\mu)$, the functions $p(s)$ have a form (see (6.2.22))

$$1 - p(s) = \int_s^1 x^2 dx \int_0^{\sqrt{1-s^2/x^2}} \rho(x,\mu)\,d\mu/M', \tag{6.2.30}$$

where

$$M' = M/4\pi\rho_c R_0^3 = \int_0^1 x^2 dx \int_0^1 \rho(x,\mu)\,d\mu. \tag{6.2.31}$$

The inner integral in (6.2.30) may be evaluated by use of the expansion (6.2.27) in the form $\left(h(s,x) = \sqrt{1-s^2/x^2}\right)$

---

[4] We use hereinafter an unprimed notation for the dimensionless functions ρ', χ', B'_s.

$$\int_0^{h(s,x)} \rho(x,\mu)\,d\mu = \sum_l (4l+1)\rho_l(x) \int_0^{h(s,x)} P_{2l}(\mu)\,d\mu$$

$$= \sum_l \rho_l(x) \left\{ P_{2l+1}\left[h(s,x)\right] - P_{2l-1}\left[h(s,x)\right] \right\}. \quad (6.2.32)$$

If the function $p(s)$ is evaluated at the points $s = s_i (i = 1, \ldots, N_x)$, and the values of $\rho_l(x)$ are specified at the points $x = x_k (k = 1, \ldots, N_x)$, then it is convenient to create an array of $\frac{1}{2}N_x(N_x - 1)$ quantities $h(s_i, x_k)$, $x_k > s_i$, in order to avoid evaluating, at all steps, the same quantities $\sqrt{1 - s_i^2/x_k^2}$. The outer integral in (6.2.30) is evaluated by applying Simpson's formula, similarly to (6.2.29).

Upon evaluating $p(s)$, with a specified $j(p)$ we have to find $\Omega(s)$ from (6.2.22), and, by use of (6.2.23), the centrifugal potential $\chi(s) = \chi(x\sqrt{1 - \mu^2})$. For a solid-body rotation we have

$$\chi = \beta x^2 (1 - \mu^2), \quad \beta = \Omega^2/8\pi G\rho_c. \quad (6.2.33)$$

If the rotation has a form $\Omega = \Omega(s)$, then χ may be evaluated from (6.2.23) without finding $p(s)$.

The next step is to calculate the enthalpy H (in the same units as the potentials φ and χ in (6.2.25)). The constant B_s in (6.2.24) is determined as a maximized value of $|\chi - \varphi|$ at $x = 1$. This causes the new configuration to be always closed inside a sphere of radius $x = 1$. Calculations [65, 284] use the value of B_s at the equator: $B_s = \varphi - \chi\big|_{x=1,\mu=0}$. With dimensional enthalpy distribution $H = 4\pi G\rho_c R_0^2 H'(x,\mu)$ we can find from (6.2.18) a next approximation to the density distribution $\rho_1(x,\mu)$. If we fix ρ_c in the constructing model, then $H_c(\rho_c)$ is known, and the maximum radius R_0 is obtained from the relation

$$R_0^2 = H_c/4\pi G\rho_c H_c'. \quad (6.2.34)$$

If, on the contrary, the stellar mass M is fixed, then R_0 together with ρ_c are obtained after each iteration from the simultaneous solution of equations (6.2.31) and (6.2.34). The stellar surface can be determined from the condition $H' = 0$. Generally, the surface does not contain the nodes of the grid (x_k, μ_j) and should be consequently found by interpolation. Outside the surface the inequality $H < 0$ is valid, and we put $\rho_1(x,\mu) = 0$ at these points. Iterations continue until

$$|\rho_0 - \rho_1|/\rho_1 < \epsilon \quad (6.2.35)$$

in all points except those where $\rho_1 < 10^{-6}$. For weakly oblate figures with $R_e/R_p \leq 3$ ($R_e = R_0$ is the equatorial, R_p the polar radii) $\epsilon = 10^{-5}$ has been taken in [64]. For rapid rotation and strong oblateness the accuracy criterion has been lowered to $\epsilon = 10^{-3}$. The integral relative error Δ in the solution may be estimated with the aid of the virial theorem [145, 437]:

$$\Delta = (W + 2T + 3 \int P \, dV)/|W|, \tag{6.2.36}$$

where W is the gravitational and T the kinetic energies.

At a low compressibility of matter, there can exist equilibrium configurations with a high oblateness. For these, the expansion of the density (6.2.37) in Legendre polynomials in a spherical coordinate system is not sufficiently accurate. For constructing equilibrium models by the self-consistent field method, we make use of the spheroidal coordinates (ζ, η) related to the cylindrical (r, z) by

$$z = R_k \zeta \eta, \quad r = R_k \left[(\zeta^2 + 1)(1 - \eta^2) \right]^{1/2}. \tag{6.2.37}$$

In calculations, the parameter of spheroidal coordinates R_k is set equal to R_0. The density written in these coordinates reads (instead of (6.2.27) and (6.2.26)) [163]

$$(\zeta^2 + \eta^2) \, \rho(\zeta, \eta) = \sum_n d_{2n}(\zeta) \, P_{2n}(\eta), \tag{6.2.38}$$

$$d_{2n}(\zeta) = (4n + 1) \int_0^1 (\zeta^2 + \eta^2) \, \rho(\zeta, \eta) \, P_{2n}(\eta) \, d\eta, \tag{6.2.39}$$

and the potential (instead of (6.2.38) and (6.2.39))

$$\Phi(\zeta, \eta) = \sum_n \varphi_{2n}(\zeta) \, P_{2n}(\eta), \tag{6.2.40}$$

$$\varphi_{2n}(\zeta) = 4\pi G R_k^2 \left[q_{2n}(\zeta) \int_0^\zeta p_{2n}(x) \, d_{2n}(x) \, dx \right.$$

$$\left. + p_{2n}(\zeta) \int_\zeta^\infty q_{2n}(x) \, d_{2n}(x) \, dx \right]. \tag{6.2.41}$$

Here, $p_n(x)$, $q_n(x)$ are the Legendre functions of imaginary arguments of the first and second kind [140]. Rigidly rotating models in spheroidal coordinates have been investigated in [38]. Another approach to constructing highly flattened models, using the matched asymptotic expansion method, was suggested in [261a].

Calculations performed in [38, 64] have supplied the ratio R_p/R_e for rotational limit models given in Table 6.2 from [38]. The results in the last line of this table are obtained by use of spheroidal coordinates. The modification of the self-consistent field method suggested in [383a] has proved to be stable for arbitrarily oblate and even for unbound figures. This method uses, similarly to [64], an expansion in Legendre polynomials in a spherical system (6.2.28) but, instead of specifying in (6.2.22) the total mass M and angular momentum distribution j as in [64], it fixes in constructing the model the central density and either the semiaxes ratio or, for a toroidal figure, the

Table 6.2. The polar to equatorial radii ratio $\dfrac{R_p}{R_e}$, the energy ratio $\dfrac{E_{\text{rot}}}{E_{\text{grav}}}$ and parameter $\beta = \dfrac{\Omega^2}{8\pi G\rho_c}$ for polytropic models with limiting rotation ($\Omega(r) = \text{const.}$)

n	γ	R_p/R_e	$E_{\text{rot}}/E_{\text{grav}}$	β	References
3	1.333	0.6516	0.00900	0.001020	[64]
1.5	1.6667	0.6149	0.0595	0.01091	[437,64]
1	2	0.5580	–	0.02093	[437]
0.808	2.238	0.5215	~0.135	0.02651	[437]
0.6	2.6667	0.465	0.145	0.03375	[64,430a]
			(0.1346)	(0.03125)	
0.5	3	0.4378	0.152	0.0375	[38,430a]
			(0.1364)	(0.03367)	

location of the outer and inner boundaries. This method is generalized in [383b] for constructing equilibrium triaxial figures. These involve expansion over the spherical harmonics $P_l^m(\mu) \cos m\phi$, instead of Legendre polynomials in (6.2.26). Modification of the SFC method with inclusion of grid adaptation was developed in [372a].

6.2.4 Methods Based on the Newton–Raphson Iteration Scheme

In the self-consistent method the convergence is poor for configurations with high density contrast (core-halo type) or with high flatness. The variant of this method, suggested in [383a] has a good convergence, but its application could be complicated by the restriction of fixing of the semiaxis ratio (not M and $j(p)$), especially for calculation of the evolution of rotating stars. Methods based on the Newton–Raphson iteration scheme have a good convergence in a wide range of cases.

The approach discussed below was first used in the construction of arbitrary form bodies for incompressible fluids in [352], and was extended in [351] for a general case of a compressible fluid.

Here we shall formulate this method for structures with axial and equatorial symmetry [351] in spherical coordinates (R, θ, φ). The angle θ is broken into N_T equidistant points

$$\theta_i = \frac{\pi}{2} \frac{i-1}{N_T - 1}, \quad (i = 1, \ldots, N_T). \tag{6.2.42}$$

The radius along each angle θ_i is divided into N_R points

$$R_{ij} = R_j(\theta_i) = R_{\text{surf}}(\theta_i) \frac{i}{N_R}, \quad (j = 1, \ldots, N_R). \tag{6.2.43}$$

Using (6.2.25–29), we can find the values of the gravitational potential Φ_{ij} in terms of the density

$$\rho_{ij} = \rho(R_{ij}, \theta_i). \tag{6.2.44}$$

We have (in the centre)

$$\Phi_{00} = -4\pi G \sum_l \sin\theta_l \Delta\theta_l \sum_m R_{lm} \Delta R_{lm} \rho_{lm} \qquad \text{(in the centre),} \tag{6.2.45}$$

$$\Phi_{ij} = -4\pi G \sum_l \sin\theta_l \Delta\theta_l \sum_m R_{lm}^2 \Delta R_{lm}$$
$$\times \sum_n f_{2n,lm}^{ij} P_{2n}(\cos\theta_i) P_{2n}(\cos\theta_l) \rho_{lm}, \tag{6.2.46}$$
$$(i = 1,\ldots,N_T; \quad j = 1,\ldots,N_R),$$

where

$$f_{2n}(R, R') = \begin{cases} R'^{2n}/R^{2n+1} & \text{for } R \geq R', \\ R^{2n}/R'^{2n+1} & \text{for } R < R', \end{cases} \tag{6.2.47}$$
$$f_{2n,lm}^{ij} = f_{2n}(R_{ij}, R_{lm}).$$

At the surface, i.e. for $j = N_R$ and $1 \leq i \leq N_T$, we have

$$\rho_{i,N_R} = \rho(R_{i,N_R}, \theta_i) = 0. \tag{6.2.48}$$

Here, $\Delta\theta_l$ and ΔR_{lm} are the angular and radial spacings multiplied by relevant weight factors which depend on the adopted numerical integration scheme (see, for example, (6.2.26)). In [351] Simpson's method has been used for integration in the radial direction and the trapezoidal method in the angular direction. Substituting (6.2.45–47) into the integral of the equilibrium equation (6.2.24) written for all points (θ_i, R_{ij}) and for the centre, we obtain, in common with (6.2.48), a non-linear system of equations of the order $(N_R + 1)N_T + 1$ with unknown values of the constant B_s, $N_T \cdot N_R + 1$ densities ρ_{ij} including ρ_c, N_T surface radii R_{i,N_R}, and one parameter characterizing the angular velocity Ω_c. The total number of unknowns exceeds the number of equations by two, so two parameters, say, ρ_c or M, and Ω_c or $R_{\text{surf}}(\pi/2)/R_{\text{surf}}(0)$, may be chosen arbitrarily to close the system. It is also assumed here that the enthalpy $H(\rho)$ and the single-parameter law $\Omega = \Omega(\Omega_c, r)$ have been substituted in (6.2.24).

In [351] an explicit form of (6.2.24) is given for the polytrope

$$P = K\rho^\gamma, \quad H = \frac{\gamma}{\gamma - 1} K\rho^{\gamma-1} \tag{6.2.49}$$

and for the angular velocity

$$\Omega(r) = \Omega_c/(1 + r^2/A^2), \quad \chi = -\frac{A^2 \Omega_c^2}{2(1 + r^2/A^2)}. \tag{6.2.50}$$

We then obtain a set of non-linear equations

$$\frac{\gamma}{\gamma - 1} K \rho_c^{\gamma - 1} + \Phi_{00} + \frac{A^2 \Omega_c^2}{2} = B_s, \tag{6.2.51}$$

$$\frac{\gamma}{\gamma - 1} K \rho_{ij}^{\gamma - 1} + \Phi_{ij} + \frac{A^2 \Omega_c^2}{2(1 + R_{ij}^2 \sin^2 \theta_i / A^2)} = B_s. \tag{6.2.52}$$

The non-linear set of equations (6.2.51), (6.2.52), and (6.2.48) is solved in [351] by the Newton–Raphson iteration scheme.

Applying the general integral method requires a large amount of computer time. For $n = 10$ in (6.2.46), $N_s = 15$, $N_r = 40$ with specified ρ_c and Ω_c, we obtain 616 non-linear equations requiring ~ 2.5 s of CPU time per iteration on a CRAY-1 computer with account of internal structure of its vector processor. A relative accuracy of 10^{-4} requires about ten iterations.

In [248a] application of a Newton iterative scheme was carried out after obtaining explicit solutions of (6.2.51) and (6.2.52) relative to ρ_{ij} and ρ_c, and substitution of this solution into (6.2.3). The Poisson equation was solved by using a uniform grid in R (1/R outside the star) and θ coordinates. Contrary to (6.2.46), no expansion in Legendre polynomials was used. The resulting non-linear finite difference system was solved by a Newton–Raphson relaxation scheme, where a corresponding system of linear equations with sparse matrices of coefficients was solved using a special elimination technique. This method is faster than the method from [351], but still rather slow on large grids. For a grid with $N_s = 51$, $N_R = 402$ it takes about 100 s CPU time per iteration on a CONVEX 220 computer without a vectorization.

6.3 Evolution of Rotating Stars

In real stars there is no barotropy, so $P = P(\rho, T)$, and the equilibrium equation should be supplemented with the energy and heat transfer equations. We write out hydrodynamic equations analogous to (6.2.1) and (6.2.2) in Eulerian spherical coordinates (r, θ, φ), in the absence of a magnetic field, retaining terms linear in velocities v_r and v_θ which are taken to be small. Similarly to Sect. 6.2, the star is assumed to be axisymmetric, i.e. $\partial / \partial \varphi = 0$. We have

$$\frac{\partial v_r}{\partial t} + \frac{1}{\rho} \frac{\partial P}{\partial r} + \frac{\partial \Phi}{\partial r} - \Omega^2 r \sin^2 \theta = 0, \quad v_\phi = \Omega r \sin \theta, \tag{6.3.1}$$

$$\frac{\partial v_\theta}{\partial t} + \frac{1}{\rho r} \frac{\partial P}{\partial \theta} + \frac{1}{r} \frac{\partial \Phi}{\partial \theta} - \Omega^2 r \sin \theta \cos \theta = 0, \tag{6.3.2}$$

$$\frac{\partial v_\varphi}{\partial t} + v_r \frac{\partial v_\varphi}{\partial r} + \frac{v_\theta}{r} \frac{\partial v_\varphi}{\partial \theta} + \frac{v_r v_\varphi}{r} = 0 \quad \text{(radiative zone)} \tag{6.3.3}$$

$$\frac{\partial \rho}{\partial t} + \frac{1}{r^2} \frac{\partial}{\partial r} (r^2 \rho v_r) + \frac{1}{r \sin \theta} \frac{\partial}{\partial \theta} (\rho v_\theta \sin \theta) = 0. \tag{6.3.4}$$

An equation for Φ is given in (6.2.18). For the convective zone, a convective viscosity equalizing the angular velocity (see Chap. 8, Vol. 2) must be added to the right-hand side of (6.3.3). For the convective core we may take, approximately, because of a high convective (turbulent) viscosity,

$$\Omega = \text{const.} \quad \text{(adiabatic convection zone).} \tag{6.3.3a}$$

The equation with viscosity should be used in the outer shells of convective envelopes with a small density, where circulation velocities may become high (see Chap. 8, Vol. 2). In this evolutionary calculation scheme v_φ is obtained from (6.3.3) or (6.3.3a).

For a rotating star, the energy and heat transfer equations have the form

$$\nabla \cdot \boldsymbol{l} = \frac{1}{r^2} \frac{\partial}{\partial r} (r^2 l_r) + \frac{1}{r \sin\theta} \frac{\partial}{\partial \theta} (\sin\theta l_\theta) =$$

$$\epsilon_n - \epsilon_\nu - T \left(\frac{\partial S}{\partial t} + v_r \frac{\partial S}{\partial r} + \frac{v_\theta}{r} \frac{\partial S}{\partial \theta} \right), \tag{6.3.5}$$

$$\boldsymbol{l} = -\frac{4}{3} \frac{acT^3}{\kappa\rho} \nabla T \quad \text{(radiation heat transfer)}, \tag{6.3.6}$$

$$\nabla T = \frac{T}{\rho} \gamma_2 \nabla P \quad \text{(adiabatic convection).} \tag{6.3.7}$$

We write also the equation for evolutionary changes in chemical composition

$$\frac{\partial x_i}{\partial t} + v_r \frac{\partial x_i}{\partial r} + \frac{v_\theta}{r} \frac{\partial x_i}{\partial \theta} =$$

$$= \begin{cases} -4m_p \left(\dfrac{\epsilon_{\text{CNO}}}{Q_{\text{CNO}}} + \dfrac{\epsilon_{\text{pp}}}{Q_{\text{pp}}} \right), \quad i = H, \\[2ex] 4m_p \left(\dfrac{\epsilon_{\text{CNO}}}{Q_{\text{CNO}}} + \dfrac{\epsilon_{\text{pp}}}{Q_{\text{pp}}} \right) - \dfrac{3m_\alpha \epsilon_{3\alpha}}{Q_{3\alpha}} \\ \quad -m_\alpha \epsilon_{12C\alpha}/Q_{12C\alpha}, \quad i = {}^4\text{He}, \\[2ex] \dfrac{3m_\alpha \epsilon_{3\alpha}}{Q_{3\alpha}} - \dfrac{m_{12C} \epsilon_{12C\alpha}}{Q_{12C\alpha}}, \quad i = {}^{12}\text{C}, \end{cases} \tag{6.3.8}$$

and analogously for other reactions. Here $\boldsymbol{l} = (l_r, l_\theta, 0)$ erg cm^{-2} s^{-1} is the heat flux vector, ϵ_n is determined in (6.1.5–6), and physical properties of matter are the same as those used in Sect. 6.1 for a non-rotating star.

Contrary to the equations of Sect. 6.2 for a barotropic medium, the evolutionary equations (6.3.1–8) require the inclusion of the velocities v_r and v_θ which inevitably arise in the presence of heat flows.

6.3.1 Meridional Circulation

The assumption that a rotating star is in a stationary state with zero velocities $v_r = v_\theta = 0$ may sometimes lead to the contradiction known as von Zeipel's paradox [199]. We consider its derivation for the case of a solid-body rotation. At $v_r = v_\theta = 0$ the equilibrium equation reduces to (6.2.17).

According to Poincaré's theorem, a barotropic medium in equilibrium must have the centrifugal potential (6.2.23). On the other hand, the existence of a centrifugal potential in (6.2.17) causes the quantity $dH = dP/\rho$ to be a total differential. It follows from (6.2.17) and (6.2.23) that ∇P is everywhere parallel to $\nabla(\chi - \Phi)$, which is possible only when the constant pressure surfaces coincide with equipotential surfaces $W = \chi - \Phi = \text{const.}$, and $P = P(W)$ in the star. Since dP/ρ is a total differential, $\rho = \rho(W)$, and so the equation of state leads to $T = T(W)$. We see, then, that for a solid-body rotation, and also for another form of centrifugal potential from (6.2.23), the equilibrium with $v_r = v_\theta = 0$ is possible only at

$$P = P(W), \quad \rho = \rho(W), \quad T = T(W). \tag{6.3.9}$$

With (6.3.9), (6.3.6) becomes

$$l = \left(-\frac{4acT^3}{3\kappa\rho}\frac{dT}{dW}\right)\nabla W. \tag{6.3.10}$$

We find from (6.3.10)

$$\nabla \cdot l = \frac{d}{dW}\left(-\frac{4acT^3}{3\kappa\rho}\frac{dT}{dW}\right)|\nabla W|^2 + \left(-\frac{4acT^3}{3\kappa\rho}\frac{dT}{dW}\right)\nabla^2 W. \tag{6.3.11}$$

Using (6.2.18) and (6.2.23), we have for a solid-body rotation

$$\nabla^2 W = 2\Omega^2 - 4\pi G\rho = f(W). \tag{6.3.12}$$

The coefficient before $|\nabla W|^2$ is a function of W. Nevertheless, ∇W changes along equipotentials, since their separation along the polar direction is less than in the equatorial plane because of oblateness of the rotating star. Hence, it follows from (6.3.11) that ∇l also changes along equipotentials W. On the other hand, in thermal equilibrium at $\partial/\partial t = v_r = v_\theta = 0$ the relation (6.3.5) yields a constancy of ∇l at equipotentials, and so leads to von Zeipel's paradox. Thus, the assumption of thermal equilibrium at $v_r = v_\theta = 0$ is not valid. In a rotating star, departures from the thermal equilibrium state cause the cooling regions to fall and the heating regions to rise, hence, a state with $v_r, v_\theta \neq 0$ arises. These slow motions are called meridional circulation. For the general case of rotation with a centrifugal potential, a proof of the absence of thermal equilibrium and the onset of meridional circulation is given in [199].

Many papers deal with meridional circulation (see reviews in [157, 75]), using various approximations. In the stationary circulation approximation, in the presence of a solid-body rotation in deep stellar regions the matter rises at the poles and falls at the equator, and vice versa in outer layers with low density ($\rho < \Omega^2/2\pi G$); for some rotation laws the direction of circulation remains the same through all the star [75, 199]. In the absence of viscosity and magnetic field, the solution for stationary rotation diverge near the surface. For example, in a star with a slow solid-body rotation and a point energy source (Cowling's model) the radial circulation velocity in the radiative zone is [199]

$$v_r = \frac{LR^2}{GM^2} Y(r, \theta), \tag{6.3.13}$$

where

$$
\begin{aligned}
Y(r, \theta) = \{&\alpha\beta_0(r) P_2(\theta) \\
&+ \alpha^2 (\overline{\rho}/\rho_0) [\beta_2(r) P_2(\theta) + \beta_4(r) P_4(\theta)]\}/(n - 3/2), \\
\end{aligned}
$$
$$\alpha = \frac{\Omega^2 R^3}{GM}, \quad n = \frac{d \ln \rho_0}{d \ln T_0}. \tag{6.3.14}$$

Here, ρ_0, T_0 are the density and temperature distributions in a spherical star, $\overline{\rho}$ is the mean stellar density, $\beta_i(r)$ are smooth functions of radius ~ 1. It should be noted that, as $\rho_0 \to 0$, v_r diverges not only at the surface, but also near the convective core boundary, where $n \to 3/2$. In differentially rotating stars the divergence of v_r near the surface appears in the first term of the expansion in α. As pointed out in [75, 199], taking into account the viscosity, magnetic field, or removing the stationary condition leads to elimination of the singularity. A stationary circulation with account of viscosity has been investigated in detail by Tassoul and Tassoul (see [604] and references therein).

The meridional circulation may influence the stellar evolution via mixing and smoothing the chemical composition. The estimates made in [229] with chemical inhomogeneities taken into account, led to the conclusion that mixing caused by circulation is unessential even for rapidly rotating stars near the upper end of the main sequence. These estimates have to be quantitatively confirmed.

6.3.2 Evolutionary Calculations Including Circulation

In calculations of rotating star evolution the following physical processes must generally be taken into account: 1) the evolution of chemical composition; 2) the thermal non-stationarity; 3) the non-stationary circulation; 4) the angular momentum exchange resulting from circulation, viscosity, magnetic field; 5) the chemical mixing; 6) the mass loss. Taking into account the first two factors does not involve any serious difficulties and are similar to calculations for a non-rotating star (see Sect. 6.1). The problems associated with circulation are usually avoided in evolutionary calculations to date (see review [199]) by fixing restrictions to the angular velocity distribution, say, a constancy of Ω at the effective potential W surfaces [436]. As shown in [445], this assumption is not valid, and Ω does not tend to be constant at these equipotential surfaces during evolution. Non-physical divergences near the surface in (6.3.13) and (6.3.14) arise from the assumption that the circulation is stationary, so correct evolutionary calculations for rotating stars make it necessary to consider the non-stationary circulation. The angular momentum exchange and chemical mixing over the star are caused, apart from circulation motions, by various transport properties. The effect of microscopic transport mechanisms, such as viscosity, diffusion and thermodiffusion, is usually negligible over evolutionary lifetime. Turbulent transport processes in regions of intense convection may be much more effective. The chemical composition in these regions may be treated as uniform because of a rapid mixing, while the angular velocity in the same regions may be taken as constant because of a high turbulent viscosity (see Chap. 8, Vol. 2). The effect of instabilities arising from rotation on the momentum transfer and mixing is studied in [445]. The unstable state of a star may result from changes in stability conditions which arise from the chemical evolution. It is shown in [445] that the time of development for such instabilities is comparable with the circulation time scale, hence, the calculation scheme described below which implies retaining terms linear in velocity, will automatically include the unstable axisymmetric modes.

It was suggested in [646b], that even in slowly rotating stars a small differential rotation leads to rotational-induced turbulence, which occurs in stellar radiative zones with motion more vigorous in the horizontal than in the vertical direction. This week anisotropic turbulence may be important for chemical mixing and angular momentum transfer in radiative zones, giving a small input in the heat transfer. Viscosity and diffusivity coefficients connected with such turbulence are estimated in [646b].

Another mechanism for transport of angular momentum and diffusion, based on the action of internal waves was considered in [561b]. It was shown that the process is very efficient and can explain the quasi-solid-body rotation of the internal radiative zone of the Sun, and diffusion generated by non-linear effects of the wave propagation can very likely explain the characteristics of lithium depletion in stars. The influence of this process on solar neutrino flux was discussed in [561c], where better prediction of the solar neutrino flux without being in disagreement with helioseismology data is expected. The magnetic field takes part in the momentum transfer both explicitly and through the effect on meridional circulation [199, 492, 493, 157]. The magnetic field is not included in the present scheme.

Self-consistent methods for evolutionary calculations with mass loss are not yet established even for non-rotating stars (see Sect. 6.1.4). The problem is even more complicated for rotating stars because of the effect of centrifugal and magnetic forces on the outflow [471].

We take the stellar mass to be constant. Equations (6.3.1–8) and (6.2.18) or (6.2.3) are basic for evolutionary calculations. The method described incorporates properties of the self-consistent field method (see Sect. 6.2.3) and Henyey method (Sect. 6.2.2), and extends the method from [436] to the presence of circulation. We have to solve $8 + i$ equations, where i is the number of calculated chemical elements, for the same number of unknowns

$$\rho, T, v_r, v_\theta, v_\varphi, l_r, l_\theta, \Phi, x_i. \qquad (6.3.15)$$

For constructing an initial model, we may specify an entropy distribution over the star determined from, say, a non-rotating chemically homogeneous model with the same stellar mass, and find a mechanically equilibrium model by one of the methods of Sect. 6.2. All time derivatives of $D^{(i)}$ in equations (6.3.1–5) and (6.3.8) should be replaced by finite differences

$$\frac{\partial D^{(i)}}{\partial t} = \frac{D^{(i)}(t) - D^{(i)}(t - \Delta t)}{\Delta t}. \tag{6.3.16}$$

The star is divided into N_T points θ_j along the angle, so that $0 < \theta_j < \pi/2$, $j = 1, \ldots, N_T$, and into N_R points r_{ij}, $i = 1, \ldots, N_R$ along the radius. The differential equations are replaced by finite difference equations. Besides the unknowns in the node points, except the centre, the central values ρ_c, T_c, and also N_T values of the outer radii represent additional unknowns. To determine these, we can use the condition of mass constancy in the form (6.2.31), for example, for obtaining ρ_c, and the condition (6.1.15) giving the pressure in N_T boundary points, to determine the outer radii $r_j^{\text{out}} \equiv r_{i_{\text{out}}}$, $i_{\text{out}}(j) \leq N_R$, $j = 1, \ldots, N_T$. For a rotating star, instead of g in (6.1.15), we have to use the effective value

$$g_{\text{ef}} = (|\nabla(\Phi| - \Omega^2 r)_j^{\text{out}}. \tag{6.3.17}$$

The existence of the axis and plane of symmetry $\theta = 0$ and $\theta = \pi/2$ causes the derivatives $\partial/\partial\theta$ of all quantities there (6.3.15) to become zero. Besides, at $\theta = 0$ and $\theta = \pi/2$ we have $l_\theta = v_\theta = 0$, and on the axis of symmetry, i.e. at $\theta = 0$, we have $v_\varphi = 0$. We have also to take into account that at the outer surface, r_j^{out},

$$(l_n)_j^{\text{out}} = \frac{ac}{4} \left(r_j^{\text{out}}\right)^2 \left(T_j^{\text{out}}\right)^4$$

$$l_n = \frac{\sqrt{l_r^2 + \left(\frac{dr_s}{d\theta}\right)^2 l_\theta^2}}{\sqrt{1 + \left(\frac{dr_s}{d\theta}\right)^2}} \quad \begin{array}{l} \text{is the heat flux component} \\ \text{normal to the surface} \end{array} \tag{6.3.18}$$

$$r_s(\theta_j) \equiv r_j^{\text{out}},$$

similarly to (6.1.11). As a result, the number of unknowns obviously coincides with that of the difference equations. A block diagram for the above method of evolutionary calculations is shown in Fig. 6.1. All quantities on the diagram in Fig. 6.1a are determined by an implicit time scheme (see (6.1.41) and (6.1.42)). In the hydrogen-burning phase, an explicit scheme, as in Fig. 6.1b, is likely to be sufficient for finding the new distribution of chemical composition. Shown also is a possible breaking of the iterative process into parts. The dimension of the set of difference equations is about $(7+i) \times N_T \times N_R$ for the scheme shown in Fig. 6.1a, and $\sim 4 \times N_T \times N_R$ and $\sim 3 \times N_T \times N_R$ for the internal and external iteration blocks, respectively, for the scheme in Fig. 6.1b. This method is capable of giving results of sufficiently high accuracy only if supercomputers are used. Any block for finding

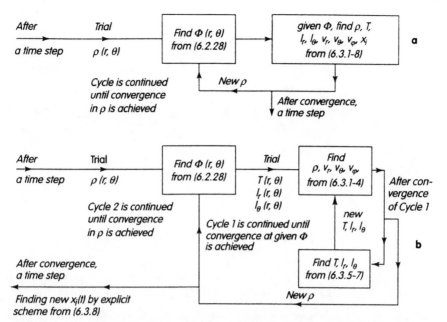

Fig. 6.1. Possible block-diagrams for evolutionary calculations of rotating star including circulation. Everywhere difference analogies of corresponding equations are implied.

the quantities $\rho, T, l_r, l_\theta, v_r, v_\theta, v_\varphi, x_i$ in Fig. 6.1 reduces to solving a set of difference equations by an iteration scheme similar to the Henyey method. In the most general formulation, one may renounce all iteration cycles and, using the difference form of Poisson's equation (6.2.46), find all the quantities (6.3.15), including Φ, in a general iterative Newton–Raphson procedure similar to the method from Sect. 6.2.4. The problem is reduced to a set of $\sim (8+i) \times N_T \times N_R$ non-linear difference equations.

References

Abbreviations

AsJ – Astronomicheskii Jurnal (Soviet Astronomy)
PAJ – Pis'ma v Astronomicheskii Jurnal (Soviet Astronomy Letters)
A – Astrofizika (Astrophysics)
UFN – Uspekhi Fizicheskikh Nauk (Soviet. Phys. Uspekhi)
JETF – Jurnal Eksperimental'noi i Teoreticheskoi Fiziki
  (Journal of Experimental and Theoretical Physics, JETP)
NI – Nauchnyie Informatzii Astronomicheskogo Soveta AN SSSR (Scientific Information of the Astonomical Council of the Academy of Sciences of the USSR)
PMTF – Jurnal Prikladnoi Mekhaniki i Tekhnicheskoi Fiziki (Journal of Applied Mechanics and Physics of Engineering)
JVMiMF – Jurnal Vychislitel'noi Matematiki i Matematicheskoi Fiziki (Journal of Applied Mathematics and Mathematical Physics)
M – Moscow
L – Leningrad

1 Abramovitz, M., Stegun, I. (eds.) (1968): *Handbook of Mathematical Functions* (Dover, New York)
2 Abrikosov, A. A. (1954): On the internal structure of hydrogen planets. Vopr. Kosmogonii **3** 12–19
3 Abrikosov, A. A. (1960): Some properties of highly compressed matter. JETF **39**, 1797–1805
4 Aleksandrov, A. F., Bogdankevich, L. S., Rukhadze, A. A. (1978): *Grounds of Plasma Electrodynamics* (M, Visschaya Schkola)
5 Allen, K. U. (1973): *Astrophysical Quantities* (Athlone Press, Univ. of London)
6 Alfven, G., FältHammar, K. G. (1963): *Cosmical Electrodynamics. Fundamental Principles* (Oxford, Clarendon)
7 Ambartzumian, V. A. (1968): *Problems of the Evolution of the Universe* (Erevan, AN ArmSSR)
8 Ambartzumian, V. A., Fuors, A. (1971): **7**, 557–572
9 Ambartzumian, V. A., Saakyan, G. S. (1960): On degenerate superdense gas of elementary particles. AsJ **37**, 193–209
10 Ardelyan, N. V. (1983): Divergence of difference schemes for two-dimensional equations of acoustics and Maxwell equations. JVMMF **23**, 1168–76
11 Ardelyan, N. V. (1983): On applying iterative methods for the realization of implicit difference schemes of two-dimensional magnetohydrodynamics, JVMMF **23**, 1417–26
12 Ardelyan, N. V., Bisnovatyi-Kogan, G. S., Popov, Yu. P. (1979): Investigation of magnetorotational supernova explosion. Cylindrical model. AsJ **56**, 1244–55

13 Ardelyan, N. V., Bisnovatyi-Kogan, G. S., Popov, Yu. P., Chernigovskii, S. V. (1987): Calculation of collapse of rotating gas cloud on Lagrangian grid. AsJ **64**, 495–508

14 Ardelyan, N. V., Bisnovatyi-Kogan, G. S., Popov, Yu. P., Chernigovskii, S. V. (1987): Core collapse and the formation of a rapidly rotating neutron star. AsJ **64**, 761–772

15 Ardelyan, N. V., Guzshin, I. S. (1982): On an approach to constructing conservative difference schemes. Vest. MGU, Vych. Mat. i Kib., N 3, pp. 3–10

16 Ardelyan, N. V., Chernigovskii, S. V. (1984): On the divergence of difference schemes for two-dimensional equations of gas dynamics with the inclusion of gravity to the acoustical approximation. Differential'nyie Uravneniya **20**, 1119–27

17 Akhiyezer, A. I., Berestetzkii, V. B. (1969): *Quantum Electrodynamics* (M, Nauka)

18 Basko, M. M. (1982): Two-group approximation in spherically symmetric problems of transfer theory. Preprint ITEP, N 7

19 Basko, M. M., Imshennik, V. S. (1975): The stability loss by low-mass stars under conditions of neutronization. AsJ **52**, 469–80

20 Basko, M. M., Rudzskii, M. A., Seidov, Z. F. (1980): Low-mass star collapse A. **16**, 321–335

20a Berezin, I. S., Zhidkov, N. P. (1960): *Calculation methods. Vol. 2* (M, Fiz-mat)

21 Berestetzkii, V. B., Lifshitz, E. M., Pitaevskii, A. P. (1968) *Relativistic Quantum Theory, part I* (M, Nauka)

22 Bethe, H. A. (1971): Theory of nuclear matter. Ann. Rev. Nucl. Sci. **21**, 93–244

23 Bethe, H. A, Morrison, F. (1956): *Elementary Nuclear Theory* (New York, Wiley)

23a Bisnovatyi-Kogan, G. S. (1965): Isotropic corrections to Maxwellian distribution functions and the rate of energy exchange. PMTF No. 3, 74–76

23b Bisikalo, D. V., Boyarchuk, A. A., Kuznetzov, O. A., Chechetkin, V. M. (1997): Influence of a common envelope of a binary system on a mass transfer through the inner Lagrange point. AsJ **74**, 889–897

23c Bisikalo, D. V., Boyarchuk, A. A., Kuznetzov, O. A., Khruzina, T. S., Cherepashchuk, A. M., Chechetkin, V. M. (1998): Evidences to absence of a shock interaction between jet and disk ("hot spot") in semi-detached binary systems. AsJ **75**, 40–53

24 Bisnovatyi-Kogan, G. S. (1966): Critical mass of hot isothermal white dwarf with the inclusion of general relativity effects. AsJ **43**, 89–95

25 Bisnovatyi-Kogan, G. S. (1967): Perfect gas flow in a spherically symmetric gravitational field with the inclusion of radiative heat transfer and radiation pressure. Prikl. Mat. Mech. **31**, 762–769

26 Bisnovatyi-Kogan, G. S. (1968): Explosions of massive stars. AsJ **45**, 74–80

27 Bisnovatyi-Kogan, G. S. (1968): On the role of radiation in the solar wind formation. Mekh. Jidk. Gaza No 4, 182–183

28 Bisnovatyi-Kogan, G. S. (1968): The mass limit of hot superdense stable configurations. A. **4**, 221–238

29 Bisnovatyi-Kogan, G. S. (1968): Late evolutionary stages of stars. Cand. Thes., Inst. Prikl. Mat.

30 Bisnovatyi-Kogan, G. S. (1970): Pulsar as a neutron star and weak interactions. Radiofiz. **13**, 1868–1872

31 Bisnovatyi-Kogan, G. S. (1970): On the mechanism of explosion of rotating star as a supernova. AsJ **47**, 813–816

32 Bisnovatyi-Kogan, G. S. (1977): Equilibrium and stability of stars and stellar systems. Doct. Thes., M. Space Res. Inst.

32a Bisnovatyi-Kogan, G. S., Moiseenko, S. G. (1992): The breaking of mirror symmetry of magnetic field in rotating stars and its possible astrophysical effects. AsJ **69**, 563–571

33 Bisnovatyi-Kogan, G. S. (1985): X-ray sources in close binaries: theoretical aspects. Abastuman. Obs. Biull. No. 58, 175–210

34 Bisnovatyi-Kogan, G. S. (1987): The cooling of white dwarfs with the inclusion of non-equilibrium beta-processes. PAJ **13**, 1014–1018

34a Bisnovatyi-Kogan, G. S. (1989, 1990): Two generations of low-mass X-ray binaries and recycled radiopulsars. A. **31**, 567–577; A. **32**, 193–194

35 Bisnovatyi-Kogan, G. S., Blinnikov, S. I. (1976): Hot corona around a disk accreting onto black hole and Cyg X-1 model. PAJ **2**, 489–493

36 Bisnovatyi-Kogan, G. S., Blinnikov, S. I. (1978): Propagation of waves in medium with high radiative density. preprint of IKI Pr 421

37 Bisnovatyi-Kogan, G. S., Blinnikov, S. I. (1978): Propagation of waves in a medium with high radiative density. I. Basic equations - homogeneous medium case. A. **14**, 563–577

38 Bisnovatyi-Kogan, G. S., Blinnikov, S. I. (1981): Equilibrium of rotating gas disks of a finite thickness. AsJ **58**, 312–323

39 Bisnovatyi-Kogan, G. S., Blinnikov, S. I., Zakharov, A. F. (1984): Numerical model of explosion near the neutron star surface and gamma-ray bursts. AsJ **61**, 104–111

40 Bisnovatyi-Kogan, G. S., Blinnikov, S. I., Kostyuk, N. D., Fedorova, A. V. (1979): Evolution of rapidly rotating stars on the stage of gravitational contraction. AsJ **56**, 770–780

41 Bisnovatyi-Kogan, G. S., Blinnikov, S. I., Fedorova, A. V. (1977): [The Role of Rotation in Young Star Evolution], in *Early Stages of Stellar Evolution* (Kiev, Naukova Dumka), pp. 40–46

42 Bisnovatyi-Kogan, G. S., Blinnikov, S. I., Schnol', E. E. (1975): Stability of stars in the presence of a phase transition. AsJ **52**, 920–929

43 Bisnovatyi-Kogan, G. S., Zel'dovich, Ya. B. (1966): Adiabatic Outflow and equilibrium states with energy excess, AsJ **43**, 1200–1206

44 Bisnovatyi-Kogan, G. S., Zel'dovich, Ya. B. (1967): The Matter outflow out of stars under the effect of high opacity in the envelope. Report at 35 Comm. XIII Gen. As. IAU, preprint of IPM

45 Bisnovatyi-Kogan, G. S., Zel'dovich, Ya. B. (1968): The matter outflow out of stars under the effect of high opacity in the atmosphere, AsJ **45**, 241–250

46 Bisnovatyi-Kogan, G. S., Kajdan, Ya. M. (1966): Critical parameters of stars, AsJ **43**, 761–771

47 Bisnovatyi-Kogan, G. S., Kajdan, Ya. M., Clypin, A. A., Shakura, N. I. (1979): Accretion onto rapidly moving gravitating centre, AsJ **56**, 359–67

47a Bisnovatyi-Kogan, G. S., Komberg, B. V. (1974): Pulsars and close binary systems. Sov. Astron. **18**, 217–221

47b Bisnovatyi-Kogan, G. S., Komberg, B. V. (1976): Binary radiopulsar as an old object with small magnetic field: possible evolutionary scheme. Sov. Astron. Lett. **2**, 130–132

48 Bisnovatyi-Kogan, G. S., Lamzin, S. A. (1976): Outflow out of stars on the stage of gravitational contraction. AsJ **53**, 742–749

49 Bisnovatyi-Kogan, G. S., Lamzin, S. A. (1977): The matter outflow out of stars on early evolutionary stages, in *Early Stages of Stellar Evolution* (Kiev, Naukova Dumka), pp. 107–118

50 Bisnovatyi-Kogan, G. S., Lamzin, S. A. (1977): Models for outflowing envelopes of T Tauri stars. AsJ **54**, 1268–1280

51 Bisnovatyi-Kogan, G. S., Lamzin, S. A. (1980): The chromosphere, corona and X-ray emission of RU Lupi, a T Tauri star. PAJ **6**, 34–38

51a Bisnovatyi-Kogan, G. S., Lamzin, S. A. (1984): Stars with neutron cores. The possibility of the existence of objects with low neutrino luminosity. Sov. Astron. **28**, 187–193

51b Bisnovatyi-Kogan, G. S., Moiseenko, S. G. (1992): Violation of mirror symmetry of the magnetic field in a rotating star and possible astrophysical manifestations. Sov. Astron. **36**, 285–289

52 Bisnovatyi-Kogan, G. S., Nadyozhin, D. K. (1969): A Calculation method for evolution of stars with mass loss. NI, Issue 11, 27–39

53 Bisnovatyi-Kogan, G. S., Romanova, M. M. (1982): Diffusion and neutron heat conduction in the neutron star crust. JETF **83**, 449–459

54 Bisnovatyi-Kogan, G. S., Rudzskii, M. A., Seidov, Z. F. (1974): Non-equilibrium beta-processes and the role of excited nuclear states. JETF **67**, 1621–1630

55 Bisnovatyi-Kogan, G. S., Seidov, Z. F. (1969): On the relationship between white dwarfs and type I supernovae. A. **5**, 243–247

56 Bisnovatyi-Kogan, G. S., Seidov, Z. F. (1970): Non-equilibrium beta-processes as a source of thermal energy of white dwarfs. AsJ **47**, 139–144

57 Bisnovatyi-Kogan, G. S., Seidov, Z. F. (1984): On the pulsation of star with a phase transition. A. **21**, 563–571

58 Bisnovatyi-Kogan, G. S., Syunyaev, R. A. (1971): Galactic nuclei and qusars as sources of infrared emission. AsJ **48**, 881–893

59 Bisnovatyi-Kogan, G. S., Fridman, A. M. (1969): On the mechanism of X-ray emission of a neutron star. AsJ **46**, 721–724

60 Bisnovatyi-Kogan, G. S., Chechetkin, V. M. (1973): Non-equilibrium composition of neutron star envelopes and nuclear sources of energy. Pis'ma v JETF **17**, 622–626

61 Bisnovatyi-Kogan, G. S., Chechetkin, V. M. (1979): Non-equilibrium envelopes of neutron stars, their role in supporting X-ray emission and nucleosynthesis. UFN **127**, 263–296

62 Bisnovatyi-Kogan, G. S., Chechetkin, V. M. (1981): Gamma-ray bursts as an effect of neutron star activity. AsJ **58**, 561–568

63 Blatt, J. M., Weiskopf, V. F. (1952): *Theoretical Nuclear Physics* (New York, Wiley)

64 Blinnikov, S. I. (1975): Self-consistent field method in the theory of rotating stars. AsJ **52**, 243–254

64a Blinnikov, S. I. (1987): On the thermodynamics of relativistic Fermi-gas. PAJ **13**, 820–823

64b Blinnikov, S. I., Dunina-Barkovskaya, N. V., Nadyozhin, D. K. (1996): The equation of state for Fermi-gas: approximations for various degrees of relativism and degeneracy. ApJ Suppl. **106**, 171–203

65 Blinnikov, S. I., Lozinskaya, T. A., Chugai, N. N. (1987): Supernovae and their remnants. Itogi Nauki i Tekhniki, Astronomiya **32**, 142–200

66 Blinnikov, S. I., Rudzskii, M. A. (1984): Collapse of low-mass iron core of star. PAJ **10**, 363–369

67 Blinnikov, S. I., Rudzskyi, M. A. (1989): Annihilation neutrino losses in dense stellar matter. Sov. Astron. **33**, 377–380

67a Blinnikov, S. I., Rudzskii, M. A. (1988): New representations of the thermodynamical functions of Fermi gas. A **29**, 644–651

68 Blinnikov, S. I., Khokhlov, A. M. (1986): The formation of detonation in degenerate cores. PAJ **12**, 318–324

69 Bogolyubov, N. N., Shirkov, D. V. (1980): *Quantum Fields* (M, Nauka)

70 Bronshtein, I. N., Semendyaev, K. A. (1953): *Handbook on Mathematics* (M, Gostekhteorizd)

71 de Broglie, L. (1936): *The Magnetic Electron. Dirac's Theory* (Khar'kov, Gos. Nauch. -Tekh. Izd. Ukrainy)

72 Bugaev, E. V., Kotov, Yu. D., Rozental, I. L. (1970): *Cosmic muons and neutrinos* (M, Atomizdat)

73 Bugaev, E. V., Rudzskii, M. A., Bisnovatyi-Kogan, G. S., Seidov, Z. F. (1978): The interaction of intemediate-energy neutrinos with nuclei. Preprint of IKI AN SSSR Pr-403

74 Vaisenberg, A. O. (1964): *Mu-Meson* (M, Nauka)

75 Vandakurov, Yu. V. (1976): *Convection in the Sun and 11-year cycle* (L, Nauka)

76 Vartanyan, Yu. L., Ovsepyan, A. V., Adzhjyan, G. S. (1973): On the stability and radial pulsations of rotating neutron stars. AsJ **50**, 48–59

77 Vartanyan, Yu. L., Ovakimova, N. K. (1972): Neutron-rich nuclei in Fermi-gas. AsJ **49**, 87–95

78 Vartanyan, Yu. L., Ovakimova, N. K. (1976): Cold neutron evaporartion from nuclei in a superdense matter. Soobzsh. Burokan Obs. No. 49, 87–95

79 Vartanyan, Yu. L., Ovsepyan, A. V. (1971): Evolution and radial pulsations of isothermal white dwarfs with the inclusion of rotation, neutronisation and GR effects. A **7**, 107–119

80 Varshavskii, V. I. (1970): The Schwarzschild method with logarithmic variables. NI, Issue 16, 77–82

81 Varshavskii, V. I. (1972): Evolution of massive stars with the inclusion of semiconvection. NI, Issue 21, 25–45

82 Varshavskii, V. I., Tutukov, A. V. (1975): Evolution of massive stars. AsJ **52**, 227–233

83 Vlasiuk, M. P., Polezhaev, V. I. (1970): Numerical study of convective motions in a horizontal gas layer heated from below. Preprint of IPM AN SSSR 1–74

83a Vorontzov, S. V., Zharkov, V. N.: Helioseismology, Itogi Nauki i Tekhniki, Astronomiya **38**, 253–338

84 Wu, C. S., Moszkowski, S. A. (1966): *Beta-Decay* (Interscience, New York)

85 Gavrichenko, K. V., Nadyozhin, D. K. (1980): Simple formulae for the rates of electron and positron capture by nuclei. Preprint of ITEF No. 123

86 Gayar, M. K., Nikolich, M. (eds.) (1977): *Weak Interactions* (Inst. Nat. de Phys. Nucl. et de Phys. de Particuler, Paris)

87 Gershtein, S. S., Zel'dovich, Ya. B. (1955): On meson corrections to the beta-decay theory. JETF **29**, 698–699

88 Gershtein, S. S., Zel'dovich, Ya. B. (1966): The rest-mass of the muon neutrino and cosmology. Pis'ma v JETF **4**, 174–177

89 Ginzburg, V. L. (1975): *Theoretical Physics and Astrophysics* (M, Nauka)

90 Ginzburg, V. L., Rukhadze, A. A. (1970): *Waves in Magnetoactive Plasma* (M, Nauka)

91 Godnev, I. N. (1956): *Calculation of Thermodynamic Functions on the Base of Molecular Data* (M, Gostekhizdat)

92 Godunov, S. K., Riaben'ki, V. S. (1962): *Introduction in the Theory of Difference Schemes* (M, Fizmatgiz)

93 Gradshtein, I. S., Ryzhik, I. M. (1971): *Tables for Integrals, Sums, Series, Productions* (M, Nauka)

94 Greenspan, H. P. (1968): *The Theory of Rotating Fluids* (Cambridge, Cambridge University Press)

94a Grinfel'd, M. A. (1981): Gravitational stability and proper oscillations of liquid bodies with phase transition surfaces. DAN SSSR **258**, 1342–1346

95 Gurevich, L. E., Libedinskii, A. I. (1955): On the causes of stellar bursts, *Proc of 4th Conf. on Cosmogony Problems*, M., AN SSSR, 143–171

96 Dungey, J. W. (1958): *Cosmic Electrodynamics* (Cambridge, Cambridge University Press)

97 Deuterium, *Chemical Dictionary.* (M., SE., 1983), p. 149

98 De Jager, C. (1980): *The Brightest Stars* (Dordrecht, Reidel)

99 Dzbelepov, B. S., Peker, L. K. (1966): *Schemes of Decay of Radioactive Nuclei* (M, Nauka)

100 Dolgov, A. D., Zel'dovich, Ya. B. (1980): Cosmology and elementary particles. UFN **130**, 559–614

101 Dorofeev, O. F., Rodionov, V. N., Ternov, I. M. (1985): Anisotropic emission of neutrinos arising in beta-processes under the effect of a strong magnetic field. PAJ **11**, 302–309

102 Zel'dovich, Ya. B. (1957): On nuclear reactions in a superdense cold gas. JETF **33**, 991–993

103 Zel'dovich, Ya. B. (1961): Equation of state at a superhigh density and relativistic restrictions. JETF **41**, 1609–1615

104 Zel'dovich, Ya. B. (1962): Static solutions with energy excess in general relativity. JETF **42**, 1667–1671

105 Zel'dovich, Ya. B. (1963): Hydrodynamical stability of star. Voprosy Kosmogonii **9**, 157–170

106 Zel'dovich, Ya. B. (1984): *Chemical Physics and Hydrodynamics, Selected Works* (M, Nauka) 279–280

107 Zel'dovich, Ya. B., Blinnikov, S. I., Shakura, N. I. (1981): *Physical Grounds of Stellar Structure and Evolution* (M, MSU)

108 Zel'dovich, Ya. B., Gusseinov, O. Kh. (1965): The neutronization of matter during collapse and neutrino spectrum. DAN SSSR **162**, 791–793

108a Zel'dovich, Ya. B., Ivanova, L. N., Nadyozhin, D. K., (1972): Non-stationary hydrodynamical accretion onto neutron star. AsJ **49**, 253–264

109 Zel'dovich, Ya. B., Novikov, I. D. (1965): Relativistic astrophysics. II. UFN **86**, 447–536

110 Zel'dovich, Ya. B., Novikov, I. D. (1971): *Gravitation Theory and Stellar Evolution* (M, Nauka)

111 Ivanova, L. N., Imshennik, V. S., Nadyozhin, D. K. (1969): A study of the dynamics of supernova explosion. NI, Issue 13, 3–78

111a Imshennik, V. S., Kalyanova, N. L., Koldoba, A. V., Chechetkin, V. M. (1999): Is detonation burning possible in a degenerate carbon-oxigen core of pre-supernova ? PAJ **25**, 250–259

112 Imshennik, V. S., Kotok, E. V., Nadyozhin, D. K. (1965): Calculation of homogeneous models by the progonka method. NI, Issue 1, 48–54

113 Imshennik, V. S., Morozov, Yu. I. (1969): Relativistically covariant equations for interaction of radiation with matter. AsJ **46**, 800–809

114 Imshennik, V. S., Nadyozhin, D. K. (1965): Thermodynamic properties of matter at high temperatures and densities. AsJ **42**, 1154–1167

115 Imshennik, V. S., Nadyozhin, D. K. (1972): Heat conductivity by neutrinos in collapsing stars. JETF **63**, 1548–1561

116 Imshennik, V. S., Nadyozhin, D. K. (1982): Final stages of stellar evolution and supernova bursts. Itogi Nauki i Tekhniki, Astronomiya **21**, 63–129

116a Imshennik, V. S., Nadyozhin, D. K. (1988): Supernova 1987a in great magellanic cloud: observations and theory. UFN **156**, 561–651

117 Imshennik, V. S., Nadyozhin, D. K., Pinaev, V. S. (1966): The kinetic equilibrium of beta-processes in stellar interiors. AsJ **43**, 1215–1225

118 Imshennik, V. S., Nadyozhin, D. K., Pinaev, V. S. (1967): Neutrino emission of energy in beta-interactions of electrons and positrons with nuclei. AsJ 44, 768–777

119 Imshennik, V. S., Chechetkin, V. M. (1970): Thermodynamics under conditions of hot neutronization of matter and hydrodynamical stability of stars on late evolutionary stages. AsJ 47, 929–941

120 Ioffe, B. L., Lipatov, L. N., Khoze, V. A. (1983): Deeply Inelastic Processes. Phenomenology. Quark-Parton Model (M, Energoatomizdat)

121 Kaplan, S. A. (1949): 'Superdense' stars. Uchyonye Zapiski L'vovskogo gosuniversiteta, ser. fiz-mat. 15, 109–116

122 Kaplan, S. A. (1950): Cooling of white dwarfs. AsJ 27, 31–33

123 Kardashev, N. S. (1964): Magnetic collapse and the nature of powerful sources of cosmic radio emission. AsJ 41, 807–813

124 Cowling, T. G. (1957): Magnetohydrodynamics. (New York, Interscience)

125 Kirzhnitz, D. A. (1960): On the internal structure of superdense stars. JETF 38, 503–508

126 Kirzhnitz, D. A., Lozovik, Yu. E., Shpatakovskaya, G. V. (1975): A statistic model of matter. UFN 117, 3–47

127 Kozik, V. S., Liubimov, V. A., Novikov, Ye. G., Nozik, V. Z., Tretyakov, E. F. (1980): On the estimate of n mass by spectrum of beta-decay of Tritium in Valin. Yadernaya Fizika 32, 301–303

128 Cox, A. N. (1965): Absorption coefficients and the opacity of stellar matter, in Stellar Structure, ed. by L. H. Aller, D. B. McLaughlin. (Chicago, University of Chicago Press)

129 Cox, A., Stuart, G. (1969): Radiative absorption and heat conductivity: opacities for 25 stellar mixtures. NI, Issue 15, 3–103

130 Cox, J. P. (1980): Stellar Pulsation Theory. (Princeton, Princeton University Press)

131 Kolesnik, I. G. (1973): Gravitational contraction of protostars. I. Bulk losses of energy. Astrometriya i Astrofizika, No. 18, 45–58

132 Kolesnik, I. G. (1979): The hydrodynamics of protostar collapse. Preprint of ITF No. 79–44P, Kiev

133 Kolesnik, I. G. (1980): The numerical simulation of protostar collapse. Physical and mathematical formulation of the problem. I. Basic equations. Astrometriya i Astrofizika No. 40, 3–18; II. Computational algorithms, Ibid. No. 41, 40–58

133a Koldoba, A. V., Tarasova, E. V., Chechetkin, V. M. (1994): On the instability of the detonation wave in thermonuclear model of supernova. PAJ 20, 445–449

134 Kopysov, Yu. S. (1975): Neutrinos and the internal structure of the Sun. Preprint of IYaI No. P-0019

135 Kravtzov, V. A. (1974): Atomic Masses and Binding Energies of Nuclei (M, Atomizdat)

136 Krasnov, N. F. (1971): Aerodynamics (M, Vysshaya Shkola)

137 Krylov, V. I. (1967): The Approximate Evaluation of Integrals (M, Nauka)

138 Krylov, N. S. (1950): Investigations on the Statistical Physics Foundation (M, L, AN SSSR)

139 Lavrentyev, M. A., Shabat, B. V. (1958): Methods of the Theory of Functions of a Complex Variable (M, Nauka)

140 Lamb, G. (1932): Hydrodynamics (Cambridge, Cambridge University Press)

141 Landau, L. D. (1969): On the Theory of Stars. Works I (M, Nauka) 86–89

142 Landau, L. D., Lifshitz, E. M. (1953): Mechanics of Continuum Media (M, Gostekhteorizdat)

143 Landau, L. D., Lifshitz, E. M. (1962): The Theory of Fields (M, Nauka)

144 Landau, L. D., Lifshitz, E. M. (1963): *Quantum Mechanics* (M, Nauka)
145 Landau, L. D., Lifshitz, E. M. (1976): *Statistical Physics* **I** (M, Nauka)
146 Landau, L. D., Lifshitz, E. M. (1982): *Electrodynamics of Continua* (M, Nauka)
147 Ledoux, P. (1957): To the conditions of equilibrium in the centre of stars and their evolution. *Nuclear Processes in Stars*, Proc. of Lirge Coll. (M, IL), pp. 152–162
148 Luyten, W. G. (1975): The white dwarfs: discovery and observation. *White Dwarfs*, ed. by Imshennik, (M, Mir), pp. 13–22
148a Lifshitz, E. M., Pitaevskii, L. P. (1979): *Physical Kinetics* (M, Nauka)
149 Lozinskaya, T. A. (1986): *Supernovae and Stellar Wind. The Interaction with Gas of Galaxy* (M, Nauka)
150 Loytzianskii, L. G. (1987): *Fluid Mechanics* (M, Nauka)
151 Lyutostanskii, Yu. S., et al. (1984): A kinetic model for beta-process. Preprint of IPM, No. 95
152 Mazetz, E. P., et al. (1979): A bursting X-ray pulsar in Dorado. PAJ **5**, 307–312
153 Manchester, R. N., Taylor, G. H. (1980): *Pulsars* (M, Mir)
154 Massevich, A. G., Ruben, G. V., Lomnev, S. P., Popova, E. I. (1965): Calculation of homogeneous 4 $M\odot$, 8 $M\odot$, 16 $M\odot$ models for various chemical compositions and various absorption laws. NI, Issue 1, 2–47
155 Massevich, A. G., Tutukov, A. V., (1974): The evolution of massive stars and semiconvection problem, NI, Issue 29, 3–26
156 Massevich, A. G., Tutukov, A. V. (1988): *Stellar evolution: Theory and Observations* (M, Nauka)
157 Mestel, L. (1965): Meridional circulation in stars, in *Stellar Structure*, ed. by L. H. Aller, D. B. McLaughlin (Chicago, University of Chicago Press)
158 Migdal, A. B. (1978): *Fermions and bosons in strong fields* (M, Nauka)
159 Mikhalas, D. (1978): *Stellar Atmospheres*, Parts I and II (San Francisco, Freeman)
160 Mikheevs, S. P., Smirnov, A. Yu. (1985): The resonant enhancement of oscillations in matter and spectroskopy of solar neutrinos. Yadernaya Fizika **42**, 1441–1448
161 Mikheevs, S. P., Smirnov, A. Yu. (1986): Neutrino oscillations in a medium with varying density and bursts caused by gravitational collapses of stars. JETF **91**, 7–13
162 Morozov, Yu. I. (1970): Radiative transfer equations in non-inertial coordinate systems. PMTF No. 1, 3–11
163 Morse, Ph. M., Feschbach, H., (1953): *Methods of theoretical physics* (New York, McGraw-Hill)
163a Muslimov, A. (1995): The system PSR 1718-19 and a hypothesis about latent magnetism in neutron stars. AA. **295** L27-L29
164 Mukhin, K. N. (1983): *Experimental nuclear physics* **1** *Physics of atomic nucleus* (M, Energoatomizdat)
165 Nadyozhin, D. K. (1966): Evolution of a star with M=30 $M\odot$ in hydrogen-burning phase., NI, Issue 4, 37–64
166 Nadyozhin, D. K. (1974): Asymptotic formulae for the equation of state of electron–positron gas. NI, Issue 32, 3–72
167 Nadyozhin, D. K. (1974): Tables for the equation of state of electron–positron gas. NI, Issue 33, 117–142
168 Nadyozhin, D. K., Chechetkin, V. M. (1969): Neutrino emission by URCA-process at high temperatures. AsJ **46**, 270–279
169 Novikov, I. D., Frolov, V. P. (1986): *Physics of Black Holes* (M, Nauka)

170 Audoze, J., Reeves, H. (1982): *The Origin of Light Elements*, in *Essays in Nuclear Astrophysics* (Cambridge, Cambridge University Press), pp. 340–358

171 Okun', L. B. (1963): *The Weak Interaction of Elementary Particles* (M, Nauka)

172 Okun', L. B. (1981): *Leptons and Quarks*, (M, Nauka)

173 Okun', L. B., (1984): *The Physics of Elementary Particles* (M, Nauka)

174 Parker, E. N. (1963): *Interplanetary Dynamical Processes* (New York, Interscience)

175 Petrov, P. P. (1977): T Tauri stars: modern observational data, in *Early Stages of Stellar Evolution* (Kiev, Naukova Dumka) pp. 66–100

176 Pikel'ner, S. B. (1961): *Fundamentals of Cosmic Electrodynamics* (M, Nauka)

177 Porfiryev, V. V., Redkoborodyi, Yu. N. (1969): The electron screening effect in nuclear reactions at high densities. A **5**, 393–413

178 Pskovskii, Yu. P. (1985):*Novae and Supernovae Stars* (M, Nauka)

179 Ptitzyn, D. A., Chechetkin, V. M. (1982): To the problem of formation of elements beyond the iron peak in supernovae. PAJ **8**, 600–606

180 Radtzig, A. A., Smirnov, B. M. (1980): *Handbook on Atomic and Molecular Physics* (M, Atomizdat)

181 Redkoborodyi, Yu. N. (1976): The quantum theory of screening effects in thermonuclear reactions. I. Relativistic electron plasma. A **12**, 495–510

182 Reeves, H., (1965): Sources of stellar energy in stellar structure, in *Stellar Structure*, ed. by L. H. Aller, D. B. McLaughlin (Chicago, University of Chicago Press)

183 Rosseland, S., (1949): *Pulsation Theory of Variable Stars* (Oxford, Clarendon)

184 Rublyov, S. V., Cherepazshuk, A. M. (1974): [Wolf–Rayet Stars], in *Nonstationary Phenomena and Stellar Evolution* (M, Nauka) pp. 47–124

185 Ruderman, M. (1965): *Astrophysical Neutrino*, Reports on Progress in Physics **61**, 411–462

186 Rudzskii, M. A., Seidov, Z. F. (1974): Thermal effects in beta-processes. Izv. AN Azerb. SSR, ser. fiz. mat. i tekh. Nauk **4**, 98–106

187 Rudzskii, M. A., Seidov, Z. F. (1974): Low-mass star collapse and heating by beta-processes. AsJ. **51**, 936–939

188 Saakyan, G. S. (1972): *Equilibrium Configurations of Degenerate Gas Masses*, (M, Nauka)

189 Samarskii, A. A., Popov, Yu. P. (1975): *Difference Schemes in the Gas Dynamics* (M, Nauka)

190 Sedrakyan, D. N., Shakhbazyan, K. M., Movsisyan, A. G. (1984): Magnetic momenta of neutron stars of real baryon gas. A **21**, 547–561

191 Seidov, Z. F. (1967): Equilibrium of a star with a phase transition. A **3**, 189–201

191a Seidov, Z. F. (1970): *Stars With a Phase Transition*. (PhD Thesis, Erevan)

192 Seidov, Z. F. (1971): Polytropes with a phase transition. II., The polytrope $n = 1$. Soobzsh. Shemakh. Obs., Issue 5, 58–69

193 Seidov, Z. F. (1984): The equilibrium, stability and pulsation of stars with a phase transition. Preprint of IKI, No. Pr-889

194 Sears, R. L., Brownlee, R. R. (1965): The Evolution of Stars and Determination of Their Age, in *Stellar Structure*, ed. by L. H. Aller, D. B. McLaughlin (Chicago, University of Chicago Press)

195 Smirnov, V. I. (1962): *Course of High Mathematics, v. 2* (M, Nauka)

196 Smirnov, V. I. (1962): *Course of High Mathematics, v. 4* (M, Gostekhizdat)

197 Sobolev, V. V. (1967): *Course of Theoretical Astrophysics* (M, Nauka)

198 Swinney, H. L., Gollub, J. P. (eds.) (1981):*Hydrodynamic Instabilities and the Transition to Turbulence* (Berlin Heidelberg, Springer)

199 Tassoul, J. L. (1978): *Theory of Rotating Stars* (Princeton, Princeton University Press)

200 Tutukov, A. V., Fadeev, Yu. A. (1981): The formation of an extended envelope around pulsating star. NI, Issue 49, 48–63

201 Harrison, B. K., Thorne, K. S., Wakano, M., Wheeler, J. A., (1965): *Gravitational Theory and Gravitational Collapse* (Chicago, University of Chicago Press)

201a Ulrich, R. K. (1982): *S-Process*. In: Essays in Nuclear Astrophysics. (Cambridge, Cambridge University Press)

202 Urpin, V. A., Yakovlev, D. G. (1980): The heat conductivity caused by inter-electron collisions in degenerate relativistic electron gas. AsJ **57**, 213–215

203 Urpin, V. A., Yakovlev, D. G. (1980): Thermogalvanomagnetic phenomena in white dwarfs and neutron stars. AsJ **57**, 738–748

203a Ustyugov, S. D., Chechetkin, V. M., (1999): Supernova explosion in presence of large scale convective instability in rotating protoneutron star. AsJ **76**, 816–824

204 Uus, U. (1969): Calculations of the structure of thin nuclear-burning shells in stellar models. NI, Issue 13, 126–144

205 Uus, U. (1970): The evolution of 1. 5, 2. 3 and 5 $M_\odot$ stars on the stage of carbon core growth. NI, Issue 17, 3–24

206 Uus, U. (1971): The penetration of the convective envelope of a star in the nuclear-burning zone. NI, Issue 20, 60–63

207 Uus, U. (1979): On the possibility of a consistent treatment of stellar turbulent convection. Publ. of Tartus. Astrofis. Obs. im. V. Ya. Struve **47**, 103–119

208 Fadeev, Yu. A. (1981): On the possibility for dust particles to form in the FG sagittae atmosphere. NI, Issue 50, 3–9

209 Fyodorova, A. V. (1978): The effect of magnetic field on the minimum mass of a main-sequence star. NI, Issue 42, 95–109

210 Fyodorova, A. V. Blinnikov, S. I. (1978): The effect of accretion and rotation on the minimum mass of a main-sequence star. NI, Issue 42, 75–94

211 Feygenbaum, M. (1980): Universal behaviour in nonlinear systems. Los Alamos Sci. **1**, No. 1 pp. 4–27

212 Fermi, E. (1950): *Nuclear Physics.* (Chicago, University of Chicago Press)

213 Fermi, E. (1934): Versuch einer Theorie der β-Strahlen. Zt. f. Phys. **88**, 161–171

214 Fok, V. A. (1976): *Fundamentals of Quantum Mechanics* (M, Nauka)

215 Frank-Kamenetzkii, D. A. (1959): *Physical Processes in Stellar Interiors* (M, Fizmatgiz)

216 Frank-Kamenetzkii, D. A. (1962): Neutronization kinetics at superhigh densities. JETF **42**, 875–879

217 Tsuruta, S., Cameron, A. G. W. (1970): URCA shells in dense stellar interiors. ApSS **7**, 374–395

218 Chandrasekhar, S. (1939): *An Introduction to the Study of Stellar Structure* (Chicago, University Chicago Press)

219 Chandrasekhar, S. 1950: *Radiative Transfer* (Oxford, Clarendon)

220 Chandrasekhar, S. (1969): *Ellipsoidal Equilibrium Figures* (New Haven, Yale University Press)

221 Chandrasekhar, S. (1983): *The Mathematical Theory of Black Holes* (Oxford, Clarendon)

222 Chapman, S., Cowling, T. G. (1952): *Mathematical Theory of Inhomogeneous Gases* (Cambridge, Cambridge University Press)

222a Cherepashchuk, A. M. (ed.) (1988): *Catalog of close binary systems at late evolutionary stages.* (M., MGU Press)

222b Cherepashchuk, A. M. (1996): Masses of black holes in binary systems. UFN **166**, 809–832

222c Cherepashchuk, A. M. (2000): X-ray nova binary systems. Space Sci. Rev. **93**, No.3-4, Nov.

223 Chechev, V. P., Kramarovski, Ya. M. (1981): Theory of nuclear synthesis in stars: s-process. UFN **134**, 431–467

224 Chechetkin, V. M.(1969): Equilibrium state of matter at high temperatures and densities. AsJ **46**, 202–206

224a Chechetkin, V. M., Ustyugov, S. D., Gorbunov, A. A., Polezhaev, V. I. (1997): On the neutrino mechanism of supernovae explosion. PAJ **23**, 34–40

225 Chugai, N. N.(1984): Neutrino helicity and space velocities of pulsars. PAJ **10**, 210–213

226 Shakura, N. I.(1972): Accretion disk model around relativistic star in a close binary system. AsJ **49**, 921–929

227 Shapiro, S. L., Teukolsky, S. A. (1983): *Black holes, white dwarfs, and neutron stars: The physics of compact objects.* (NY, Wiley-Interscience)

228 Shwartzman, V. F. (1971): Halos around black holes. AsJ **48**, 479–488

229 Schwarzschild, M. (1958): *Structure and Evolution of the Stars* (Princeton, Princeton University Press)

230 Schweber, S. S. (1961): *An Introduction to Relativistic Quantum Field* (Evanston, Row, Peterson and Co.)

231 Shekhter, V. M. (1976): Weak interaction with neutral currents. UFN **119**, 593–632

232 Shirokov, Yu. M., Yudin, N. P. (1980): *Nuclear Physics* (M, Nauka)

233 Shklovski, I. S. (1956): On the nature of planetary nebulae and their nuclei. AsJ **33**, 315–329

234 Shklovski, I. S. (1976): *Supernovae Stars* (M, Nauka)

235 Schlichting, H. (1962): *The Appearance of Turbulence*, (M, IL)

236 Shnol', E. E. (1974): On the stability of stars with a density jump caused by phase transition. Preprint of IPM, No. 93

237 Shustov, B. M. (1978): An algorithm for calculation of gasodynamic evolution of the envelope of a massive protostar. NI, Issue 42, 60–74

238 Shustov, B. M. (1979): The evolution of protostar envelopes I. The stage of cocoons. NI, Issue 46, 63–92

239 Shustov, B. M. (1979): The evolution of protostar envelopes II. spectra of outgoing radiation of protostars and compact zones N II. NI, Issue 46, 93–110

240 Ebeling, W., Kraeft, W. D., Kremp, D. (1979): *Theory of Bound States and Ionization Equilibrium in Plasmas and Solids* (Academic-Verlag, Berlin)

241 Ergma, E. V. (1969): Models for 4 $M\odot$ and 20 $M\odot$ envelopes. NI, Issue 12

242 Ergma, E. V. (1972): A non-local model of convection for stellar envelopes. NI, Issue 23, 33–46

243 Ergma, E. V. (1982): Thermonuclear bursts in neutron star envelopes. Itogi Nauki i Tekhniki, Astronomiya **21**, 130–150

244 Ergma, E. V. (1986): Thermonuclear processes in accreting white dwarfs (novae, simbiotic novae and type I supernovae). Itogi Nauki i Tekhniki, Astronomiya **31**, 228–257

245 Yakovlev, D. G. (1980): *Phenomena of Heat and Charge Transfer in Neutron Stars and White Dwarfs* (PhD Thesis, L, LFTI)

246 Yakovlev, D. G., Urpin, V. A. (1980): On the heat and charge conductivity in neutron stars and white dwarfs. AsJ **57**, 526–536

247 Yakovlev, D. G., Shalybkov, D. A. (1987): The effect of electron screening on thermonuclear reaction rates. PAJ **13**, 730–736

248 Yakovlev, D. G., Shalybkov, D. A. (1988): The degenerate cores of white dwarfs and neutron star envelopes: thermodynamics and plasma screening in thermonuclear reactions. Itogi Nauki i Tekhniki, Astronomiya **38**, 191–252

248a Aksenov, A. G., Blinnikov, S. I. (1994): A Newton iteration method for obtaining equilibria of rapidly rotating stars. AA **290**, 674–681

248* Abdurashitov, J.N. et al. (1999): Measurement of the solar neutrino capture rate with Gallium metal. Astro-ph/9907113

248** Abdurashitov, J.N. et al. (1999): Measurement of the solar neutrino rate by SAGE and implications for neutrino oscillations in vacuum. Astro-ph/9907131

249 Alcock, Ch., Illarionov, A. F. (1980): The surface chemistry of stars. II. Fractionated accretion of interstellar matter. ApJ **235**, 541–553

250 Alexander, D. (1975): Low temperature Rosseland opacity tables. ApJ Suppl **29**, 363–374

250* Alexander, D. R., Ferguson, J. W. (1994): Low-temperature Rosseland opacities. ApJ. **437**, 879–891

250a Alpar, M. A., Cheng, A. F., Ruderman, M. A., Shaham, J. (1982): A new class of radio pulsars. Nature **300**, 728–730

250b Andersen, J. (1991): Accurate masses and radii of normal stars. AA Review. **3**, 91–126

250c Anita, H. M., Basu, S. (1999): High-frequency and high-wave-number Solar oscillations. ApJ **519**, 400–406

251 Appenzeller, I. (1970): The evolution of a vibrationally unstable main sequence star of 130 $M_\odot$. AA **5**, 355–371

252 Appenzeller, I. (1970): Mass loss rates for vibrationally unstable very massive main–sequence stars. AA **9**, 216–220

253 Appenzeller, I., Tscharnuter, W. (1974): The evolution of a massive protostar. AA **30**, 423–430

253a Ardelyan, N. V., Bisnovatyi-Kogan, G. S., Kosmachevskii, K. V., Moiseenko, S. G. (1996): An implicit Lagrangian code for the treatment of nonstationary problems in rotating astrophysical bodies. AA Suppl. **115**, 573-594

253b Ardeljan, N. V., Bisnovatyi-Kogan, G. S., Moiseenko S. G. (1996): 2D calculations of the collapse of rotating magnetized gas cloud. ApSS **239**, 1–13

253c Ardeljan, N. V., Bisnovatyi-Kogan, G. S., Moiseenko, S. G. (2000): Nonstationary magnetorotational processes in a rotating magnetized cloud. AA **355**, 1181–1190

254 Arnett, D., (1967): Mass dependence in gravitational collapse of stellar cores. Canad. J. Phys **45**, 1621–1641

255 Arnett, D., (1969): A possible model of supernovae: detonation of ^{12}C. ApSS **5**, 180–212

255a Arnett, D., (1977): Neutrino trapping during gravitational collapse of stars. ApJ **218**, 815–833

256 Arponen, J., (1972): Internal structure of neutron stars NP **A191**, 257–284

256a Artemova, J. V., Bisnovatyi-Kogan, G. S., Björnsson, G., Novikov, I. D. (1996): Structure of accretion discs with optically thin-thick transitions. ApJ **456**, 119–123

256b Audard, N., Provost, J. (1994): Sesmological properties of intermediate-mass stars. AA **282**, 73–85

257 Auman, J., Bodenheimer, P. (1967): The influence of water-vapor opacity and the efficiency of convection on models of late-tape stars. ApJ **149**, 641–648

258 Baade, W., Zwicky, F. (1934): Supernovae and cosmic rays. PR **45**, 138–139

259 Bahcall, J. (1978): Solar neutrino experiment. Rev. Mod. Phys. **50**, 881–903

259a Bahcall, J. N. (1989): *Neutrino Astrophysics* (Cambridge, Cambridge University Press)

259b Bahcall, J. N., Barnes, C. A., Christensen-Dalsgaard, J. et. al. (1994): Has a standard model solution to the solar neutrino problem been found? Preprint astro-ph 9404002

260 Bahcall, J., Cleveland, B. T., Davis, R. Jr., Rowley, J. K. (1985): Chlorine and gallium solar neutrino experiment. ApJ Lett **292**, L79–L82

260a Bahcall, J. N., Lande, K., Lanou, R. E., Learned, J. G., Robertson, R. G. H, Wolfenstein, L. (1995): Progress and prospects in neutrino astrophysics. Nature **375**, 29–34

260b Bahcall, J. N., Pinsonneault, M. (1992): Standart solar models with and without helium diffusion and the solar neutrino problem. Rev. Mod. Phys. **64**, 885–926

261 Baker, N., Gough, D. (1979): Pulsation of model RR lyrae stars. ApJ **234**, 232–244

261a Balmforth, N. J., Howard, L. N., Spiegel, E. A. (1993): Equilibria of rapidly rotating polytropes. MN **260**, 253–272

262 Bardeen, J. (1965): Binding energy and stability of spherically symmetric masses in general relativity. Preprint OAP No. 36

263 Burrows, A., Lattimer, J. (1986): The birth of neutron stars. ApJ **307** 178–196

263a Basu, S., Christensen-Dalsgaard, J., Schon, J., Thompson, M.J., Tomczyk, S. (1996): The Sun's hydrostatic structure from LOWL data. ApJ **460**, 1064–1070

264 Baud, B. et al (1984): High-sensitivity IRAS observations of the Chamaeleon I dark cloud. ApJ Lett **278**, L53–L55

265 Baym, G., Bethe, H., Pethick, Ch. (1971): Neutron star matter. NP **A175**, 255–271

266 Baym, G., Pethick, Ch. (1975): Neutron stars. Ann. Rev. Nucl. Sci. **25**, 27–77

267 Baym, G., Pethick, Ch., Sutherland, P. (1971): The groun state of matter at high densities: equation of state and slellar model. ApJ **170**, 306–315

268 Beaudet, G., Petrosian, V., Salpeter, E. E. (1967): Energy losses due to neutrino processes. ApJ **150**, 979–999

269 Beaudet, G., Salpeter, E. E. Silvestro, M. L. (1972): Rates for URCA neutrino processes. ApJ **174**, 79–90

270 Becker, S., Iben, I. (1979): The asymptotic giant branch evolution of intermediate-mass stars as a function of mass and composition. I. Through the second dredge-up phase. ApJ **232**, 831–853

271 Becker, S., Iben, I., Tuggle, R. (1977): On the frequency–period distribution of Cepheid variables in galaxies in the local group ApJ **218**, 633–653

272 Beichman, C. A. et al. (1984): The formation of Solar-type stars: IRAS observations of the dark cloud Barnard 5. ApJ Lett **278**, L45–L48

273 Beichman, C. et al. (1986): Candidate Solar-type protostars in nearby molecular cloud cores ApJ **307**, 337–349

274 Bethe, H. (1986): Possible explanation of Solar neutrino puzzle. PR Lett **56**, 1305–1308

275 Bethe, H., Johnson, M. (1974): Dense barion matter calculations with realistic potentials. NP **A230**, 1–58

276 Bethe, H., Wilson, J. (1985): Revival of a stalled supernova shock by neu-
     trino heating. ApJ **295**, 14–23

276a Bethe, H. (1991): The supernova shock. ApJ **449**, 714–726

276a* Bhattacharya, D. van den Heuvel, E. P. J. (1991): Formation and evolution
     of binary and millisecond radio pulsars. Phys. Rep. **203**, 1–124

276b Bhattacharya, D., Wijers, R., Hartman, J and Verbunt, F. (1992): On the
     decay of the magnetic fields of single radio pulsars. AA **254**, 198–212

277 Bisnovatyi-Kogan, G. S. (1973): Stellar envelopes with supercritical lumi-
     nosity. ApSS **22**, 307–320

278 Bisnovatyi-Kogan, G. S. (1975): Gamma-ray bursts from neutron stars.
     Report on COSPAR, Varna. Preprint IKI No. D203

279 Bisnovatyi-Kogan, G. S. (1979): Magnetohydrodynamical processes near
     compact objects Rivista Nuovo Cimento **2**, No. 1

280 Bisnovatyi-Kogan, G. S. (1980): Magnetohydrodynamical model of super-
     novae explosion. Ann. New–York Acad. Sci. **336**, 389–394

281 Bisnovatyi-Kogan, G. S. (1981): Pre-main-sequence stellar evolution, in
     Proc. Symp. IAU Mo. 93: *Fundamental problems of stellar evolution*, ed.
     by D. Sugimoto, D. Lamb, D. Schramm (Dordrecht, Reidel) pp. 87–97.

282 Bisnovatyi-Kogan, G. S. (1982): Physical processes in stars on late stages
     of stellar evolution. Astron. Nachrichten **203**, 131–137

282a Bisnovatyi-Kogan, G. S. (1990): Angular velocity distribution in convective
     regions and the origin of solar differential rotation. Solar Phys. **128**, 299–
     304

282b Bisnovatyi-Kogan, G. S. (1992): Recycled pulsars and LMXB. Proc. Coll.
     IAU No. 128 (Zelena Gora, Pedagog. University Press), 209–212.

282c+ Bisnovatyi-Kogan, G. S. (1993): A self-consistent solution for an accretion
     disk structure around a rapidly rotating non-magnetized star. AA **274**,
     796–806

282c Bisnovatyi-Kogan, G. S. (1993): Asymmetric neurino emission and forma-
     tion of rapidly moving pulsars. Astron. Ap. Transact. **3**, 287–294

282d Bisnovatyi-Kogan, G. S. (1994): Analytical self-consistent solution for the
     structure of polytropic accretion discs with boundary layers. MN **269**, 557–
     562

282e Bisnovatyi-Kogan, G. S. (2000): Oscillations and convective motion in stars
     with URCA shells. Preprint astro-ph/0004281

282f Bisnovatyi-Kogan, G. S. (1995): Close galactic origin of gamma ray bursts.
     ApJ Suppl. **97**, 185-187

283 Bisnovatyi-Kogan, G. S., Blinnikov, S. I. (1972): The equilibrium, stability
     and evolution of a rotating magnetized gaseous disk. ApSS **19**, 119–144

284 Bisnovatyi-Kogan, G. S., Blinnikov, S. I. (1974): Static criteria for stability
     of arbitrary rotating stars. AA **31**, 391–404

285 Bisnovatyi-Kogan, G. S., Blinnikov, S. I. (1977): Disk accretion onto a black
     hole at subcritical luminosity. AA **59**, 111–125

286 Bisnovatyi-Kogan, G. S., Blinnikov, S. I. (1980): Spherical accretion onto
     compact X-ray source with preheating: no thermal limits for the luminosity.
     MN **191**, 711–719

287 Bisnovatyi-Kogan, G. S., Chechetkin, V. M. (1974): Nucleosynthesis in su-
     pernova outbursts and the chemical composition of the envelopes of neutron
     stars. ApSS **26**, 25–46

288 Bisnovatyi-Kogan, G. S., Chechetkin, V. M. (1983): Nuclear fission in the
     neutron stars and gamma-ray bursts. ApSS **89**, 447–451

288a Bisnovatyi-Kogan, G. S., Dorodninsyn, A. V. (1999): On modeling radia-
     tion-driven envelopes at arbitrary optical depths. AA **344**, 647–654

288b Bisnovatyi-Kogan, G. S., Dorodninsyn, A. V. (1998): Application of Galerkin method to the problem of stellar stability, gravitational collapse and black hole formation. Grav. and Cosmology **4**, 174–182

289 Bisnovatyi-Kogan, G. S., Imshennik, V. S., Nadyozhin, D. K., Chechetkin, V. M. (1975): Pulsed gamma-ray emission from neutron and collapsing stars and supernovae. ApSS **35**, 23–41

289a Bisnovatyi-Kogan, G. S., Lovelace, R. V. E. (1997): Influence of Ohmic heating on advection-dominated accretion flows. ApJ Lett. **486**, L43–L46

289b Bisnovatyi-Kogan, G. S., Lovelace, R. V. E. (2000): Magnetic field limitations on advection dominated flows. ApJ **529**, 978–984

290 Bisnovatyi-Kogan, G. S., Nadyozhin, D. K., (1972): The evolution of massive stars with mass loss. ApSS **15**, 353–374

291 Bisnovatyi-Kogan, G. S., Popov, Yu. P., Samochin, A. A. (1976): The magnetohydrodynamical rotational model of supernovae explosion. ApSS **41**, 321–356

291a Bisnovatyi-Kogan, G. S., Postnov, K. A. (1993): Geminga braking index and pulsar motion. Nature **366**, 663–665

292 Bisnovatyi-Kogan, G. S., Ruzmaikin, A. A. (1973): The stability of rotating supermassive stars. AA **27**, 209–221

293 Bisnovatyi-Kogan, G. S., Ruzmaikin, A. A. (1974): The accretion of matter by a collapsing star in the presence of a magnetic field. ApSS **28**, 45–59

294 Bisnovatyi-Kogan, G. S., Ruzmaikin, A. A. (1976): The accretion of matter by a collapsing star in the presence of a magnetic field. II. Selfconsistent stationary picture. ApSS **42**, 401–424

295 Bisnovatyi-Kogan, G. S., Vainshtein, S. I. (1971): Generation of magnetic fields in rotating stars and quasars. Astrophys. Lett **8**, 151–152

296 Black, D. C., Bodenheimer, P. (1975): Evolution of rotating interstellar clouds. I. Numerical techniques. ApJ **199**, 619–632

297 Black, D. C., Bodenheimer, P. (1976): Evolution of rotating interstellar clouds. II. The collapse of protostars of 1, 2, and 5 $M_\odot$. ApJ **206**, 138–149

298 Blake, J., Schramm, D. (1976): A possible alternative to the r-process. ApJ **209**, 846–849

299 Blake, J., Woosley, S., Weaver, T., Shramm, D. (1981): Nucleosynthesis of neutron-rich heavy niclei during explosive helium burning in massive stars. ApJ **248**, 315–320

300 Blandford, R. D., Znaeck, R. L. (1977): Electromagnetic extraction of energy from Kerr black holes. MN **179**, 433–456

300a Blinnikov, S. I., Sasorov, P. V., Woosley S. E. (1995): Self-acceleration of nuclear flames in supernovae. Space Sci. Rev. **74**, 299–311

301 Bludman, S., Van Riper, K. (1978): Diffusion approximation to neutrino transport in dense matter. ApJ **224**, 631–642

302 Bodansky, D., Clayton, D., Fowler, W. (1968): Nuclear quasi-equilibrium during silicon burning. ApJ Suppl. **16**, 299–371

303 Bodenheimer, P., Ostriken, J. (1970): Rapidly rotating stars. IV. Pre-main sequence evolution of massive stars. ApJ **161**, 1101–1115

304 Bodenheimer, P., Ostriken, J. (1974): Do the pulsars make supernovae? II. Calculations of light curves for type II events. ApJ **191**, 465–471

305 Bodenheimer, P., Tscharnuter, W. (1979): A comparison of two independent calculations of the axysimmetric collapse of a rotating protostar. AA **74**, 288–293

306 Bohm-Vitenze, E. (1958): Uber die Wasserstoffkonvektionzone in Sternen verschiedener Effectivtemperaturen und Leuchtkrafte. Z. Astrophys. **46**, 108–143

306* Bonazzola, S., Gourgoulhon, E., Mark, J.-A. (1999): Spectral methods in general relativistic astrophysics. Journ. Comput. Appl. Math. **109**, 433–473

306a Bonazzola, S. and Gourgoulhon, E. (1994): A virial identity applied to relativistic stellar models. Classical and quantum gravity **11**, 1775–1784

306b Bondi, H. (1952): On spherically symmetrical accretion. MN **112**, 195–204

306c Bondi, H., Hoyle, F. (1944): On the mechanism of accretion by stars. MN **104**, 273–282

307   Boss, A. P. (1980): Protostellar formation in rotating interstellar cloud. I. Numerical methods and tests. ApJ **236**, 619–627

308   Boss, A. P. (1980): Protostellar formation in rotating interstellar cloud. II. Axially symmetric collapse. ApJ **237**, 563–573

309   Boss, A. P. (1980): Protostellar formation in rotating interstellar cloud. II. Nonaxysimmetric collapse. ApJ **237**, 866–876

310   Boss, A. P. (1980): Collapse and equilibrium of rotating adiabatic clouds. ApJ **242**, 699–709

311   Boss, A. P., Haber, J. G. (1982): Axysimmetric collapse of rotating isothermal clouds. ApJ **255**, 240–244

311a Braaten, E., Segel, D. (1993): Neutrino energy loss from the plasma process at all temperatures and densities. Phys. Rev. **D48**, 1478–1491

312   Bray, R. J., Loughhead, R. E. (1967): *The solar granulation* (L, Chapman and Hall)

313   Buchler, J-R., Barkat, Z. (1971): Properties of low density neutron star matter. PR Lett **27**, 48–51

314   Buchler, J-R., Datta, B. (1979): Neutron gas: temperature dependence of the effective interaction. PR **C19**, 494–497

315   Buchler, J-R., Yuen, W. R. (1976): Compton scattering opacities in a partially degenerate electron plasma at high temperatures. ApJ **210**, 440–446

316   Bugaev, E. V., Bisnovatyi-Kogan, G. S., Rudzskyi, M. A., Seidov, Z. F. (1979): The interaction of intermediate energy neutrinos with nuclei. NP **A324**, 350–364

317   Burbidge, E. M., Burbidge, G. R., Fowler, W., Hoyle, F. (1957): Synthesis of the elements in stars. Rev. Mod. Phys. **29**, 547–650

317a Burrows, A. (1987): Convection and the mechanism of Type II supernovae. ApJ Lett. **318**, L57-L61

317b Burrows, A., Fryxell, B. A. (1993): A convective trigger for supernova explosion. ApJ Lett. **418**, L33–L35

317c Burrows, A., Hayes, J., Fryxell, B. A. (1995): On the nature of core collapse supernova explosion. ApJ **450**, 830–850

318   Cabrit, S., Bertout, C. (1986): CO lines formation in bipolar flows. I. Accelerated outflows. ApJ **307**, 313–323

318* Camilo, F., Lorimer, D. R., Freire, P., Lyne, A.G., Manchester, R. N. (2000): Observations of 20 millisecond pulsars in 47 Tucanae at 20 centimeters. ApJ **535**, 975–990

318a Cannon, R. C., Eggleton, P. P., Zytkow, A. N., Podsiadlowski P. (1992): The structure and evolution of Thorne–Zytkow objects. ApJ **386**, 206–214

319   Carson, T. R., Huebner, W. F., Magee, N. H., Merts, A. L. (1984): Discrepancy in the CNO opacity bump resolved. ApJ **283**, 466–468

320   Castor, J. L. (1972): Radiative transfer in spherically symmetric flows. ApJ **178**, 779–792

320* Castor, J. I., Abbott, D. C., Klein, R. I. (1974): Radiation-driven winds in Of stars. ApJ **195**, 157–174

320** Castro-Tirado, A. J., Gorosabel, J. (1999): Optical observations of GRB afterglows: GRB 970508 and GRB 980326 revisited. AA Suppl. **138**, 449–450

320a Caughlan, G. R. and Fowler, W. A. (1988): Thermonuclear reaction rates
      V. At. data and nucl. data tabl. **40**, 283–334

 321 Caughlan, G., Fowler, W., Harris, M., Zimmerman, B. (1985): Tables of
      thermonuclear reaction rates for low-mass nuclei ($1 \leq Z \leq 14$). At. data
      and nucl. data tabl. **32**, 197–234

321a Cavalerie, A., Isaak, G., et al. (1981): Structure of 5-minute solar oscilla-
      tions 1976–1980. Solar Phys. **74**, 51–57

321b Chabrier, G. (1993): Quantum effects in dense Coulombic matter: applica-
      tion to the cooling of white dwarfs. ApJ **414**, 695–700

321c Chabrier, G., Schatzman, E. (eds) (1994): *The equation of state in astro-
      physics.* Proc. IAU Coll. (Cambridge, Cambridge University Press)

 322 Chandrasekhar, S. (1931): The highly collapsed configuration of a stellar
      mass. MN **91**, 456–466

 323 Chandrasekhar, S. (1964): The dynamical instability of gaseous masses ap-
      proching the Schwarzschild limit in general relativity. ApJ **140**, 417–433

323a Chandrasekhar, S. (1981) *Selected Papers* (Chicago, Chicago University
      Press), **1**, p. 185

 324 Chan, K., Sofia, S., Wolff, L. (1982): Turbulent compressible convection
      in a deep atmosphere. I. Preliminary two-dimensional results. ApJ **263**,
      935–943

324a Charbonnel, C., Meynet, G., Maeder, A., Schaller, G., Schaerer, D. (1993):
      Grids of stellar models. III. From 0. 8 to 120 $M_\odot$ at Z=0. 004. AA Suppl.
      101, 415–419

 325 Chechetkin, V. M., Gershtein, S. S., Imshennik, V. S., Ivanova, L. N.,
      Khlopov, M. Yu. (1980): Supernovae of types I and II and the neutrino
      mechanism of thermonuclear explosion of degenerated carbon-oxygen stel-
      lar cores. ApSS **67**, 61–98

 326 Chechetkin, V. M., Kowalski, M. (1976): Production of heavy elements in
      nature. Nature **259**, 643–644

326a Chevalier, R., Dwarkadas, V. V. (1995): The presupernova H II region
      around SN 1987A. ApJ Lett. **452**, L45–L48

 327 Chiosi, C., Stalio, R. (eds.) (1981): Effects of mass loss in stellar evolution,
      in *Proc. IAU Coll.* No. 59 (Dordrecht, Reidel)

327a Christensen-Dalsgaard, J., Däppen, W. (1992): Solar oscillations and the
      equation of state. AA Review **4**, 267–361

 328 Clifford, F. E., Tayler, R. J. (1965): The equilibrium distribution of nuclides
      in matter at high temperatures. Mem. Roy. Astron. Soc. London **69**, 21–81

 329 Cloutman, L., Whitaker, R. (1980): On convective and semiconvective mix-
      ing in massive stars. ApJ **237**, 900–902

 330 Cohen, M., Kuhi, L. V. (1979): Observational studies of pre-main-sequence
      evolution. ApJ Suppl. **41**, 743–843

330a Colgate, S. (1989): Hot bubbles drive explosions. Nature **341**, 489–490.

330b Colgate, S., Petschek, A. G. (1980): Explosive supernova core overturn and
      mass ejection. ApJ Lett. **236**, L115–L119

330c Colgate, S., Krauss, L. M., Schramm, D. N., Walker, T. P. (1990): Mag-
      netohydrodynamic jets, pulsar formation and SN 1987A. Astrophys. Lett.
      27, 411–418

330d Colgate, S., Herant, M. E., Benz, W. (1993): Neutron star accretion and
      the neutrino fireball. Phys. Rep. **227**, 157–174

 331 Colgate, S., White, R. (1966): The hydrodynamic behavior of supernovae
      explosions. ApJ **143**, 626–681

331* Colpi, M., Shapiro, S., Teukolsky S. (1993): A hydrodynamical model for
      the explosion of a neutron star just below the minimum mass. ApJ **414**,
      717–734

331a Commins, P. H., Bucksbaum, P. H. (1983): *Weak interactions of Leptons and Quarks* (Cambridge, Cambridge University Press)

331b Costa, E. on behalf of the BEPPOSAX GRB Team (1999): X-ray afterglows of gamma-ray bursts with BeppoSAX. AA Suppl. **138**, 425–429

332 Couch, R., Arnett, D. (1972): Advanced evolution of massive stars. I. Secondary nucleothynthesis during helium burning. ApJ **178**, 771–777

333 Couch, R., Schmiedenkamp, A., Arnett, D. (1974): S-process nucleosynthesis in massive stars: core helium burning. ApJ **190**, 95–100

334 Cox, A., Tabor, J. (1976): Radiative opacity tables for 40 stellar mixtures. ApJ Suppl. **31**, 271–312

335 Cox, D. P., Tucker, W. H. (1969): Ionization equilibrium and radiative cooling in a low-density plasma. ApJ **157**, 1157–1167

335a Däppen, W., Mihalas, D., Hummer, D. G. (1988): The equation of statefor stellar envelopes. III. Thermodynamical quantities. ApJ **332**, 261–270

335b Dar, A., Shaviv, G. (1994): A standard model solution to the solar neutrino problem? Preprint astro-ph 9401043

335c Dar, A., Shaviv, G. (1994): Has a standard physics solution to the solar neutrino problem been found?-A response. Preprint astro-ph 9404035

336 Dicus, D. (1972): Stellar energy-loss rate in a convergent theory of a weak and electromagnetic interactions. PR **D6**, 941–949

337 Downes, D. et al. (1981): Outflow of matter in the KL nebula: the role of IRc2. ApJ **244**, 869–883

338 Dravins, D., Lindegren, L., Nordlund, A. (1981): Solar granulation: influence of convection on spectral line asymmetries and wavelength shifts. AA **96**, 345–364

338a. Drobyshevskii, E. M. (1977): Outward transport of angular momentum by gas convection and the equatorial acceleration of the Sun. Solar Phys. **51**, 473–479

338b Drotleff, H. W., Denker, A., Knee, H., Soine, M., Wolf, G., Hammer, J. W., Greife, U., Rolfs, C., Trautvetter, H. P. (1993): Reaction rates of the s-process neutron sources $^{22}Ne(\alpha,n)^{25}Mg$ and $^{13}C(\alpha,n)^{16}O$. ApJ **414**, 735–739

339 Durney, B. R. (1970): The interaction of rotation with convection, in *Stellar Rotation*, ed. by A. Slettebak (Dordrecht, Reidel), pp. 30–36

340 Durney, B. R. (1976): On theories of solar rotation, in *Basic Mechanisms of Solar Activity*, ed. by Bumba, Kleczek (Dordrecht, Reidel), pp. 243–295

341 Dyck, H. H., Simon, Th., Zukerman, B. (1982): Discovery of an infrared companion to T Tauri. ApJ Lett. **255**, L103–L106

342 Dzembowski, W. (1971): Nonradial oscillations of evolved stars. I. Quasiadiabatic approximation. Acta Astron. **21**, 289–306

342a Dzytko, H., Turck-Chieze, S., Delburgo-Salvador, P., Lagrange, C. (1995): The screened nuclear reaction rates and the solar neutrino puzzle. ApJ **447**, 428–442

343 Eardley, D. M., Lightman, A. P. (1976): Inverse Compton spectra and the spectrum of Cyg X–1. Nature **262**, 196–197

344 Eggleton, P. (1967): The structure of narrow shells in red giants. MN **135**, 243–250

344a Eggleton, P. (1971): The evolution of low mass stars. MN **151**, 351–354

344b Eggleton, P. (1972): Composition changes during stellar evolution. MN **155**, 361–376

344c Eggleton, P. (1973): A numerical treatment of double shell source stars. MN **163**, 279–284

344d Eggleton, P. (1994): Programme STAR. Private communication

345 Eggleton, P. (1983): Towards consistency in simple prescriptions for stellar convection. MN **204**, 449–461

345a Eggleton, P., Faulkner, J., Flannery, B. (1973): An approximate equation of state for stellar material. AA **23**, 325-330

345b Eich, C., Zimmermann, M. E., Thorne, K. S., Zytkow, A. N. (1989): Giant and supergiant stars with degenerate neutron cores. ApJ **346**, 277–283

346[++] El Eid, M. (1995): Effect of convective mixing on the red-blue loops in the Hertzsprung–Russell diagram. MN **275**, 983–1002

346[+] El Eid, M., Hartmann D. H. (1993): Stellar models and the brightening of P Cygni. ApJ **404**, 271–275

346 El Eid, M. F., Hillebrandt, W. (1980): A new equation of state of supernova matter. AA **42**, 215–226

346[*] El Eid, M., Langer, N. (1986): The evolution of very luminous stars II. Pair creation supernova in massive Wolf–Rayet stars. AA **167**, 274–281

346a Ellis, J. (1995): Particle physics and cosmology. Ann. N. -Y. Acad. Sci. **759**, 170–187

346b Elsworth, Y., Howe, R., Isaak, G. R., McLeod, C. P., Miller, B. A., New, R., Wheeler, S. J., Gough, D. O. (1995): Slow rotation of the Sun's interior. Nature **376**, 669–672

347 Emerson, J. P. et al. (1984): IRAS observations near young objects with bipolar outflows: L1551 and HH 46–47. ApJ Lett. **278**, L49–L52

348 Epstein, R. I. (1979): Lepton-driven convection in supernovae. MN **188**, 305–325

349 Ergma, E., Paczynski, B. (1974): Carbon burning with convective URCA neutrinos. Acta Astron. **24**, 1–16

350 Ergma, E., Tutukov, A. V. (1976): Evolution of carbon–oxygen dwarfs in binary systems. Acta Astron. **26**, 69–76

351 Eriguchi, Y., Müller, E. (1985): A general method for obtaining equilibria of self-gravitating and rotating gases. AA **146**, 260–268

352 Eriguchi, Y., Sugimoto, D. (1981): Another equilibrium sequence of self-gravitating and rotating incompressible fluid. PTP **65**, 1870–1875

352a Errico, L., Vittone, A. (eds.) (1993): *Stellar jets and bipolar outflows.* Proc. OAC 6, Capri 1991. (Kluwer)

353 Ezer, D., Cameron, A. G. W. (1971): Pre-main-sequence stellar evolution with mass loss. ApSS **10**, 52–70

354 Faulkner, J., Roxburgh, I. W., Strittmatter, P. A. (1968): Uniformly rotating main-sequence stars. ApJ **151**, 203–216

355 Feigelson, E. D., DeCampli, W. M. (1981): Observations of X-ray emission from T Tauri stars. ApJ Lett. **243**, L89–L94

355[*] Feroci, M., Frontera, F., Costa, E., Amati, L., Tavani, M., Rapisarda, M., Orlandini, M. (1999): A giant outburst from SRG 1900+14 observed with the BeppoSAX gamma-ray burst monitor. ApJ Lett. **515**, L9–L12

355a Fillipenko, A. V. (1991): The optical diversity of supernovae, in *Proc. ESO/EIPC Workshop on SN 1987A and other Supernovae*, ed. by Danziger, I. J., Kjär, K., pp. 343–362

356 Finzi, A., Wolf, R. (1971): Ejection of mass by radiation pressure in planetary nebulae. AA **11**, 418–430

357 Flowers, E., Itoh, N. (1979): Transport properties of dense matter II. ApJ **230**, 847–858

358 Fontaine, G., Graboske, H. C. Jr., van Horn, H. M. (1977): Equation of state for stellar partial ionization zones. ApJ Suppl. **35**, 293–358

358a Forestini, M. (1994): Low-mass stars: pre-main sequence evolution and nucleosynthesis. AA **285**, 473–488

359 Fowler, W. (1977): The solar neutrino problem. Preprint OAP, No. 507

360 Fowler, W., Caughlan, G., Zimmerman, B. (1967): Thermonuclear reaction rates. Ann. Rev. Astron. Ap. **5**, 525–570

361 Fowler, W., Caughlan, G., Zimmerman, B. (1975): Thermonuclear reaction rates II. Ann. Rev. Astron. Ap. **13**, 69–112

362 Fowler, W., Engebrecht, C., Woosley, S. (1978): Nuclear partition functions. ApJ **226**, 984–995

363 Fowler, W., Hoyle, F. (1964): Neutrino processes and pair formation in massive stars and supernovae. ApJ Suppl. **9**, 201–319

364 Fraley, G. (1968): Supernovae explosions induced by pair-production instability. ApSS **2**, 96–114

365 Friedman, B., Pandariphande, V. R. (1981): Hot and cold nuclear and neutron matter. NP **A361**, 502–520

366 Friedman, B., Ipser, J., Parker, L. (1986): Rapidly rotating neutron star models. ApJ **304**, 115–139

366a Fukasaku, K., Fujita, T. (1997): Reexamination of standard solar model to the solar neutrino problem. PTP **98**, 1251-1260

366b Fukuda, V. et al. (1998): Measurements of the Solar Neutrino Flux from Super-Kamiokaude. PRL **81**, 1158-1162

367 Fuller, G., Fowler, W., Newman, M. (1980): Stellar weak-interaction rates for sd-shell nuclei. I. Nuclear matrix element systematics with application to AL^{26} and selected nuclei of importance to the supernova problems. ApJ Suppl. **42** 445–473

368 Fuller, G., Fowler, W., Newman, M. (1982): Stellar weak-interaction rates for intermediate mass nuclei. II: $A = 21$ to $A = 60$. ApJ **252**, 715–740

369 Fuller, G., Fowler, W., Newman, M. (1982): Stellar weak-interaction rates for intermediate mass nuclei. III: rates tables for the free nucleons and $A = 21$ to $A = 60$. ApJ Suppl. **48**, 279–295

370 Fuller, G., Fowler, W., Newman, M. (1985): Stellar weak-interaction rates for intermediate mass nuclei. IV. Interpolation procedures for rapidly varying lepton capture rates using effective $\lg(ft)$ values ApJ **293**, 1–16

370a Gabriel, M. (1989): The D_{nl} values and the structure of the solar core. AA **226**, 278–283

371 Gahm, G. F. (1980): X-ray observations of T Tauri stars. ApJ Lett. **242**, L163–L166

372 Gahm, G. F., Fradga, K., Liseau, R., Dravins, D. (1979): The far UV spectrum of T Tauri star RU Lupi. AA **73**, L4–L6

372a Galkin, S. A., Denissov, A. A., Drozdov, V. V., Drozdova, O. M. (1993): A finite-difference adaptive grid method for computations of equilibria of self-gravitating rotating barotropic gas. AA **269**, 674–681

373 Gaustad, J. (1963): The opacity of diffuse cosmic matter and the early of star formation. ApJ **138**, 1050–1073

373a Geppert, U., Urpin, V. (1994): Accretion driven magnetic field decay in neutron stars. MN **271**, 490–496

374 Gillman, R. (1974): Planck mean cross-section for four grain materials. ApJ Suppl. **28**, 397–403

375 Giovanelli, F., Bisnovatyi-Kogan, G. S., Golynskaya, I. M. et al. (1984): Coordinated X-ray, ultraviolet and optical observations of T Tauri stars, in *Proc. symp. int. X-ray astronomy 84*, ed. by M. Oda, R. Giacconi (Bologna), pp. 77–80

375a Glatzel, W., Kiriakidis, M. (1993): The stability of massive main-sequence stars. MN **262**, 85–92

375b Glatzel, W., Kiriakidis, M. (1993): Stability of massive stars and Humphrey–Davidson limit. MN **263**, 375–384

376 Goldreich, P., Julian, W. (1969): Pulsar electrodynamics. ApJ **157**, 869–880

377 Golenitskii, S. V., Mazets, E. P. et al. (1985): Annihilation radiation in cosmic gamma-ray bursts. Preprint LFTI im. Ioffe, No. 959

378 Gonczi, G., Osaki, Y. (1980): On local theories of time-dependent convection in the stellar pulsation problem. AA **84**, 304–310

378* Goode, P. R., Dziembowski, W. A., Korzennik, S. G., Rhodes, E. J. (1991): What we know about the Sun's internal rotation from solar oscillations. ApJ **367**, 649–657

378a Goodman, J., Dar, A., Nussinov, Sh. (1987): Neutrino annihilation in Type II supernovae. ApJ Lett. **314**, L7–L10

378a* Gough, D. O. (1969): The anelastic approximation for thermal convection. J. Atmosph. Sci. **26**, 448–456

378a** Gough, D. O. (1987): Linear adiabatic stellar pulsation. *Proc. Les Houches 1987*, ed. by J.-P. Zahn and J. Zinn-Justin, pp. 399–560

378b Gough, D. O., Toomre, J. (1982): Single-mode theory of diffusive layers in thermohaline convection. J. Fluid Mech. **125**, 75–97

379 v. Groote, H., Hilf, E. R., Takahashi, K. (1976): A new semiempirical shell correction to the droplet model. Gross theory of nuclear magics. At. data nucl. data tables **17**, 418–427, 476–608

380 Grossman, A. (1969): The surface boundary condition and approximate equation of state for low-mass stars, in *Proc. Symp. Low-luminosity Stars*, ed. by Sh. Kumar (Gordon and Breach), pp. 247–254

381 Grossman, A. (1970): Evolution of low-mass stars. I. Contraction to the main sequence. ApJ **161**, 619–632

382 Grossman, A., Mutschlecner, J., Pauls, T. (1970): Evolution of low-mass stars. II. Effects of premodal deuterium burning and nongray surface condition during pre-main-sequence contraction. ApJ **162**, 613–619

383 Grossman, A., Graboske, H. Jr. (1971): Evolution of low-mass stars. III. Effects of nonideal thermodynamic properties during the pre-main-sequence contraction. ApJ **164**, 475–490

383* Grossman, S., Narayan, R., Arnett, D. (1993): A theory of nonlocal mixing-length convection. I. The moment formalism. ApJ **407**, 284–315

383** Grossman, S., Narayan, R. (1993): A theory of nonlocal mixing-length convection. II. Generalized smoothed particle hydrodynamics simulations. ApJ Suppl. **89**, 361–394

383a Hachisu, I. (1986): A versatile method for obtaining structures of rapidly rotating stars. ApJ Suppl. **61**, 479–508

383b Hachisu, I. (1986): A versatile method for obtaining structures of rapidly rotating stars. II. Three-dimensional self-consistent field method. ApJ Suppl. **62**, 461–500

384 Hamada, T., Salpeter, E. (1961): Models for zero-temperature stars. ApJ **134**, 683–698

384* Han, Z., Podsiadlowski, P., Eggleton, P. P. (1994) A possible criterion for envelope ejection in AGB or FGB stars. MN **270**, 121–130

384a Hansen, J. P., Torrie, G. M., Viellerfosse, P. (1977): Statistical mechanics of dense ionized matter. VII. Equation of state and phase separation of ionic mixtures in a uniform background. Phys. Rev. **A16**, 2153–2168

385 Hanson, R., Jones, B. F., Lin, D. N. C. (1983): The astrometric position of T Tauri and the nature of its companion. ApJ Lett. **270**, L27–L30

386 Harm, R., Schwarzschild, M. (1966): Red giants of population II. IV. ApJ **145**, 496–504

387 Harm, R., Schwarzschild, M. (1975): Transport from a gas giant to a blue nucleus after ejection of a planetary nebula. ApJ **200**, 324–329

388 Harm, R., Schwarzschild, M. (1964): Red giants of population II. III. ApJ **139**, 594–601

389 Harris, M., Fowler, W., Caughlan, G. Zimmerman, B. (1983): Thermonuclear reaction rates III. Ann. Rev. Astron. Ap. **21**, 165–176

389c Haselgrove, C. B., Hoyle, F. (1959): Main-sequence stars. MN **119**, 112–123

390 Hayashi, Ch. (1966): Evolution of protostars. Ann. Rev. Astron. Ap. **4**,
171–192

391 Hayashi, Ch., Hoshi, R., Sugimoto, D. (1962): Evolution of stars. Suppl.
PTP **22**, 1–183

392 Hayashi, Ch., Hoshi, R., Sugimoto, D. (1965): Advanced phases of evolution
of population II stars. Growth of the carbon core and shell helium flashes.
PTP **34**, 885–911

393 Henyey, L. G., Wilets, L., Böhm, K. -H., LeLevier, R., Levee, R. D. (1959):
A method for automatic computation of stellar evolution. ApJ **129**, 628-636

393a Herant, M., Benz, W. (1991): Hydrodynamical instabilities and mixing in
SN 1987A: two-dimensional simulations of the first 3 months. ApJ Lett.
370, L81–L84

393b Herant, M., Benz, W., Hix, W. R., Fryer, C. L., Colgate, S. A. (1994):
Inside the supernova: a powerful convective engine. ApJ **435**, 339–361

394 Herbig, G. (1957): The widths of absorption lines in T Tauri-like stars. ApJ
125, 612–613

395 Herbig, G. (1977): Eruptive phenomena in early stellar evolution. ApJ **217**,
693–715

395a Hernández, M. M., Hernández, E. P., Michel, E., Belmonte, J. A., Goupil,
M. J., Leberton, Y. (1998): Seismology of δ Scuti stars in the Praesepe
cluster. AA **338**, 511–520

396 Hillebrandt, W. (1978): The rapid neutron capture process and the synthesis of heavy and neutron-rich elements. Space Sci. Rev. **21**, 639–702

397 Hillebrandt, W. (1985): Stellar collapse and supernovae explosion, in Proc.
NATO-ASI High energy phenomena around collapsed stars,
Cargese

398 Hillebrandt, W., Nomoto, K., Wolff, R. (1984): Supernovae explosions of
massive stars. The mass range 8 to 10 $M_\odot$. AA **133**, 175–184

399 Holmes, J., Woosley, S., Fowler, E., Zimmerman, B. (1976): Tables of
thermonuclear–reaction–rate date for neutron–induced reaction on heavy
nuclei. At. data nucl. data tables **18**, 305–412

400 Hoshi, R. (1977): Basis properties of a stationary accretion disk surrounding
a black hole. PTP **58**, 1191–1204

401 Hoshi, R., Shibazaki, N. (1977): The effect of pressure gradient force on an
accretion disk surrounding a black hole. PTP **58**, 1759–1765

402 Houck, J. R. et al. (1984): Unidentified point sources in the IRAS minisur-
vey. ApJ Lett. **278**, L63–L66

403 Hoxie, D. (1970): The structure and evilution of stars of very low mass.
ApJ **161**, 1083–1099

403a Hoyle, F., Fowler, W. A. (1960): Nucleosynthesis in supernovae. ApJ **132**,
565–590

403b Hummer, D. G., Mihalas, D. (1988): The equation of state for stellar envelopes. I. An occupation probability formalism forr the truncation of internal partition functions. ApJ **331**, 794–814

404 Hunt, R. (1971): A fluid dynamical study of the accretion process. MN **154**,
141–165

405 Hurlburt, N., Toomre, J., Massaguer, J. (1984): Two-dimensional compressible convection extending over multiple scale heights. ApJ **282**, 557–573

406 Iben, I. Jr. (1965): Stellar evolution. I. The approach to the main sequence.
ApJ **141**, 993–1018

407 Iben, I. (1965): Stellar evolution. II. The evolution of a $3\,M_\odot$ star from main sequence through core helium burning. ApJ **142**, 1447–1467

408 Iben, I. (1966): Stellar evolution. III. The evolution of a $5\,M_\odot$ star from main sequence through core helium burning. ApJ **143**, 483–504

409 Iben, I. (1966): Stellar evolution. IV. The evolution of a $9\,M_\odot$ star from main sequence through core helium burning. ApJ **143**, 505–515

410 Iben, I. (1966): Stellar evolution. V. The evolution of a $15\,M_\odot$ star through core helium burning from the main sequence. ApJ **143**, 516–526

411 Iben, I. (1967): Stellar evolution. VI. Evolution from the main sequence to the red-giant branch for stars of mass $1\,M_\odot$, $1.25\,M_\odot$ and $1.5\,M_\odot$. ApJ **147**, 624–649

412 Iben, I. (1967): Stellar evolution. VII. Evolution of $2.25\,M_\odot$ star from the main sequence to the helium burning phase. ApJ **147**, 650–663

413 Iben, I. (1967): Stellar evolution within and off the main sequence. Ann. Rev. Astron. Ap. **5**, 571–626

414 Iben, I. (1971): On the specification of the blue edge of the RR Lyrae instability trip ApJ **166**, 131–151

415 Iben, I. (1974): Post main sequence evolution of single stars. Ann. Rev. Astron. Ap. **12**, 215–256

416 Iben, I. (1975): Thermal pulses; p-capture, α-capture, s-process nucleosynthesis; and convective mixing in a star of intermediate mass. ApJ **196**, 525–547

417 Iben, I. (1976): Solar oscillations as a guide to solar structure. ApJ Lett. **204**, L147–L150

418 Iben, I. (1976): Futher adventures of a thermally pulsing star. ApJ **208**, 165–176

419 Iben, I. (1982): Low-mass asymptotic giant branch evolution I. ApJ **260**, 821–837

420 Iben, I. (1984): On the frequency of a planetary nebula nuclei powered by helium burning and on the frequency of white dwarfs with hydrogen-deficient atmospheres. ApJ **277**, 333–354

421 Iben, I. (1985): The life and times of an intermediate mass star – in isolation/in a close binary. Quart. J. Roy. Astron. Soc. **26**, 1–39

421a Iben, I. (1991): Single and binary star evolution. ApJ Suppl. **76**, 55–114

422 Iben, I., Kaler, J., Truran, J., Renzini, A. (1983): On the evolution of those nuclei of planetary nebulae, that experience a final helium shell flash. ApJ **264**, 605–612

423 Iben, I., Renzini, A. (1983): Asymptotic giant branch evolution and beyond. Ann. Rev. Astron. Ap. **21**, 271–342

424 Iben, I., Renzini, A. (1984): Single star evolution I. Massive stars and early evolution of low and intermediate mass stars. Phys. Rep. **105**, 329–406

425 Iben, I., Rood, R. (1970): Metal-poor stars I. Evolution from the main sequence to the giant branch. ApJ **159**, 605–617

426 Iben, I., Tutukov, A. V. (1984): Cooling of low-mass carbon-oxygen dwarfs from the planetary nucleus stage through the crystallization stage. ApJ **282**, 615–630

427 Ichimaru, S. (1982): Strongly coupled plasma: high density classical plasmas and degenerate electron liquids. Rev. Mod. Phys. **54**, 1017–1059

428 Ichimaru, S., Utsumi, K. (1984): Enhancement of thermonuclear reaction rate due to screening by relativistic degenerate electrons long range correlation effect. ApJ **286**, 363–365

428a Iglesias, C. A., Rogers, F. J., Wilson, B. J. (1992): Spin-orbit interaction effects on the Rosseland mean opacity. ApJ **397**, 711–728

428b Iglesias, C. A., Rogers, F. J. (1993): Radiative opacities for carbon and oxygen-rich mixtures. ApJ **412**, 752–760

428c Iglesias, C. A., Wilson, B. J., Rogers, F. J., Goldstein, W. H., Bar-Shalom, A., Oreg, J. (1995): Effects of heavy metals on astrophysical opacities. ApJ **445**, 855–860

429 Illarionov, A. F., Sunyaev, R. A. (1975): Why the number of galactic X-ray stars is so small? AA **39**, 185–195

430 Imshennik, V. S. Nadyozhin, D. K. (1979): Neutrino chemical potential and neutrino heat conductivity with allowance for neutrino scattering. ApSS **62**, 309–333

430a Ipser, J., Managan, R. (1981): On the existence and structure of inhomogeneous analogues of the Dedekind and Jakobi ellipsoids. ApJ **250**, 362–372

431 Itoh, N., (1981): Physics of dense plasmas and the enhancement of thermonuclear reaction rates due to strong screening. Supl. PTP **70**, 132–141

431a Itoh, N., Hayashi, H., Kohyama, Y. (1993): Electrical and thermal conductuvity of dense matter in the crystalline lattice phase. III. Inclusion of lower densities. ApJ **418**, 405–413

432 Itoh, N., Mutaku, S., Iyetomi, H., Ichimaru, S. (1983): Electrical and thermal conductivities of dense matter in the liquid metal phase I. High temperature results. ApJ **273**, 774–782

433 Itoh, N., Totsuji, H., Ichimaru, S. (1977–1978): Enhancement of thermonuclear reaction rates due to strong screening. ApJ **218**, 477–483, **220**, 742

434 Itoh, N., Totsuji, H., Ichimaru, S., De Witt, H. (1979–1980): Enhancement of thermonuclear reaction rates due to strong screening. II. Ionic mixtures. ApJ **234**, 1079–1084, **239**, 415

435 Ivanova, L. N., Imshennik, V. S., Chechetkin, V. M. (1974): Pulsation regime of the thermonuclear explosion of a star's dense carbon core. ApSS **31**, 497–514

436 Jackson, S. (1970): Rapidly rotating stars. The coupling of the Henyey and the self-consistent-fluid methods. ApJ **161**, 579–585

437 James, R. A. (1964): The structure and stability of rotating gas masses. ApJ **140**, 552–582

438 Juman, C. (1965): Baryon star models. ApJ **141**, 187–194

438a Janka, H. -T., Müller, E. (1994): Neutron star recoils from anisotropic supernovae. AA **290**, 496–502

438b Janka, H. -T., Müller, E. (1995): The first second of a Type II supernova: convection, accretion, and shock propagation. ApJ Lett. **448**, L109–L113

439 Kamija, Y. (1977): The collapse of rotating gas clouds. PTP **58**, 802–815

439a Kato, M., Iben, I. (1992): Self-consistent models of Wolf–Rayet stars as helium stars with optically thick winds. ApJ **394**, 305–312

440 Keene, J. et al. (1983): Far-infrared detection of low-luminosity star formation in the Bok globule B335. ApJ Lett. **274**, L43–L47

441 Kellman, S., Gaustad, J. (1963): Rosseland and Planck mean absorption coefficients for particles of ice, graphite and silicon dioxide. ApJ **138**, 1050–1073

441a Khokhlov, A. M. (1989): The structure of detonation waves in supernovae. MN **239**, 785–808

441b Khokhlov, A. M. (1991): Mechanisms for the initiation of detonation in degenerate matter of Supernovae. AA **246**, 383–396

441c Khokhlov, A. M. (1991): Delayed detonation model for type Ia Supernova. AA **245**, 114–128

442 Kippenhahn, R. (1963): Differential rotation in stars with convective envelopes. ApJ **137**, 664–678

443 Kippenhahn, R., Thomas, H. C., Weigert, A. (1965): Sternentwicklung IV. Zentrales Wasserstoff und Heliumbrenner bei einen Stern von 5 Sonnenmassen. Zeit. Astrophys. **61**, 241–267

444 Kippenhahn, R., Thomas, H. C., Weigert, A. (1966): Sternentwicklung V. Der Kohlenstoff-Flash bei einen Stern von 5 Sonnenmassen. Zeit. Astrophys. **64**, 373–394

445 Kippenhahn, R., Thomas, H. C. (1981): Rotation and stellar evolution, in *Proc. IAU Symp. No. 93, Fundamental Problems of the Theory of Stellar Evolution*, ed. by D. Sugimoto, D. Lamb, D. Shramm. (Dordrecht, Reidel), pp. 237–256

445a Kippenhahn, R., Weigert, A. (1990): *Stellar structure and evolution* (Berlin, Heidelberg, Springer)

446 Kippenhahn, R., Weigert, A., Hofmeister, E. (1967): Methods for calculating stellar evolution. Meth. Comput. Phys. **7**, 129–190

447+ Kitamura, H., Ichimaru, S. (1995): Pycnonuclear reaction rates in stellar interiors. ApJ **438**, 300–307

447 Kohijama, Y., Itoh, N., Munakata, H. (1986): Neutrino energy losses in stellar interiors II. Axial-vector contribution to the plasma neutrino energy loss rate. ApJ **310**, 815–819

447a Koide, H., Matzuda, T., Shima, E. (1991): Numerical simulations of axisymmetric accretion flows. MN **252**, 473–481

447b Kosovichev, A. G., Christensen-Dalsgaard, J., Däppen, W., Dziembovski, W., Gough, D. O., Thompson, M. J. (1992): Sources of uncertainity in direct seismological measurements of the solar helium abundance. MN **259**, 536–547

447c Kraft, R. P., Greenstein, J. L. (1969): A new method for finding faint members of the Pleiads. In *Low luminosity stars* (New York, Gordon and Breach), pp. 65–82

448 Kuan, P. (1975): Emission envelopes of T Tauri stars. ApJ **202**, 425–432

449 Kuhi, L. V. (1964): Mass loss from T Tauri stars. ApJ **140**, 1409–1433

450 Kulkarni, S. R. (1986): Optical identification of binary pulsars implications for magnetic field decay in neutron stars. ApJ Lett. **306**, L85–L90

450a Kulkarni, S.R. et al. (29 authors) (1999): The afterglow, redshift and extreme energetics of the γ-ray burst of 23 January 1999. Nature **398**, 389–399

451 Kundt, W. (1976): Are supernova explosions driven by magnetic strings? Nature **261**, 673–674

452 Kutter, G. S., Savedoff, M. P., Schuerman, D. W. (1969): A mechanism for the production of planetary nebulae. ApSS **3**, 182–197

453 Kutter, G. S., Sparks, W. (1974): Studies of hydrodynamic events in stellar evolution III. Ejection of planetary nebulae. ApJ **192**, 447–455

454 Kwok, S. (1982): From red giants to planetary nebulae. ApJ **258**, 280–288

454a Lafon, J. -P. J., Beruyer, N. (1991): Mass loss mechanisms in evolved stars. AA Rev. **2**, 349–389

455 Lamb, D. Q., Pethick, C. J. (1976): Effects of neutrino degeneracy in supernova models. ApJ Lett. **209**, L77–L82

456 Lamb, D. Q., Lattimer, J. M., Pathick, C. J., Ravenhall, D. G. (1978): Hot dense matter and stellar collapse. PR Lett. **41**, 1623–1626

457 Lamb, D. Q., Van Horn, H. M. (1975): Evolution of crystallizing pure C^{12} white dwarfs. ApJ **200**, 306–323

458 Lamb, D. Q. (1981): Neutron star binaries, pulsars and burst sources. Preprint (Urbana University)

459 Lamb, S., Iben, I., Howard, M. (1976): On the evolution of massive stars through the core carbon-burning phase. ApJ **207**, 209–232

459a Lambert, D. L. (1992): The p-nuclei: abundances and origin. AA Rev. **3**, 201–256

460 Lamers, H. (1981): The dependence of mass loss on the basic stellar parameters, in *Effects of mass loss on stellar evolution*, ed. by C. Chiosi, R. Stalio (Dordrecht, Reidel), pp. 19–23

460a Lamers, H. J. G. L., Leitherer, C. (1993): What are the mass-loss rates of O stars? ApJ **412**, 771–791

461 Lampe, M. (1968): Transport coefficients of degenerate plasma. PR **170**, 306–319

462 Langanke, K., Wiescher, M., Fowler, W., Gorres, J. (1986): A new estimate of the $Ne^{19}(p,\gamma)Na^{20}$ and $O^{15}(\alpha,\gamma)Ne^{19}$ reaction rates at stellar energies. ApJ **301**, 629–633

462a Langer, N., El Eid, M. F., Fricke, K. J. (1985): Evolution of massive stars with semiconvective diffusion. AA **145**, 179–191

462b Langer, N., Hamann, W. -R., Lennon, M., Najarro, F., Pauldrach, A. W. A., Puls, J. (1994): Towards an understanding of very massive stars. A new evolutionary scenario relating O stars, LBVs and Wolf–Rayet stars. AA **290**, 819–833

463 Larson, R. (1973): The evolution of protostars – theory. Found. Cosm. Phys. **1**, 1–70

464 Lattimer, J., Mazurek, T. (1981): Leptonic overturn and shocks in collapsing stellar cores. ApJ **246**, 955–965

465 Lattimer, J. (1981): The equation of state of hot dense matter and supernovae. Ann. Rev. Nucl. Part. Sci. **31**, 337–374

466 Lattimer, J., Mazurek, T. (1980): Stellar implosion shocks and convective overturn, in Proc. DUMAND-1980

467 Le Blank, L. M., Wilson, J. R. (1970): A numerical example of the collapse of a rotating magnetized star. ApJ **161**, 541–551

468 Ledoux, P. (1974): Non-radial oscillations, in *Proc. IAU Symp. No. 59*, ed. by P. Ledoux, A. Noels, A. W. Rodgers (Dordrecht, Reidel), pp. 135–173

469 Lewellyn-Smith, C. H. (1972): Neutrino reactions at accelerator energies. Phys. Rep. **3C**, 261–379

470 Lighthill, H. J. (1950): On the stability of small planetary cores (II). MN **110**, 339–342

471 de Loore, C. (1981): The influence of mass loss on the evolution of binaries, in *Effects of mass loss on stellar evolution*, ed by C. Chiosi, R. Stalio (Dordrecht, Reidel), pp. 405–427

471* Lopez, J. L., Nanopoulos, D. V. and Zichichi, A. (1994): The top-quark mass in $SU(5)\times U(1)$ supergravity. Preprint ASC-31/94

471a Lovelace, R. V. E. (1976): Dynamo model of double radio sources. Nature **262**, 649–652

471b Lovelace, R.V.E., Bisnovatyi-Kogan, G. S., Romanova, M. M. (1995): Spin-up/spin-down of magnetized stars with accretion discs and outflows. MN **275**, 244–254

472 Lucy, L. (1967): Formation of planetary nebulae. AJ **72**, 813

473 Lucy, L. B. (1967): Gravity-darkening for stars with convective envelopes. Zeit. Astrophys. **65**, 89–92

473* Lucy, L. B. (1986): Radiatively-driven stellar winds. Preprint No. 419 European Southern Observatory

473** Lynden-Bell, D. (1969): Galactic nuclei as collapsed old quasars. Nature **223**, 690–694

473a Lyne, A. G. et al. (1987): The discovery of a millisecond pulsar in the globular cluster M28. Nature **328**, 399–401

473b Lyne, A. G., Lorimer, D. R. (1994): High birth velocities of radio pulsars. Nature **369**, 127–129

474 Mac Donald, J. (1980): The effect of a binary companion on a nova outburst. MN **191**, 933–949

475 Maeder, A. (1975): Stellar evolution III: the overshooting from convective cores. AA **40**, 303–310

476 Maeder, A. (1981): The most massive stars evolving to red supergiants: evolution with mass loss, WR stars, as post-red supergiants and pre-supernovae. AA **99**, 97–107

477 Maeder, A. (1981): Grid of evolutionary models for upper part of the HR diagram, mass loss and the turning of some red supergiants into WR stars. AA **102**, 401–410

478 Makashima, K. et al. (1986): Simultaneous X-ray and optical observations of GX 339–4 in an X-ray high state. ApJ **308**, 635–643

479 Malone, R., Johnson, M., Bethe, H. (1975): Neutron star model with realistic high-density equations of state. ApJ **199**, 741–748

480 Massaguer, J. M., Latour, J., Toomre, J., Zahn, J.-P. (1984): Penetrative cellular convection in a stratified atmosphere. AA **140**, 1–16

481 Mathews, G., Dietrich, F. (1984): The $N^{13}(p, \gamma)O^{14}$ thermonuclear reaction rate and the hot CNO cycle. ApJ **287** 969–976

481* Matsuda, T, Sekino, N., Sawada, K, Shima, E., Livio, M., Anzer, U., Börner, G. On the stability of wind accretion. AA **248**, 301–314

481a Mayle, R., Wilson, J. R., Schramm, D. N. (1987): Neutrinos from gravitational collapse. ApJ **318**, 288–306

482 Mazurek, T. (1974): Degeneracy effects of neutrino mass ejection in supernovae. Nature **252**, 287–289

482a Menard, F., Monin, J. -L., Angelucci, F., Rougan, D. (1993): Disks around pre-main-sequence binary systems: the case of Haro 6-10. ApJ Lett. **414**, L117–L120

482b Merryfield, W. J. (1995): Hydrodynamics of semiconvection. ApJ **444**, 318–337

483 Mestel, L. (1952): On the theory of white dwarf stars. I. The energy sources of white dwarfs. MN **112**, 583–594

484 Mestel, L. (1952): On the theory of white dwarf stars. II. The accretion of interstellar matter by white dwarfs. MN **112**, 598–605

485 Mestel, L., Ruderman, M. A. (1967): The energy content of a white dwarf and its rate of cooling. MN **136**, 27–38

485* Meszáros, P. (1999): Gamma-ray bursts afterglows and their implications. AA Suppl. **138**, 533–536

485a Mewe, R. (1991): X-ray spectroscopy of stellar coronae. AA Rev. **3**, 127–168

485b Meier, D. L., Epstein, R. I., Arnett, W. D., Schramm, D. N. (1976): Magnetohydrodynamic phenomena in collapsing stellar cores. ApJ **204**, 869–878

486 Meyers, W., Swiatecki, W. (1966): Nuclear masses and deformations. NP **81**, 1–60

487 Migdal, A. B., Chernoutsan, A. I., Mishustin, I. N. (1979): Pion condensation and dynamics of neutron stars. Phys. Lett. **83B**, 158–160

487a Mihalas, D., Däppen, W., Hummer, D. G. (1988): The equation of state for stellar envelopes. II. Algorithm and selected results. ApJ **331**, 815–825

487b Miller, D. S., Wilson, J. R., Mayle, R. W. (1993): Convection above the neutrinosphere in Type II supernovae. ApJ **415**, 278–285

488 Mitler, H. (1977): Thermonuclear ion–electron screening at all densities. I. Static solution. ApJ **212**, 513–532

489 Morton, D. (1967): Mass loss from three OB supergiants in Orion. ApJ **150**, 535–542

490 Miller, E., Hillebrandt, W. (1979): A magnetohydrodynamical supernova model. AA **80**, 147–154

491 Moss, D. (1973): Models for rapidly rotating pre-main-sequence stars. MN **161**, 225–237

492 Moss, D. (1977): Magnetic star models: toroidal fields and circulation. MN **178**, 51–59

493 Moss, D. (1984): Time-dependent models of rotating magnetic stars. MN **209**, 607–639

494 Muchotrzeb, B. (1983): Transonic accretion flow in a thin disk around a black hole II. Acta Astron. **33**, 79–87

495 Muchotrzeb, B., Pachinski, B. (1982): Transonic accretion flow in a thin disk around a black hole. Acta Astron. **32**, 1–11

495a Muller, R. A., Newberg, H. J. N., Pennypacker, C. R., Perlmutter, S., Sasseen, T. P., Smith, C. K. (1992): High rate for type Ic supernovae. ApJ Lett. **384**, L9–L13

496 Munakata, H., Kohyama, Y., Itoh, N. (1985): Neutrino energy loss in stellar interior. ApJ **296**, 197–203

497 Myra, E. S., Bludman, S. A., Hoffman, Y., Lichenstadt, I., Sack, N., van Riper, K. A. The effect of neutrino transport on the collapse of iron stellar cores. ApJ **318**, 744–759

498 Nadyozhin, D. K. (1977): The collapse of iron-oxygen stars: physical and mathematical formulation of the problem and computational methods. ApSS **49**, 399–425

499 Nadyozhin, D. K. (1977): Gravitational collapse of iron cores with masses 2 and 10 $M_\odot$. ApSS **51**, 284–302

500 Nadyozhin, D. K. (1978): The neutrino radiation for a hot neutron star formation and the envelope outburst problem. ApSS **53**, 131–153

501 Nakazawa, K. (1973): Effect of electron capture on temperature and chemical composition in collapsing dense stars. PTP **49**, 1932–1946

502 Nakazawa, K., Hayashi, C., Takahara, M. (1976): Isothermal collapse of rotating gas clouds. PTP **56**, 515–530

503 Nakazawa, K., Murai, T., Hoshi, R., Hayashi, C. (1970): Effect of electron capture on the temperature in dense stars. PTP **44**, 829–830

503* Narayan, R., Yi, I. (1995): Advection - dominated accretion: underfed black holes and neutron stars. ApJ **452**, 710–735

503a Nayfeh, A. (1981): *Perturbation methods* (New York, Wiley)

504 Negele, J., Vautherin, D. (1973): Neutron star matter at sub-nuclear densities. NP **A207**, 298–320

504a Neuforge, C. (1993): Low temperature Rosseland mean opacities. AA **274**, 818–820

505 Neugebauer, G. et al. (1984): The infrared astronomical satellite (IRAS) mission. ApJ Lett. **278**, L1–L6

505a Neuhäuser, R., Sterzik, M. F., Schmitt, J. H. M. M., Wichmann, R., Krautter, J. (1995): ROSAT survey observation of T Tauri stars in Taurus. AA **297**, 391–417

506 Newman, M. (1978): S-process studies: the exact solution. ApJ **219**, 676–689

506a Niemeyer, J. C., Hillebrandt, W. (1995): Microscopic instabilities of nuclear flames in Type Ia supernovae. ApJ **452**, 779–784

506b Niemeyer, J. C., Hillebrandt, W. (1995): Turbulent nuclear flames in Type Ia supernovae. ApJ **452**, 762–778

507 Nomoto, K. (1982): Accreting white dwarf models for type I supernovae II. Off-center detonation supernovae. ApJ **257**, 780–792

508 Nomoto, K. (1986): Neutron star formation in theoretical supernovae – low mass stars and white dwarfs, in *Proc. symp. IAU No. 125 The origin and evolution of neutron stars*, ed. by D. Helfand, J. Huang. (Dordrecht, Reidel)

508a Nomoto, K., Iwamoto, K., Suzuki, T., Yamaoka, H., Hashimoto, M., Höfflich, P., Pols, O. R., van den Heuvel, E. P. J. (1995): Type Ib/Ic/IIb/II-L supernovae. Preprint No. 95-13, University of Tokyo

509 Nomoto, K., Thielemann, F. -K., Wheller, J. C. (1980): Explosive nucleosynthesis and type I supernovae. ApJ Lett. **279**, L23–L26

510 Nomoto, K., Thielemann, F. -K., Miyaji, S. (1985): The triple alpha reaction at low temperatures in accreting white dwarfs and neutron stars. AA **149**, 238–245

511 Nomoto, K., Tsuruta, S. (1981): Cooling of young neutron stars and the Einstein X-ray observations. ApJ Lett. **250**, L19–L23

512 Nomoto, K., Tsuruta, S. (1987): Cooling of neutron stars: effects of finite scale of thermal conduction. ApJ **312**, 711–726

513 Nordlung, A. (1974): On convection in stellar atmospheres. AA **32**, 407–422

514 Norman, M. L., Wilson, J. R., Barton, R. T. (1980): A new calculation on rotating protostellar collapse. ApJ **239**, 968–981

515 Novikov, I. D., Thorne, K. S. (1973): Astrophysics of black holes. In Black holes, ed. by B. and C. De Witt (Gordon and Breach), pp. 343–561

515* O'Connell, R. F., Matese, J. J. (1969): Effect of a constant magnetic field on the neutron beta decay rate and its astrophysical implications. Nature **222**, 649–650

515a Ogata, Sh., Ichimaru, S., Van Horn, H. M. (1993) Thermonuclear reaction rates for dense binary–ionic mixtures. ApJ **417**, 265–272

516 Ohnishi, T. (1983): Gravitational collapse of rotating magnetized stars. Tech. Rep. Inst. At. En. Kyoto Univ. No. 198

517 Oppenheimer, J., Volkoff, G. (1939): On massive neutron cores. PR **55**, 374–381

518 Ostriker, J., Gunn, J. (1971): Do pulsars make supernovae? ApJ Lett. **164**, L95–L104

519 Ostriker, J., Mark, J. (1968): Rapidly rotating stars. I. The self-consistent-fluid method. ApJ **151**, 1075–1088

519a Owocki, S. P., Castor, J. I., Rybicki, G. B. (1988): Time-dependent models of radiatively driven stellar winds. I. Nonlinear evolution of instabilities for a pure absorption model. ApJ **335**, 914–930

520 Paczynski, B. (1969): Envelopes of red supergiant. Acta Astron. **19**, 1–22

521 Paczynski, B. (1970): Evolution of single stars. I. Stellar evolution from main sequence to white dwarf or carbon ignition. Acta Astron. **20**, 47–58

522 Paczynski, B. (1970): Evolution of single stars. II. Core helium burning in population I stars. Acta Astron. **20**, 195–212

523 Paczynski, B. (1970): Evolution of single stars. III. Stationary shell source. Acta Astron. **20**, 287–309

524 Paczynski, B. (1971): Evolution of single stars. V. Carbon ignition in population I stars. Acta Astron. **21**, 271–288

525 Paczynski, B. (1971): Evolution of single stars. VI. Model nuclei of planetary nebulae. Acta Astron. **21**, 471–435

526 Paczynski, B. (1972): Carbon ignition in degenerate stellar cores. Ap. Lett. **11**, 53–55

527 Paczynski, B. (1972): Linear series of stellar models. I. Thermal stability of stars. Acta Astron. **22**, 163–174

528 Paczynski, B. (1974): Evolution of stars with $M \leq 8\,M_\odot$, in *Proc. Symp IAU No. 66 Late Stages of Stellar Evolution*, ed. by R. Tayler (Dordrecht, Reidel) pp. 62–69

529 Paczynski, B. (1974): Helium flush in population I stars. ApJ **192**, 483–485

530 Paczynski, B. (1975): Core mass-interflash period relation for double-shell source stars. ApJ **202**, 558–560

531 Paczynski, B. (1977): Helium shell flashes. ApJ **214**, 812–818

532 Paczynski, B. (1983): Models of X-ray bursters with radius expansion. ApJ **267**, 315–321

533 Paczynski, B., Bisnovatyi-Kogan, G. S. (1981): A model of a thin accretion disk around a black hole. Acta Astron. **31**, 283–291

534 Paczynski, B., Schvarzenberg-Czerny, A. (1980): Disk accretion in U Geminorum. Acta Astron. **30**, 127–141

535 Paczynski, B., Wiita, P. (1980): Thick accretion disks and supercritical luminosities. AA **88**, 23–31

536 Paczynski, B., Ziolkovski, J. (1968): On the origin of planetary nebulae and Mira variables. Acta Astron. **18**, 255–266

536a Palla, F., Stahler, S. W. (1993): The pre-main-sequence evolution of intermediate-mass stars. ApJ **418**, 414–425

536b Pallavicini, R. (1989): X-ray emission from stellar coronae. AA Rev. **1**, 177-207

537 Pandharipande, V. (1971): Dense neutron matter with realistic interaction. NP **A174**, 641–656

538 Pandharipande, V., Pines, D., Smith, R. (1976): Neutron star structure: theory, observation and speculation. ApJ **208**, 550–566

539 Papaloizou, J. C. B. (1973): Nonlinear pulsations of upper main sequence stars. I. A perturbation approach. MN **162**, 143–168

540 Papaloizou, J. C. B. (1973): Nonlinear pulsations of upper main sequence stars. II. Direct numerical investigations. MN **162**, 169–187

541 Papaloizou, J. C. B., Whelan, J. A. J. (1973): The structure of rotating stars: the J^2 method and results for uniform rotation. MN **164**, 1–10

541a Park, M. -G. (1990): Self-consistent models of spherical accretion onto black holes. I. One-temperature solutions. ApJ **354**, 64–82

541b Park, M. -G. (1990): Self-consistent models of spherical accretion ontoblack holes. II. Two-temperature solutions with pairs. ApJ **354**, 83–97

542 Patterson, J. (1984): The evolution of cataclysmic and low-mass X-ray binaries. ApJ Suppl. **54**, 443–493

543 Petrosian, V., Beaudet, G., Salpeter, E. E. (1967): Photoneutrino energy loss rates. PR **154**, 1445–1454

543a Poe, C. H., Owocki, S. P., Castor, J. I. (1990): The steady state solutions of radiatively driven stellar winds for a non-Sobolev, pure absorption model. ApJ **358**, 199–213

543b Pogorelov, N. V., Ohsugi, Y., Matsuda, T. (2000): Towards steady-state solutions for supersonic wind accretion on to gravitating objects. MN **313**, 198–208

544 Pollock, E. L., Hansen, J. P. (1973): Statisticsl mechanics of dense ionized matter. II. Equilibrium properties and melting transition of the crystallized one-component plasma. PR **8A**, 3110–3122

544*+ Podsiadlowski, P., Cannon, R. C., Rees, M. J. (1995): The evolution and final fate of massive Thorne–Zytkow objects. MN **274**, 485–490

544* Pols, O. R., Tout, Ch. A., Eggleton, P. P., Han, Zh. (1995): Approximate input physics for stellar modeling. MN **274**, 964–974

544a Pontecorvo, B., Bilenky, S. (1987): Neutrino today. Preprint JINR, Dubna, No. E1, 2-87-567

544b Pringle, J. E., Rees, M. J. (1972): Accretion disc models for compact X-ray sources. AA **21**, 1–9

545 Radhakrishnan, V. (1982): On the nature of pulsars. Contemp. Phys. **23**, 207–231

546 Raikh, M. E., Yakovlev, D. G. (1982): Thermal and electrical conductivities of crystals in neutron stars and degenerate dwarfs. ApSS **87**, 193–203

547 Ramsey, W. H. (1950): On the stability of small planetary cores (I). MN **110**, 325–338

548 Ravenhall, D., Bennett, C., Pechick, C. (1972): Nuclear surface energy and neutron-star matter. PR Lett. **28**, 978–981

549 Reimers, D. (1981): Winds in red giants, in *Physical processes in red giants*, ed. by I. Iben, A. Renzini (Dordrecht, Reidel), 269–284

550 Regev, O., Lilio, M. (1985): X-ray bursters – hot way to chaos, in *Chaos in astrophysics*, ed. by J. M. Perdang, J. R. Buchler, E. A. Spiegel. (Dordrecht, Reidel)

550* Regev, O. (1983): The disk-star boundary layer and its effect on the accretion disk structure. AA **126**, 146–151

550** Riffert, H., Herold, H. (1995): Relativistic accretion disk structure revisited. ApJ **408**, 508–511

550a Robnik, M., Kundt, W. (1983): Hydrogen at high pressures and temperatures. AA **120**, 227–233

551 Rosenfeld, L. (1974): Astrophysics and gravitation, in *Proc. 16 Solvay conf. on phys*, Univ. de Bruxells, p. 174

551a Rogers, F. J., Iglesias, C. A. (1992): Rosseland mean opacities for variable compositions. ApJ **401**, 361–366

552 Rose, W., Smith, R. (1970): Final evolution of a low-mass star I. ApJ **159**, 903–912

553 Roth, M., Weigert, A. (1972): Example of multiple solutions for equilibrium stars with helium cores. AA **20**, 13–18

553* Roxburgh, I. W. (1989): Integral constraints on convective overshooting. AA **211**, 361–364

553** Roxburgh, I. W. (1992): Limits on convective penetration from stellar cores. AA **266**, 291–293

553*** Roxburgh, I. W., Simmons, J. (1993): Numerical studies of convective penetration in plane parallel layers and the integral constraint. AA **277**, 93–102

553[4*] Roxburgh, I. W., Vorontsov, S. V. (1994): Seismology of the stellar cores: a simple theoretical discription of the 'small frequency separations'. MN **267**, 297–302

553[5*] Roxburgh, I. W., Vorontsov, S. V. (1994): The asymptotic theory of stellar acoustic oscillations: fourth-order approximation for low-degree modes. MN **268**, 143–158

553[6*] Roxburgh, I. W., Vorontsov, S. V. (1996): An asymptotic description of solar acoustic oscillations of low and intermediate degree. MN **278**, 940–946

553[7*] Roxburgh, I. W., Vorontsov, S. V. (1995): An asymptotic description of solar acoustic oscillations with an elementary exitation source. MN **272**, 850–858

553a Rubbia, C., Jacob, M. (1990): The Z^0. American Scientist **78**, 502–519

553b Rüdiger, G. (1989): *Differential rotation and stellar convection, Sun and solar-type stars* (Berlin, Akademie-Verlag)

553b* Rybicki, G. V., Owocki, S. P., Castor, J. I. (1990): Instabilities in line-driven stellar winds. IV. Linear perturbations in three dimensions. ApJ **349**, 274–285

553c Sager, R., Piskunov, A. E., Myakutin, V. I., Joshi, U. C. (1986): Mass and age distributions of stars in young open clusters. MN **220**, 383–403

554 Sakashita, S., Hayashi, C. (1959): Internal structure and evolution of very massive stars. PTP **22**, 830–834

554a Salgado, M., Bonazzola, S., Gourgoulhon, E., Haensel, P. (1994): High precision rotating neutron star models. AA **291**, 155–170

555 Salpeter, E. E. (1961): Zero temperature plasma. ApJ **134**, 669–682

556 Salpeter, E. E. (1964): Accretion of interstellar matter by massive objects. ApJ **140**, 796–799

557 Salpeter, E. E., Van Horn, H. M. (1965): Nuclear reaction rates at high densities. ApJ **155**, 183–202

558 Salpeter, E. E., Zapolsky, H. S. (1967): Theoretical high-pressure equations of state including correlation energy. PR **158**, 876–886

559 Sampson, D. H. (1959): The opacity at high temperatures due to Compton scattering. ApJ **129**, 734–751

560 Sato, K. (1979): Nuclear composition in the inner crust of neutron stars. PTP **62**, 957–968

561 Scalo, J. (1981): Observations and theories of mixing in red giants, in *Physical processes in red giants*, ed. by I. Iben, A. Renzini. (Dordrecht, Reidel) pp. 77–114 PTP **62**, 957–968

561a Schaller, G., Schaerer, G., Meynet, G., Maeder, A. (1992): New grids of stellar models from 0. 8 to 120 $M_\odot$ at Z=0. 02 and Z=0. 001. AA Suppl. **96**, 269–331

561b Schatzman, E. (1993): Transport of angular momentum and diffusion by the action of internal waves. AA **279**, 431–446

561c Schatzman E. (1995): Solar neutrino and transport processes. Proc. Symp. *Physical Processes in Astrophysics*, ed. by I. W. Roxburgh, J. -L. Masnou (Berlin, Heidelberg, Springer), pp. 171–184

561d Schatzman, E., Praderie, F. (1993):*The stars* (Berlin, Heidelberg, Springer)

562 Schinder, P. et al. (1987): Neutrino emission by the pair, plasma and photoprocesses in the Weinberg–Salam model. ApJ **313**, 531–542

562a Schmidt-Kaler, Th. (1982): Physical parameters of the stars. Landholt-Börnstein. Astronomy and Astrophysics **2**, 1–35; 449–456

563 Schonberner, D. (1979): Asymptotic giant branch evolution with steady mass loss. AA **79**, 108–114

564 Schonberner, D. (1981): Late stages of stellar evolution: central stars of planetary nebulae. AA **103**, 119–130

565 Schonberner, D. (1981): Late stages of stellar evolution II. Mass loss and the transition of asymptotic giant branch into hot remnant. ApJ **272**, 708–714

566 Schonberner, D. (1986): Late stages of stellar evolution III. The observed evolution of central stars of planetary nebulae. AA **169**, 189–193

567 Schramm, D., Wagoner, R. (1977): Element production in the early Universe. Ann. Rev. Nucl. Sci. **27**, 37–74

567a Schwarzschild, B. (1994): Anomalous cosmic-ray data suggest oscillation between neutrino flavors. Physics Today **No. 10**, 22–24

567b Schwarzschild, B. (1995): Chromium surrogate Sun confirms that solar neutrinos really are missing. Physics Today **No. 4**, 19–21

568 Schwarzschild, M., Harm, R. (1959): On the maximum mass of stable stars. ApJ **129**, 637–646

569 Schwarzschild, M., Harm, R. (1962): Red giants of population II. II. ApJ **139**, 158–165

570 Schwarzschild, M., Harm, R. (1965): Thermal instability in non-degenerate stars. ApJ **142**, 855–867

571 Schwarzschild, M., Harm, R. (1967): Hydrogen mixing by helium-shell flashes. ApJ **150**, 961–970

572 Schwarzschild, M., Harm, R. (1973): Stability of the Sun against spherical thermal perturbations. ApJ **184**, 5–8

573 Schwarzschild, M., Sebberg, H. (1962): Red giants of population II. I. ApJ **136**, 150–157

574 Seeger, P., Fowler, W., Clayton, D. (1965): Nucleosynthesis of heavy elements by neutron capture. ApJ Suppl. **11**, 121–166

575 Shakura, N. I., Sunyaev, R. A. (1973): Black holes in binary systems. Observational appearance. AA **24**, 337–355

575a Shapiro, S. L., Lightman, A. P., Eardley, D. M. (1976): A two-temperature accretion disc model for Cygnus X-1: structure and spectrum. ApJ **204**, 187-199

576 Shaviv, G., Salpeter, E. E. (1973): Convective overshooting in stellar interior models. ApJ **184**,191–200

576a Shi, X., Schramm, D. N., Dearborn, D. S. P. (1994): On the solar model solution to the solar neutrino problem. Phys. Rev. **D50**, 2414–2420

577 Shima, E., Matsuda, T., Takeda, H., Sawada, K. (1985): Hydrodynamic calculations on axisymmetric accretion flow. MN **217**, 367–386

577a Shimizu, T., Yamada, S., Sato, K. (1994): Axisymmetric neutrino radiation and the mechanism of supernova explosions. ApJ Lett. **432**, L119–L122

578 Slattery, W. L., Doolen, G. D., De Witt, H. E. (1982): N dependence on the classical one-component plasma Monte-Carlo calculations. PR A **A26**, 2255–2258

579 Smak, J. (1976): Eruptive binaries VI. rediscussion of U Geminorum. Acta Astron. **26**, 277–300

580 Smak, J. (1984): Accretion in cataclysmic binaries IV. Accretion disks in dwarf novae. Acta Astron. **34**, 161–189

581 Sofia, S., Chau, K. (1984): Turbulent compressible convection in a deep atmosphere II. Two-dimensional results for main sequence A5 and F0 type envelopes. ApJ **282**, 550–556

581⁺ Spiegel, E. A. (1963): A generalization of the mixing-length theory of turbulent convection. ApJ **138**, 216–225

581a Spiegel, E. A., Zahn, J.-P. (eds.) (1976): *Problems of stellar convection.* Proc. IAU Coll. **No. 38** (Berlin, Springer)

581b Spruit, H. (1992): The rate of mixing in semiconvective zones. AA **253**, 131–138

582 Sramek, R., Panagia, N., Weiler, K. (1984): Radio emission from type I supernova SN 1983. 51 in NGC 5236. ApJ Lett. **285**, L59–L62

583 Stahler, S. (1983): The birthline for low-mass stars. ApJ **274**, 822–829

584 Stahler, S., Shu, F., Taam, R. (1980): The envelope of protostars. I. Global formulation and results. ApJ **241**, 637–654

585 Stahler, S., Shu, F., Taam, R. (1980): The envelope of protostars. II. The hydrostatic core. ApJ **242**, 226–241

586 Stahler, S., Shu, F., Taam, R. (1981): The envelope of protostars. III. The accretion envelope. ApJ **248**, 727–737

586* Stein, J., Barkat, Z., Wheeler, J. C. (1999): The role of kinetic energy flux in the convective URCA process. ApJ **523**, 381–385

586a Stothers, R. B. (1992): Upper limit to the mass of pulsationally stable stars with uniform chemical composition. ApJ **392**, 706–709

587 Stothers, R., Chin, C. (1973): Stellar evolution at high mass based on the Ledoux criterion for convection. ApJ **179**, 555–568

588 Stothers, R., Chin, C. (1979): Stellar evolution at high masses including the effects of a stellar wind. ApJ **233**, 267–279

589 Stothers, R., Chin, C. (1985): Stellar evolution at high mass with convective core overshooting. ApJ **292**, 222–227

589* Stothers, R. B., Chin, C. (1993): Dynamical instability as the cause of the massive outburst in Eta Carinae and other luminous blue variables. ApJ Lett. **408**, L85–L88

589a Stothers, R., Chin, C. (1993): Iron and molecular opacities and the evolution of population I stars. ApJ **412**, 294–300

589b Stothers, R., Chin, C. (1994): Galactic stars applied to tests of the criterion for convection and semiconvection in an inhomogeneous star. ApJ **431**, 797–805

590 Strom, S. E., Strom, K., Rood, R. T., Iben, I. (1970): On the evolutioonary status of stars above the horizontal branch in globular clusters. AA **8**, 243–250

591 Sigimoto, D. (1964): Helium flash in less massive stars. PTP **32**, 703–725

592 Sigimoto, D. (1970): On the numerical stability of computation of stellar evolution. ApJ **159**, 619–628

593 Sigimoto, D., Nomoto, K. (1980): Presupernova models and supernovae. Space Sci. Rev. **25**, 155–227

594 Sigimoto, D., Nomoto, K., Eriguchi, Y. (1981): Stable numerical method of computations of stellar evolution. Suppl. PTP **70**, 115–131

595 Sigimoto, D., Yamamoto, Y. (1966): Second helium flash and an origin of carbon stars. PTP **36**, 17–36

596 *Supernova 1987A*, in *Proc. workshop ESO*, July 1987.

597 Sweigart, A. (1971): A method for suppression of the thermal instability in helium-shell burning stars ApJ **168**, 79–97

598 Sweigart, A. (1973): Initial asymptotic branch evolution of population II stars. AA **24**, 459–464

599 Sweigart, A., Mengel, J., Demarque, P. (1974): On the origin of the blue halo stars. AA **30**, 13–19

600 Sunyaev, R. A., Titarchuk, L. G. (1980): Comptonization of X-rays in plasma clouds. Typical radiation spectra. AA **86**, 121–138

601 Sztajno, M., et al. (1986): X-ray bursts from GX 17+2, a new approach. MN **222**, 499–511

602 Talbot, R. J. (1971): Nonlinear pulsations of unstable massive main-sequence stars I. Small-amplitude tests of an approximation technique. ApJ **163**, 17–27

603 Talbot, R. J. (1971) Nonlinear pulsations of unstable massive main-sequence stars II. Finite-amplitude stability. ApJ **165**, 121–138

603a Tassoul, M. (1980): Asymptotic approximations for stellar nonradial pulsations. ApJ Suppl. **43**, 469–490

604 Tassoul, J. -L., Tassoul, M. (1984): Meridional circulation in rotating stars VIII. The solar spin-down problem. ApJ **286**, 350–358

604a Taylor, J. H. (1992): Pulsar timing and relativistic gravity. Philos Trans. R. Soc. Lond. A. **341**, 117–134

604b Taylor, J. H., Manchester, R. N., Lyne, A. G. (1993): Catalog of 558 pulsars. ApJ Suppl. **88**, 529–568

605 Taylor, J. H., Weisberg, J. M. (1982): A new test of general relativity: gravitational radiation and the binary pulsar PSR 1913+16. ApJ **253**, 908–920

605a Taylor, J. H., Weisberg, J. M. (1989): Further experimantal tests of relativistic gravity using the binary pulsar PSR 1913+16. ApJ **349**, 434–450

605a* Terebey, S., Chandler, C. J., Andre, P. (1993): The contribution of disks and envelopes to the millimeter continuum emission from very low-mass stars. ApJ **414**, 759–772

605b Thorne, K. S., Zytkow, A. N. (1977): Stars with degenerate neutron cores. I. Structure of equilibrium models. ApJ **212**, 832–858

606 Tohlins, J. E. (1980) Ring formation in rotating protostellar clouds. ApJ **236**, 160–171

606a Tomczuk, S., Schou, J., Thompson, M. J. (1995): Measurement of the rotational rate in deep solar interior. ApJ Lett. **448**, L57–L60

607 Tomonaga, S. (1938): Innere Reibung und Warmeleitfahigkeit der Kernmaterie. Zeit. Phys. **110**, 573–604

607a Toomre, J. (1993): Thermal convection and penetration. *Proc. Les Houches 1987*, ed. by J.-P. Zahn and J. Zinn-Justin

607b Tooper, R. F. (1969): On the equation of state of a relativistic Fermi-Dirac gas at high temperatures. ApJ **156**, 1075–1100

608 Trautvetter, H., et al. (1983): The Ne-Na cycle and the $C^{12} + \alpha$ reaction. Proc. Second Workshop on Nuclear Astrophysics. Preprint MPA, No. 90, pp 24–33

609 Trimble, V. (1975) The origin and abundances of chemical elements. Rev. Mod. Phys. **47** 877–976

609a Trimble, V. (1991): The origin and abundances of the chemical elements revisited. AA Rev. **3**, 1–46

610 Tscharnuter, W. (1975) On the collapse of rotating protostars. AA **39** 207–212

611 Uehling, E. A., Uhlenbeck, G. E.: Transport phenomena in Einstein–Bose and Fermi–Dirac gases. I. (1933): PR **43** 552–561; II. (1934): PR **46** 917–929

612 Ulrich, R. (1976): A nonlocal mixing-length theory of convection for use in numerical calculations. ApJ **207** 564–573

612a Umeda, H., Nomoto, K., Tsuruta, S., Muto, T., Tatsumi, T. (1994): Neutron star cooling and pion condensation. ApJ **431** 309–320

613 Unno, W. (1981): Development of the stellar convection theory. Suppl. PTP No. 70, 101–114

614 Unno, W., Kondo, M. (1976): The Eddington approximation generalized for radiative transfer in spherically symmetric systems I. Basic method. Publ. Astron. Soc. Japan **28** 347–354

615 Unno, W., Kondo, M. (1977): The Eddington approximation generalized for radiative transfer in spherically symmetric systems II. Non-gray extended dust-shell models. Publ. Astron. Soc. Japan **29** 693–710

616 Unno, W., Osaki, Y., Ando, H., Shibahashi, H. (1989): *Nonradial Oscillations of Stars* (Tokyo, Tokyo University Press)

617 Upton, I. K. L., Little, S. J., Dworetsky, M. M. (1968): Dynamical stability in pre-main-sequence stars. ApJ **154**, 597–611

618 Vanbeveren, D. (1980): Evolution with mass loss: massive stars, massive binaries. PhD Thesis (Brussels, Brussels University)

619 Van den Hulst, J., et al. (1983): Radio discovery of a young supernova. Nature **306**, 566–568

620 Van Horn, H. M. (1968): Crystallization of white dwarfs. ApJ **151**, 227–238

621 Vardya, M. S. (1960): Hydrogen–Helium adiabats for late type stars. ApJ Suppl. **4**, 281–336

622 Vardya, M. S. (1965): Thermodynamics of a solar composition gaseous mixture. MN **129**, 205–213

622b Wagenhuber, J., Weiss, A. (1994): Termination of AGB-evolution by hydrogen recombination. AA **290**, 807–814

623 Wagoner, R. (1969): Synthesis of the elements within objects exploding from very high temperatures. ApJ Suppl. **18**, 247–296

624 Wallace, R. K., Woosley, S. E. (1981): Explosive hydrogen burning. ApJ Suppl. **45**, 389–420

625 Weaver, T., Woosley, S. (1980): Evolution and explosion of massive stars. Ann. New-York Acad. Sci. **336**, 335–357

626 Weaver, T., Zimmerman, G., Woosley, S. (1978): Presupernova evolution of massive stars. ApJ **225**, 1021–1029

627 Weigert, A. (1966): Sternentwicklung VI. Entwicklung mit neutrinoverlusten und thermische pulse der Helium-Schalenquelle bei einem Stern von 5 sonnenmassen. Zeit. Astrophys. **64**, 395–425

628 Weir, A. D. (1976): Axisymmetric convection in rotating sphere. Part I. Stress-free surface. J. Fluid Mech. **75**, 49–79

629 Wendell, C. E., Van Horn, H. M., Sargent, D. (1987): Magnetic field evolution in white dwarfs. ApJ **313**, 284–297

630 Westbrook, Ch., Tarter, B. (1975): On protostellar evolution. ApJ **200**, 48–60

631 Wiedemann, V. (1981): The initial/final mass relation for stellar evolution with mass loss, in *Effects of Mass Loss on Stellar Evolution*, ed. C. Chiosi, R. Stalio (Dordrecht, Reidel) pp. 339–349

632 Wilson, L. A. (1975): Fe I fluorescence in T Tauri stars II. Clues to the velocity field in the circumstellar envelopes. ApJ **197**, pp. 365–370

633 Wilson, J. R., Mayle, R., Woosley, S., Weaver, T. (1986): Stellar core collapse and supernova. Ann. New York Acad. Sci. **470**, 267–293

633* Wilson, J. R., Mayle, R. W. (1993): Report on the progress of supernova research by the Livermore group. Phys. Rep. **227**, 97–111

633a Wolszczan, A. (1991): A nearby 39.7 ms radio pulsar in a relativistic binary system. Nature **350**, 688–690

634 Wood, P. R. (1974): Dynamical models of asymptotic-giant-branch stars, in *Proc. IAU Symp. No. 59: Stellar Instability and Evolution*, ed. by P. Ledoux, et al. (Dordrecht, Reidel), pp. 101–102

635 Wood, P. R. (1979): Pulsation and mass loss in Mira variables. ApJ **227**, 220–231

636 Wood, P. R., Faulkner, D. J. (1986): Hydrostatic evolutionary sequences for the nuclei of planetary nebula. ApJ **307**, 659–674

637 Woodward, P. (1978): Theoretical models of star formation. Ann. Rev. Astron. Ap. **16**, 555–584

638 Woosley, S., Fowler, W., Holmes, J., Zimmerman, B. (1978): Semiempirical thermonuclear reaction rate data for intermediate mass nuclei. At. Data Nucl. Data Tables **22**, 371–441

639 Woosley, S., Fowler, W., Holmes, J., Zimmerman, B. (1975): Tables of thermonuclear reaction rate data for intermediate mass nuclei. Preprint OAP, No. 422, pp. 1-1-5, A1–A179

639a Woosley, S. E., Langer, N., Weaver, T. A (1995): The presupernova evolution and explosion of helium stars that experience mass loss. ApJ **448**, 315-338

640 Woosley, S., Weaver, T. (1986): Theoretical models for type I and type II supernovae, in *Nucleosynthesis and its implications on nuclear and particle physics* ed. J. Audouze, N. Mathieu (Dordrecht, Reidel), pp. 145–166

641 Wynn-Williams, C. (1982): The search for infrared protostar. Ann. Rev. Astron. Ap. **20**, 587–618

641a Young, E. J., Chanmugam, G. (1995): Postaccretion magnetic field evolution of neutron stars. ApJ Lett. L53–L56

642 Yorke, H. (1979): The evolution of protostellar envelopes of masses $3\,M_\odot$ and $10\,M_\odot$. I. Structure and hydrodynamic evolution. AA **80**, 308–316

643 Yorke, H. (1979): The evolution of protostelar envelopes of masses $3\,M_\odot$ and $10\,M_\odot$. II. Radiation transfer and spectral apppearance. AA **85**, 215–220

644 Yorke, H. (1981): Protostars and their evolution, in *Proc. ESO Conf. on Scientific importance of high angular resolution of infrared and optical wavelength* (Garching), pp. 319–340

645 Yorke, H., Krugel, H. (1977): The dynamical evolution of massive proto-stellar clouds. AA **54**, 183–194

646 Yorke, H., Shustov, B. M. (1981): The spectral appearance of dusty proto-stellar envelopes. AA **98**, 125–132

646a Zahn, J.-P. (1991): Convective penetration in stellar interiors. AA **252**, 179–188

646b Zahn, J.-P. (1992): Circulation and turbulence in rotating stars. AA **265**, 115–132

647 Zapolsky, H. S., Salpeter, E. E. (1969): The mass-radius relations for cold spheres of low mass. ApJ **158**, 809–813

648 Zhevakin, S. A. (1963): Physical basis of the pulsation theory of variable stars. Ann. Rev. Astron. Ap. **1**, 367–400

648a Zhou, Sh., Evans, N. J., Kömpe, C. and Walmsley, C. M. (1993): Evidence for protostellar collapse in B335. ApJ **404**, 232–246

649 Ziebarth, K. (1970): On the upper mass limit for main sequence stars. ApJ **162**, 947–962

650 Zimmerman, B., Fowler, W., Caughlan, G. (1975): Tables of thermonuclear reaction rates. Preprint OAP, No. 399, pp. 1–35

651 Ziolkowski, J. (1972): Evolution of massive stars. Acta Astron. **22**, 327–374

652 Żytkov, A. (1972): On the stationary mass outflow from stars I. The computational method and results for 1 $M_\odot$ star. Acta Astron. **22**, 103–139

653 Żytkov, A. (1973): On the stationary mass outflow from stars II. The results for 30 $M_\odot$ star. Acta Astron. **23**, 121–134

List of Symbols and Abbrevations

Here, only global symbols used throughout the book are indicated. Some of the notations may be used for other variables in different parts of the book, where they are defined.

Latin Symbols

A, A_i: atomic mass of a nucleus
$\mathbf{A}$, A_i: vector potential of electromagnetic field
a: constant of radiation energy density
a: average distance between ions
a_0: Bohr radius
a_Z: atomic radius in the Thomas–Fermi model
$\mathbf{B}$, B_i: vector of the magnetic field strength
$B_c = m_e^2 c^3 / e\hbar$: critical value of the magnetic field
$B_{A,Z}$: binding energy of a nucleus with atomic mass A and charge Z
B_n: binding energy per nucleon
$B(T)$: energy density of an equilibrium radiation
$B_\nu(T)$: spectral intensity of equilibrium Planck radiation
c: speed of light in vacuum
c_p, C_p: heat capacity at constant pressure
c_v, C_v: heat capacity at constant volume
d, d_{over}: overshooting length
d_{pc}: distance to a star in parsecs
ds: interval of space-time
dV, dv: element of a volume
$d\Omega$: element of a solid angle
$\mathcal{D}(x)$: Debye function
E: specific energy of matter
E_b: binding energy per nucleon
E_e, E_{e^-}: specific energy of electrons in matter
E_{e^+}: specific energy of positrons in matter
E_n: specific energy of neutrons in matter
E_N: specific energy of nuclei in matter

E_{conv}: specific energy of convective motion

E_{zp}: zero-point energy of three-dimensional oscillator

E_{ν_e}: specific energy of electron neutrinos in thermodynamic equilibrium

$E_{\bar{\nu}_e}$: specific energy of electron antineutrinos in thermodynamic equilibrium

$\mathbf{E}$, E_i: electrical field strength

E_M: magnetic energy per unit volume

$\tilde{E}$: internal energy per baryon

e: electrical charge of the electron

$e \equiv e(r)$: energy of a star within radius r

F: specific free energy of matter

F_i: vector of radiative energy flux density

F_{conv}: convective energy flux density

F_{rad}: radiative energy flux density

$F_0(u)$: Fermi function of beta decay

f: particle distribution function

f_e: electron distribution function

f_ν: photon distribution function

f_ν: rate of neutrino emission losses per unit mass

G: gravitational constant

G_W: coupling constant of the weak interaction

$G_V = G_W$: constant of weak vector-type interaction

G_A: constant of weak axial-type interaction

$g_A = G_V/G_A$: relative constant of axial weak interaction

g: gravitational acceleration

$g_\odot$: gravitational acceleration at the surface of the Sun

g_{ef}: effective gravitational acceleration, including centrifugal acceleration

g_{ik}: metric tensor

$g_{A,Z}$: statistical weight of a nucleus with atomic mass number A and charge Z

g_0, g_1: statistical weights of nuclei entering a nuclear reaction

g_2, g_3: statistical weights of nuclei resulting from a nuclear reaction

g_{ij}: statistical weight of an element i in the ionization state j

g_n: statistical weight of a neutron

g_p: statistical weight of a proton

g_{bf}: Gaunt factor for bound–free transitions

g_{ff}: Gaunt factor for free–free transitions

H: matrix element of nuclear transformation

H: specific enthalpy of matter

H_β: matrix element of beta reaction

H_μ: matrix element of muon decay

H_n: matrix element of a neutron decay

H_p: pressure scale height

H_ρ: density scale height

$\hbar$: Planck constant divided by 2π

h: Planck constant

h: thickness of a layer; half-thickness of a disk

I: nuclear spin

I_{ij}: ionization energy (potential) of the j-th electron of element i

I_ν: spectral intensity of radiation

$I_{\nu e}$: spectral intensity of neutrino radiation

I_ϕ: azimuthal component of a surface electric current density

J_0: total angular momentum of a star

j: specific angular momentum of matter

j_i: vector of electrical current density

j_ν: spectral coefficient of emission per unit mass

K, K_1, K_2: coefficients in a polytropic equation of state

k: Boltzmann's constant

k: wavenumber of a turbulent vortex

k_s: diffusion coefficient in an ion binary gas mixture

k_T: coefficient of temperature diffusion

L: stellar luminosity

L: maximal scale of a turbulent vortex

L_B: magneto-bremstrahlung luminosity

L_{opt}: photon luminosity of a star

L_{cr}, L_c: Eddington critical stellar luminosity

L_{RG}: luminosity of a red giant star

L_r: radial energy flux from a star

L_t: stellar luminosity at the turning point off a main sequence

L_{k}: flux of kinetic energy from a star

L_m: stellar luminosity in a peak of a helium shell flash

L_ν: neutrino luminosity of a star

L_r^{rad}: radial heat flux due to radiation heat conductivity

L_{th}: thermal heat flux due to heat conductivity

L_{conv}, L_r^{conv}: radial energy flux due to convection

$L_\odot$: luminosity of the Sun

L_{H}: stellar luminosity due to hydrogen burning

L_{He}: stellar luminosity due to helium burning

L_{g}: stellar luminosity due to gravitational energy production

l: mixing length of a convective element

l: specific angular momentum of matter

l: current scale of a turbulent vortex

l_i: unit vector in the direction of photon motion

l_{in}: specific angular momentum on the inner boundary of an accretion disk

l_T: mean free path of a neutrino

l_{WZ}: radius of a Wigner–Seitz spherical cell

M: stellar mass

M: absolute stellar magnitude

M_0: rest mass of a star

M_i: initial mass of a star

M_e: mass of the hydrogen envelope of a star

M_n: mass of a neutron star

M_n: non-dimensional Lane–Emden stellar mass for polytropic index n

$M_\odot$: mass of the Sun

M_c, M_{core}: mass of the stellar core

M_{CHe}: mass of the helium core of a star

M_{CO}: mass of the carbon–oxygen stellar core

M_{Fe}: mass of the iron stellar core

M_{Ch}: Chandrasekhar limit of the stellar mass

$\dot{M}$: mass loss rate from a star, or mass flux into a star

M_{bol}: bolometric absolute stellar magnitude

M_0, M_1: masses of nuclei entering a nuclear reaction

M_2, M_3: masses of nuclei emerging from a nuclear reaction

M_{n}: matrix element of neutron decay

M_Z: matrix element of beta decay of a nucleus (A,Z)

m: Lagrangian mass coordinate in a spherically symmetric star

$m = M/M_\odot$: non-dimensional stellar mass

m_0: rest mass of a star within radius r

m: visual stellar magnitude

m_{ph}: photo-visual stellar magnitude

m_B: visual stellar magnitude in a filter B

m_U: visual stellar magnitude in a filter U

m_V: visual stellar magnitude in a filter V

$m_{A,Z}$: mass of a nucleus with atomic number Z and atomic mass A

m_{e}: electron mass

m_{n}: neutron mass

m_{p}: proton mass

m_i: mass of an atomic nuclei with mass number $A_i (\approx A_i m_{\mathrm{u}})$

m_{u}: atomic mass unit $= 1/12$ times the mass of the isotope ^{12}C

m_μ: muon mass

$m_{\nu_e 0}$, m_{ν_e}: rest mass of the electron neutrino

$m_{\nu_\mu 0}$, m_{ν_μ}: rest mass of the muon neutrino

$\dot{m} = \dot{M}c^2/L_c$: non-dimensional accretion mass flux

N: number of baryons in a star

N_{b}: total number of baryons in a neutron star

N_A: Avogadro number

N_{n}: total number density of neutrons (free and bound in nuclei)

N_{p}: total number density of protons (free and bound in nuclei)

n: number density of baryons

n: Landau level

n, n_1, n_2: polytropic indices

n_0, n_1: number densities of nuclei entering a reaction

n_2, n_3: number densities of nuclei emerging from a reaction

n_A: number density of atoms

$n_{A,Z}$: number density of nuclei with atomic number Z and atomic mass A

n_{b}: number density of baryons

n_{e}, $n_{\mathrm{e}-}$, n_-: number density of electrons

$n_{\mathrm{e}+}$, n_+: number density of positrons

n_{ij}: number density of ions of element i in ionization state j

n_{n}: number density of neutrons

n_{p}: number density of protons

P: pressure of matter

P_{e}, $P_{\mathrm{e}-}$: electron pressure

$P_{\mathrm{e}+}$: positron pressure

P_{n}: neutron pressure

P_g: gas pressure

P_r: radiation pressure

P_N: nucleus pressure

P_{ik}: pressure tensor of radiation field

P_{01}: reaction rate per unit volume

p: electron momentum

p_z: z component of electron momentum

p_{ei}: electron four-vector

$p_{\mu i}$: muon four-vector

p_{Fe}: Fermi momentum of electrons

p_{Fn}: Fermi momentum of neutrons

$\mathcal{P}$: pressure integrated over the accretion disk thickness

Q: energy release per nuclear reaction

Q_6: energy obtained as heat per nuclear reaction, expressed in MeV

Q_{pair}: energy loss rate per unit volume by neutrino emission produced due to electron–positron pair annihilation

Q_{tot}: energy release per nuclear reaction, including the energy of free outflowing neutrinos, expressed in MeV

Q_n: energy of a neutron strip from a nucleus

Q_p: energy of a proton strip from a nucleus

Q_{CN}, Q_{CNO}: heat produced during the formation of a helium nucleus in the CNO cycle of hydrogen burning

$Q_{3\alpha}$: heat produced during the formation of a ^{12}C nucleus in the 3α reaction of helium burning

$Q_{^{12}\mathrm{C}\alpha}$: heat produced during the formation of a ^{16}O nucleus in the α-capture reaction by ^{12}C

$Q_{^{16}\mathrm{O}\alpha}$: heat produced during the formation of a ^{20}Mg nucleus in the α-capture reaction by ^{16}O

q_i: heat flux density

R: radius of a star

R_g, $r_g = 2GM/c^2$: stellar gravitational radius

$R_\odot$: radius of the Sun

Re: non-dimensional Reynolds number

R_{cr}, r_{cr}: critical radius where the flow velocity equals the local sound velocity

r: radial coordinate

r_i: radius of an isothermal core

r_{in}: inner radius of an accretion disk

$r_{\mathcal{D}e}$: Debye radius for electron screening

$r_{\mathcal{D}i}$: Debye radius for ion screening

$\mathcal{R}$: gas constant

S: specific entropy of matter

S: total density of radiation energy

S_ν: spectral density of radiation energy

S_e, S_{e^-}: entropy of electrons

S_{e^+}: entropy of positrons

$s = 2S/3\mathcal{R}$: non-dimensional entropy

T: thermodynamic temperature

T: rotational energy of a star

T_c: central temperature of a star

T_9: temperature expressed in 10^9 K

T_i: temperature of an isothermal core

T_{ef}, T_e: effective stellar temperature

T_{cr}: critical temperature where the flow velocity equals the local sound velocity

T_m: temperature of crystal melting

t_{ms}: time needed for a contracting star to reach the main sequence

t_{RG}: lifetime of a star during the red giant stage

$t_{r\phi}$: $r\phi$ component of the viscosity stress tensor

t_t: lifetime of a star prior to the turning point off the main sequence

t_n: characteristic time of nuclear reactions

t_β: characteristic time of beta processes

u, $\mathbf{u}$: velocity

u_s: sound velocity in a gas

u_{cr}: critical velocity of a flow, equal to the local sound velocity

$u = \Delta/m_e c^2$: non-dimensional energy of beta decay

v, v_i: velocity

v_e: electron velocity

v_n: velocity of a neutron star

v_s: same as u_s

$\langle v_i \rangle$: average (diffusive) electron velocity

W: gravitational energy of a star

W_n: probability of neutron decay

W_e: probability of electron capture by a nucleus

W_β, $W_{A,Z}$, $W_{A,Z}^+$, $W_{A,Z}^-$: probabilities of different beta reactions with nuclei

W_μ: probability of muon decay

X_H, x_H: mass abundance of hydrogen

X_α, x_{He}: mass abundance of helium

$x = pc/kT$: non-dimensional electron momentum

$x = rc^2/GM$: non-dimensional current radius of an accretion disk

x_A, x_Z: mass abundance of an element heavier than He

x_i: mass abundance of an element with atomic number i

$x_{^{12}C}$: mass abundance of carbon

x_0, x_1: mass abundances of nuclei entering a reaction

x_2, x_3: mass abundances of nuclei emerging from a reaction

Y_e: ratio of the total number of protons to the total number of baryons in nuclear matter

Y_l: lepton charge per baryon in nuclear matter

$y = p_{Fe}/m_e c$: non-dimensional Fermi momentum of electrons

$y_n = p_{Fn}/m_n c$: non-dimensional Fermi momentum of neutrons

y_{ij}: fraction of the i-th element ionized to the j-th state

Z: electrical charge of a nucleus (nuclear number)

Greek Symbols

α: fine-structure constant

$\alpha = m_e c^2/kT$: non-dimensional inverse temperature

α: coefficient connecting the $r\phi$ components of the viscous stress tensor with pressure

α: coefficient connecting the mixing length of a convective element and the characteristic scale height

α_p, α: coefficient connecting the mixing length of a convective element and the pressure scale height

α_P: Planck-averaged absorption coefficient per unit mass

α_T: thermal expansion coefficient

α_ν: spectral coefficient of absorption per unit mass

α_ν^*: spectral coefficient of absorption per unit mass taking into account stimulated transitions

α_ν^{bf}: spectral coefficient of bound–free absorption per unit mass taking into account stimulated transitions

α_ν^{ff}: spectral coefficient of free–free absorption per unit mass taking into account stimulated transitions

α_ρ, α: coefficient connecting the mixing length of a convective element and the density scale height

$\beta = \mu_{te}/kT$: non-dimensional chemical potential of electrons

β_g: ratio of gas pressure to total pressure

Γ: relativistic adiabatic index

$\Gamma = Z^2 e^2/kT\, l_{WZ}$: non-dimensional gas parameter

$\Gamma_Z = Z^2 e^2/kT\, a$: non-dimensional gas parameter

Γ; Γ_1, Γ_2: total and partial widths of a Breit–Wigner resonance

γ: polytropic or adiabatic power index

γ_1, γ_2, γ_3: adiabatic power indices

γ_i: Dirac matrices, $i = 1, 2, 3, 4, 5$

$\gamma_{\mathrm{rad}} = d\ln T/d\ln P$: logarithmic derivative along the radius of a radiative
   star

Δ: total energy of beta decay

$\Delta\nabla T$: excess of a temperature gradient in star over the corresponding adiabatic gradient

$\Delta\nabla\rho$: excess of a density gradient in star over the corresponding adiabatic
   gradient

Δt_{ThF}: duration of a thermal helium shell flash

$\delta = \Delta/m_e c^2$: non-dimensional total energy of beta decay

δ_{ij}: Kronecker symbol

ϵ: total energy of a star (with or without its rest mass)

ϵ, ϵ_e: total energy of an electron

ϵ_G: Newtonian gravitational energy of a star

ϵ_{GR}: first-order correction to the energy of a star connected with general
   relativity

ϵ_i: internal energy of a star

ϵ_{eq}: total energy of a star in static equilibrium (without rest mass)

ϵ_M: magnetic energy of a star

ϵ_{N}: total energy of a Newtonian star (without rest mass)

ϵ_{01}: energy per unit mass released in a reaction between nuclei 0 and 1

$\epsilon_{2\gamma}$: energy produced during photo-disintegration of nucleus 2

ϵ_{ff}: total free–free emission rate of electrons per unit mass

ϵ_B: total emission rate of electron magneto-bremsstrahlung per unit mass

ϵ_{Fe}: Fermi energy of electrons

ϵ_{Fn}: Fermi energy of neutrons

ϵ_β: kinetic energy of particles produced in beta decay (absorbed in beta capture)

ϵ_ν: energy loss by a star due to neutrino emission per unit mass

ϵ_n: rate of nuclear energy production in a star per unit mass

ϵ_{gr}: rate of gravitational energy production in a star per unit mass

ϵ_{CNO}: rate of heat production per unit mass in the CNO cycle of hydrogen
   burning

$\epsilon_{3\alpha}$: rate of heat production per unit mass in the 3α reaction of helium burning

$\epsilon_{^{12}C\alpha}$: rate of heat production per unit mass in the α-capture reaction by ^{12}C

$\epsilon_{^{16}O\alpha}$: rate of heat production per unit mass in the α-capture reaction by ^{16}O

η: coefficient of (dynamic) viscosity

η_T: coefficient of turbulent viscosity

θ: Debye temperature of a Coulomb lattice

θ: Lane–Emden function

θ: angle of $e\nu$ or $e\tau$ neutrino mixing in vacuum

θ_W: Weinberg angle

θ_m: angle of $e\nu$ or $e\tau$ neutrino mixing in matter

κ: Rosseland opacity

κ_{bf}: Rosseland opacity for bound–free absorption

κ_e: opacity connected with electron heat conductivity

κ_{ff}: Rosseland opacity for free–free absorption

κ_B: Rosseland opacity for magneto-bremsstrahlung absorption

κ_{se}^r: Rosseland opacity for scattering on free relativistic electrons

$\kappa_\nu = \alpha_\nu + \sigma_\nu$: spectral coefficient of combined absorption and scattering

κ_ν: neutrino opacity

κ_L: normalized monochromatic line absorption coefficient

$\kappa_\mathcal{D}$: reciprocal radius of Debye screening

Λ: Coulomb logarithm

λ: wavelength of a photon

λ: coefficient of heat conductivity

λ_e: coefficient of electron heat conductivity

λ_n: coefficient of neutron heat conductivity in the presence of nuclei

λ_{nn}: coefficient of neutron heat conductivity of a pure neutron gas

$\lambda_1(0)$: probability of a nuclear reaction between nucleus 0 and nucleus 1

μ: molecular weight $\equiv$ number of nucleons per particle

$\mu_{A,Z}$: chemical potential of nuclei with atomic number Z and atomic mass A

μ_N: average number of nucleons per nucleus

μ_Z: number of nucleons per electron

μ_n: chemical potential of neutrons

μ_p: chemical potential of protons

μ_{te}, μ_{te^-}: chemical potential of electrons

$\mu_{te^+} = -\mu_{te^-}$: chemical potential of positrons

μ_{ν_e}: chemical potential of electronic neutrinos

$\mu_{\bar{\nu}_e} = -\mu_{\nu_e}$: chemical potential of electronic antineutrinos

ν: photon frequency

ν: coefficient of kinematic viscosity

$\nu \equiv \nu(r)$: number of baryons within radius r

ν_e: total frequency of electron collisions

ν_{ee}: frequency of collisions between electrons

ν_{ei}: frequency of collisions between electrons and ions

ξ: non-dimensional Lane–Emden radius

Π: pressure term connected with artificial viscosity

ρ: matter density

ρ_c, ρ_{c0}: central density of a star

ρ_0: rest mass density

$\rho_{c,cr}$: central density of a star at a point of loss of stability

$\bar{\rho}$: average density of a star

ρ_n: density of neutrons

ρ_{nd}: neutron drip line density

ρ_{cr}: critical density, where the flow velocity equals the local sound velocity

Σ: surface matter density of a disk

σ: cross-section of a nuclear reaction

σ: coefficient of electro-conductivity

σ_{eff}: cross-section of free–free emission

σ_{aff}: cross-section of free–free absorption

σ_{abf}: cross-section of bound–free absorption

σ_{efb}: cross-section of free–bound emission (recombination cross-section)

σ^*_{abf}: cross-section of bound–free absorption taking into account stimulated transitions

σ^*_{aff}: cross-section of free–free absorption taking into account stimulated transitions

σ_e: total cross-section of scattering by free electrons (Thomson scattering)

σ_e: electron spin

σ_{er}: total cross-section of scattering by free relativistic electrons

σ_T: Thomson coefficient of scattering by free electrons per unit mass

σ_T: coefficient of turbulent electro-conductivity

σ_ν: spectral coefficient of scattering per unit mass

$\boldsymbol{\sigma}$: vector of Pauli matrices

τ: optical depth

τ: decay time of a stellar magnetic field

τ_{H}: characteristic time of hydrogen burning in a star

τ_{He}: characteristic time of helium burning in a star

τ_h: characteristic hydrodynamic time

τ_n: characteristic time of nuclear reactions

τ_{ph}: optical depth at the level of a photosphere

τ_{th}: characteristic time of thermal processes

τ_β: characteristic time of beta processes

τ_ν: cooling time of a star due to neutrino emission

$\tau_{1/2}$: half-life in beta decay

$\tau_1(0)$: average lifetime of nucleus 0 until a reaction with nucleus 1

$\tau_\gamma(2)$: time of photo-disruption of nucleus 2

τ_μ: lifetime of a muon

Φ, ϕ_G: gravitational potential

ψ: particle psi-function (bispinor)

$\bar{\psi}$: Dirac conjugate psi-function

ψ_{e}: electron psi-function

ψ_{n}: neutron psi-function

ψ_{p}: proton psi-function

ψ_μ: muon psi-function

ψ_ν: neutrino psi-function

Ω, ω: angular velocity of matter

Ω_K: Keplerian angular velocity

ω: circular frequency of a photon

ω, ω_{nl}: frequency of stellar oscillations

ω_1, ω_2: terms with artificial viscosity

ω_B: Larmor frequency
ω_i: circular frequency of ion oscillations in a crystal
ω_{pi}: plasma frequency of ions
$\nabla = d\ln T/d\ln\rho$: as γ_{rad}
$\langle 01 \rangle \equiv \langle \sigma v \rangle_{01}$: averaging over reacting nuclei, with relative velocity v

List of Abbreviations

1-D: one-dimensional
2-D: two-dimensional
AGB: asymptotic giant branch
CAK: Castor–Abbott–Klein
CHF: core helium flash
CP: charge conjugation-spatial parity
CVC: conservation of vector current
D: degenerate
D: dipole
DH: Debye–Hückel
GR: general relativity
HB: horizontal branch
HR: Herzsprung–Russel
IMS: initial main sequence
IRZ: intermediate regime zone
LD: Landau–Darreus
LI: low-intermediate(-mass stars)
LTE: local thermodynamic equilibrium
MES: minimum energy state
MHD: magneto-hydrodynamical
MHD: Mihalas–Hummer–Däppen
MRE: magneto-rotational explosion
MS: main sequence
ND: non-degenerate
NM: nuclear matter
P: pole
PCAC: partial conservation of axial current
PN: planetary nebula
PNN: planetary nebula nuclei
PSR: pulsar
RGB: red giant branch
RT: Rayleigh–Taylor
SFC: self-consistent
SN: supernova
SNI: supernova of type I
SN Ia,b,c: supernova of types Ia, Ib, Ic
SNII, SN II: supernova of type II

SNU: aolar neutrino unit
SPH: smooth particle hydrodynamics
TF: Thomas–Fermi
ThF: thermal flash
TFDH: Thomas–Fermi Debye–Hückel
UHB: upper horizontal branch
UWI: universal weak interaction
WD: white dwarf
WKB: Wentzel-Kramers-Brillouin
WR: Wolf–Rayet
WS: Wigner–Seitz
ZAMS: zero-age main sequence
ZAHB: zero-age horizontal branch

Some Important Constants

$\pi = 3.1415926536$;
$e = 2.7182818285$;
$\log e = 0.4342944819$;
1 radian=$57.2957795131°$.

Physical Constants

Light velocity: $c = 2.997925 \times 10^{10}$ cm·s^{-1}
Gravitational constant: $G = 6.67 \times 10^{-8}$ dyn· cm^2·g^{-2}
Plank constant divided by 2π: $\hbar = 1.05459 \times 10^{-27}$ ergs·s
Electron charge: e= 4.80325×10^{-10} CGSE units
Electron mass: $m_e = 9.10956 \times 10^{-28}$ g,
$\quad m_e c^2 = 0.511003$ MeV=$k \cdot 5.93013 \times 10^9$K
Physical mass unit: $m_u = (1/12)m_{12C} = 1.660531 \times 10^{-24}$ g,
$\quad m_u c^2 = 931.481$ MeV
Proton mass: $m_p = 1.672661 \times 10^{-24}$ g $= 1.00727m_u$
Neutron mass: $m_n = 1.674911 \times 10^{-24}$ g
Muon mass: $m_\mu = 1.88357 \times 10^{-25}$ g
Boltzmann constant: $k = 1.38062 \times 10^{-16}$ erg·K^{-1}
Fine structure constant: $\alpha = e^2/\hbar c = (137.036)^{-1}$
Classical electron radius: $l_e = e^2/m_e c^2 = 2.81794 \times 10^{-13}$ cm
Compton electron wavelength: $\lambda_e = \hbar/m_e c = 3.861592 \times 10^{-11}$ cm
Photon wavelength of the energy 1 eV: λ (1 eV)= 12398.54×10^{-8} cm
Photon frequency of the energy 1 eV: ν(1 eV)= 2.417965×10^{14} c^{-1}
Energy corresponding to 1 eV: E_0(1 eV)= 1.602192×10^{-12} ergs
Temperature corresponding to 1 eV: T(1 eV)= 11604.8K=E_0/k,
$\quad (E_0/k)\log e = 5.039.9$K
Radiation energy density const.: $a = \frac{\pi^3 k^4}{15c^3\hbar^3} = 7.56464 \times 10^{-15}$erg·cm^{-3}·K^{-4}
Stephan-Boltzmann const.: $\sigma = ac/4 = 5.66956 \times 10^{-5}$erg· cm^{-2}·K^{-4}·c^{-1}

Astronomical Constants

1 astronomical unit: a.u.= 1.495979×10^{13} cm
1 parsec: pc= 3.085678×10^{18} cm
1 light year= 9.460530×10^{17} cm
Solar mass: $M_{\odot} = 1.989 \times 10^{33}$ g
Solar radius: $R_{\odot} = 6.9599 \times 10^{10}$ cm
Solar luminosity: $L_{\odot} = 3.826 \times 10^{33}$ erg $\cdot$s^{-1}
Earth mass: $M_{\oplus} = 5.976 \times 10^{27}$ g
Earth equatorial radius: $R_{\oplus,e} = 6378.164$ km
Jupiter mass: $M_{\mathrm{Jup}} = 317.83 M_{\oplus}$
Jupiter equatorial radius: $R_{\mathrm{Jup,e}} = 71300$ km
Tropical year (from equinox to equinox): 1 year= 3.1556926×10^{7} s
Connection between absolute bolometric stellar magnitude and full
    luminosity: $M_{bol} = 4.74 - 2.5 \log(L/L_{\odot})$
Connection between absolute (M) and visual (m) stellar magnitudes: $M = m + 5 \log d_{\mathrm{pc}} - A_{\mathrm{absorption}}$; d_{pc} is a distance to the star in parsec.

Subject Index

Printing: Saladruck, Berlin
Binding: Buchbinderei Lüderitz & Bauer, Berlin